Lecture Notes in Computer Science

Lecture Notes in Artificial Intelligence   15464

Founding Editor

Jörg Siekmann

Series Editors

Randy Goebel, *University of Alberta, Edmonton, Canada*
Wolfgang Wahlster, *DFKI, Berlin, Germany*
Zhi-Hua Zhou, *Nanjing University, Nanjing, China*

The series Lecture Notes in Artificial Intelligence (LNAI) was established in 1988 as a topical subseries of LNCS devoted to artificial intelligence.

The series publishes state-of-the-art research results at a high level. As with the LNCS mother series, the mission of the series is to serve the international R & D community by providing an invaluable service, mainly focused on the publication of conference and workshop proceedings and postproceedings.

Lourdes Martínez-Villaseñor ·
Gilberto Ochoa-Ruiz · Martin Montes Rivera ·
María Lucía Barrón-Estrada ·
Héctor Gabriel Acosta-Mesa
Editors

Advances in Computational Intelligence

MICAI 2024 International Workshops

HIS 2024, WILE 2024, and CIAPP 2024
Tonantzintla, Mexico, October 21–25, 2024
Proceedings, Part I

Editors
Lourdes Martínez-Villaseñor ⓘ
Universidad Panamericana
Mexico City, Distrito Federal, Mexico

Martin Montes Rivera ⓘ
Universidad Politécnica de Aguascalientes
Aguascalientes, Mexico

Héctor Gabriel Acosta-Mesa ⓘ
Universidad Veracruzana
Veracruz, Mexico

Gilberto Ochoa-Ruiz ⓘ
Instituto Tecnológico y de Estudios
Superiores de Monterrey
Zapopan, Mexico

María Lucía Barrón-Estrada ⓘ
TecNM-Instituto Tecnológico de Culiacán
Sinaloa, Mexico

ISSN 0302-9743   ISSN 1611-3349 (electronic)
Lecture Notes in Artificial Intelligence
ISBN 978-3-031-83878-1   ISBN 978-3-031-83879-8 (eBook)
https://doi.org/10.1007/978-3-031-83879-8

LNCS Sublibrary: SL7 – Artificial Intelligence

© The Editor(s) (if applicable) and The Author(s), under exclusive license
to Springer Nature Switzerland AG 2025

This work is subject to copyright. All rights are solely and exclusively licensed by the Publisher, whether the whole or part of the material is concerned, specifically the rights of translation, reprinting, reuse of illustrations, recitation, broadcasting, reproduction on microfilms or in any other physical way, and transmission or information storage and retrieval, electronic adaptation, computer software, or by similar or dissimilar methodology now known or hereafter developed.
The use of general descriptive names, registered names, trademarks, service marks, etc. in this publication does not imply, even in the absence of a specific statement, that such names are exempt from the relevant protective laws and regulations and therefore free for general use.
The publisher, the authors and the editors are safe to assume that the advice and information in this book are believed to be true and accurate at the date of publication. Neither the publisher nor the authors or the editors give a warranty, expressed or implied, with respect to the material contained herein or for any errors or omissions that may have been made. The publisher remains neutral with regard to jurisdictional claims in published maps and institutional affiliations.

This Springer imprint is published by the registered company Springer Nature Switzerland AG
The registered company address is: Gewerbestrasse 11, 6330 Cham, Switzerland

If disposing of this product, please recycle the paper.

Preface

The Mexican International Conference on Artificial Intelligence (MICAI) is a yearly international conference series that has been organized by the Mexican Society for Artificial Intelligence (SMIA) since 2000. MICAI is a major international artificial intelligence (AI) forum and the main event in the academic life of the country's growing AI community.

This year, MICAI 2024 was organized by the Mexican Society for Artificial Intelligence (SMIA, Sociedad Mexicana de Inteligencia Artificial) in collaboration with the Instituto Nacional de Astrofísica, Óptica y Electrónica (INAOE) and the Universidad de las Américas Puebla (UDLAP).

The MICAI series website is www.MICAI.org. The website of the Mexican Society for Artificial Intelligence, SMIA, is www.SMIA.mx. Contact options and additional information can be found on these websites.

The conference, as is traditional, showcased a large variety of research fields and topics. Moreover, the conference included cutting-edge keynote lectures, as well as detailed paper presentations and comprehensive hands-on tutorials. Furthermore, thought-provoking panels and niche workshops provided a rich and exciting experience that aimed to cater to a wide audience.

Moreover, we continued the legacy of announcing the José Negrete Award, the SMIA Best Thesis in Artificial Intelligence Contest's results. This year, the historic and culturally rich city of Puebla was our chosen rendezvous.

MICAI conferences publish high-quality papers in all areas of AI and its applications. The proceedings of the previous MICAI events have been published by Springer in its Lecture Notes in Artificial Intelligence (LNAI) series (volumes: 1793, 2313, 2972, 3789, 4293, 4827, 5317, 5845, 6437, 6438, 7094, 7095, 7629, 7630, 8265, 8266, 8856, 8857, 9413, 9414, 10061, 10062, 10632, 10633, 11288, 11289, 11835, 12468, 12469, 13067, 13068, 13612, 13613, 14502, 14391, and 14392). Since its foundation in 2000, the conference has grown in popularity and improved in quality.

Three workshops were held jointly with the conference. The proceedings of the MICAI 2024 workshops are published in two volumes. The first volume contains 20 papers from the 17th Workshop of Hybrid Intelligent Systems (HIS 2024). The second volume contains 18 papers from the 17th Workshop on Intelligent Learning Environments (WILE 2024) and the 6th Workshop on New Trends in Computational Intelligence and Applications (CIAPP 2024).

These volumes will be of interest for researchers in all fields of artificial intelligence, students specializing in related topics, and the general public interested in recent developments in AI.

The MICAI workshops received for evaluation 58 submissions. From these submissions, 38 papers were selected for publication in these volumes after a double-blind peer-reviewing process and three reviews per submission. It was carried out by the Program Committee of the workshops. The acceptance rate was 65%.

HIS 2024

Hybrid Intelligent Systems (HIS) offer a compelling solution to complex challenges in various application domains, including biology, medicine, logistics, management, engineering, technology, social, and humanities. The 17th Workshop of Hybrid Intelligent Systems (HIS 2024) gathered relevant research associated with HIS and their capabilities for managing all these complex processes.

HIS 2024 was hosted by the Mexican Society of Artificial Intelligence (SMIA), nested within the Mexican International Conference on Artificial Intelligence (MICAI 2024). This year, the workshop featured 20 selected research articles of 32 received on the CMT platform. All the articles underwent a rigorous double-blind peer review with an acceptance rate of 62.5%, covering topics such as Machine Learning, Fuzzy Systems, Reasoning, Intelligent Control, Computer Vision, Optimization, and Expert Systems.

HIS 2024 presented the latest advancements in the field with two track sessions: the first featured articles with finished research, while the second consisted of posters that showcased proposals and prototypes for future research. The workshop's primary objective was to present expert model systems applied to specific research topics.

We sincerely thank Patricia Melin of the Tijuana Institute of Technology, our keynote speaker, for her valuable contribution to our event with the lecture "Hybrid Intelligent Models based on Neural Networks and Fuzzy Logic." HIS 2024 was enriched by her extensive experience and remarkable achievements in the field. We greatly appreciate her presence and the insight she shared with our audience.

We would also like to thank SMIA and the MICAI organizers for their continued collaboration in developing this prestigious event. It was our honor to contribute to the MICAI event with research in Hybrid Intelligent Systems.

WILE 2024

In these times when we witness the fantastic advance of science and technology around artificial intelligence, the multiplication of knowledge has become a challenge, not only for human beings but also for the most powerful computers. The assimilation of such knowledge requires the development of more effective learning methods for both humans and machines. These new methodologies, in order to be in line with the trend towards smarter systems, need to be intelligent enough to facilitate and adequately condition the learning process. This type of learning should not be limited to individual learning or to following current trends and practices, but should break with restrictive schemes and allow learning anywhere, in all kinds of situations and under any circumstances. This is why the first mental change must occur in the perception of classroom learning or solitary reading and one must think that learning must occur in the environment in which each person finds themselves.

This is why the Workshop on Intelligent Learning Environments (WILE) is organized to promote and encourage the development of methods and technologies in the living surroundings. This workshop presents new ideas and developments in all areas of academia, education, training, development, and research, using a wide variety of learning methods and artificial intelligence.

The aim of this workshop is to bring together active researchers and students in the field of Intelligent Learning Environments so that they can present and discuss innovative theoretical work and original applications, exchange ideas, establish collaboration links, discuss important recent achievements, and talk over the significance of results in the field to AI in general. Our goal for WILE 2024 was to give researchers a platform to showcase their work while also investigating new approaches to integrate AI techniques in the creation of educational systems.

We invited authors to submit papers presenting original and unpublished research in all areas related to:

- Web-based intelligent tutoring systems
- Intelligent learning management systems
- Affective tutoring systems
- Modeling, enactment, and intelligent use of emotion and affect
- Natural language and dialogue approaches to ILE design and construction
- Authoring tools in intelligent tutoring systems
- Learning companions
- Applications of cognitive science
- Semantic Web technologies
- Student modeling
- Sentiment analysis in educational applications
- Gamification and game-based learning
- Educational data mining
- Learning analytics

The entire submission, reviewing, and selection process, as well as preparation of the proceedings, was supported by Microsoft's Conference Management Toolkit. Many people took part in WILE 2024, and we are thankful for the cooperation of the Red-ICA (Conacyt Thematic Network in Applied Computational Intelligence) members who served on the Technical Committee, as well as members of The Mexican Society for Artificial Intelligence (SMIA Sociedad Mexicana de Inteligencia Artificial). As every year, SMIA 2024 was the appropriate host for this event.

CIAPP 2024

The 6th Workshop on New Trends in Computational Intelligence and Applications (CIAPP 2024) aimed to bring together researchers, students, and end users to explore the latest advances in computational intelligence. This year's workshop focused on innovative algorithms, applications in health sciences, and interdisciplinary approaches that use machine learning, computer vision, neural networks, and evolutionary computing. Participants had the opportunity to share their findings, engage in discussions, and start collaborations, fostering a vibrant community dedicated to pushing the boundaries of computational intelligence.

We thank the Sociedad Mexicana de Inteligencia Artificial (SMIA) for organizing the 23rd Mexican International Conference on Artificial Intelligence (MICAI 2024). Their dedication and hard work created a platform for researchers and students to share knowledge, foster collaboration, and advance the field of artificial intelligence.

We also extend our most sincere thanks and recognition to the Instituto Nacional de Astrofísica Óptica y Electrónica (INAOE) and the Universidad de las Américas Puebla (UDLAP) for their outstanding support and exceptional organization as a local committee of MICAI 2024. Their commitment to organizing this event created an attractive environment for collaboration and knowledge sharing. Thank you for your invaluable contributions to the success of CIAPP 2024.

We want to thank all the people involved in the organization of this conference: the authors of the papers published in these two volumes –it is their research work that gives value to the proceedings– and the organizers for their work. We thank the reviewers for their great effort spent on reviewing the submissions and the Program and Organizing Committee members.

A special acknowledgment to the local committee led by Jose Martinez-Carranza, whose meticulous coordination was instrumental in realizing MICAI 2024 in Tonantzintla, Puebla, Mexico. Our thanks extend to INAOE's director, David Sánchez de La Llave. We are also indebted to Luis Ernesto Derbez Bautista, Rector of UDLAP, for his invaluable assistance in securing the university facilities to complement the facilites of INAOE.

In addition, the success of this conference and the breadth of its program are a testament to the collaborative efforts of our organizers and committees, sponsors, and the invaluable support of the US Office of Naval Research Global (ONRG) with the grant award number N629092412100 from the agency GRANT14222939.

The entire submission, reviewing, and selection process, as well as preparation of the proceedings, was supported by Microsoft's Conference Management Toolkit (https://cmt3.research.microsoft.com/). Last but not least, we are grateful to Springer for their patience and help in the preparation of these volumes.

In conclusion, MICAI 2024 was more than just a conference. It was a confluence of minds, a testament to the indefatigable spirit of the AI community, and a beacon for the future of Artificial Intelligence. As you navigate through these proceedings, may you find inspiration, knowledge, and connections that propel you forward in your journey.

November 2023

Lourdes Martínez-Villaseñor
Gilberto Ochoa-Ruiz
Martin Montes Rivera
María Lucía Barrón Estrada
Efrén Mezura-Montes

Conference Organization

Conference Committee

General Chair

Lourdes Martínez-Villaseñor — Universidad Panamericana, Mexico

Program Chair

Gilberto Ochoa-Ruiz — Tecnológico de Monterrey, Mexico

Workshop Chair

Hiram Ponce — Universidad Panamericana, Mexico

Tutorials Chairs

Roberto Antonio Vázquez Espinoza de los Monteros — Universidad La Salle, Mexico

Doctoral Consortium Chairs

Miguel González Mendoza — Tecnológico de Monterrey, Mexico
Juan Martínez Miranda — Centro de Investigación Científica y de Educación Superior de Ensenada, Mexico

Keynote Talks Chairs

Gilberto Ochoa Ruiz — Tecnológico de Monterrey, Mexico
Iris Méndez — Universidad Autónoma de Ciudad Juárez, Mexico

Publication Chair

Hiram Ponce — Universidad Panamericana, Mexico

Financial Chairs

Hiram Calvo Instituto Politécnico Nacional, Mexico
Lourdes Martínez-Villaseñor Universidad Panamericana, Mexico

Grant Chair

Leobardo Morales IBM, Mexico

Local Organizing Committee

Jose Martinez-Carranza (INAOE)
Jezabel Guzman-Zavaleta (UDLAP)
Caleb Rascón-Estebané (UNAM)
Gustavo Rodríguez-Gómez (INAOE)
Aldrich Alfredo Cabrera-Ponce (BUAP)
Brenda Cervantes-Cuahuey (INAOE)
Delia Irazú Hernández-Farias (INAOE)
Alejandro Gutiérrez-Giles (INAOE)
Leticia Oyuki Rojas-Perez (INAOE)

Program Committee

Alberto Ochoa-Zezzatti	Universidad Autónoma de Ciudad Juárez, Mexico
Aldo Marquez-Grajales	Instituto Tecnológico Superior de Xalapa, Mexico
Alexander Bozhenyuk	Southern Federal University, Russia
Andrés Espinal	Universidad de Guanajuato, Mexico
Angel Sánchez García	Universidad Veracruzana, Mexico
Anilu Franco	Universidad Autónoma del Estado de Hidalgo, Mexico
Antonieta Martinez	Universidad Panamericana, Mexico
Antonio Neme	UNAM, Mexico
Ari Barrera Animas	Universidad Panamericana, Mexico
Asdrúbal López Chau	Universidad Autónoma del Estado de México, Mexico
Belém Priego Sánchez	Universidad Autónoma Metropolitana Unidad Azcapotzalco, Mexico
Bella Martinez Seis	Instituto Politécnico Nacional, Mexico
Betania Hernandez-Ocaña	Universidad Juárez Autónoma de Tabasco, Mexico
Claudia Gómez	Instituto Tecnológico de Ciudad Madero, Mexico

Daniela Alejandra Ochoa	CentroGEO-CONACyT, Mexico
Dante Mújica-Vargas	CENIDET, Mexico
Diego Uribe	Tecnológico Nacional de México - ITL, Mexico
Eddy Sánchez-DelaCruz	Tecnológico Nacional de México - Campus Misantla, Mexico
Eduardo Valdez	Instituto Politécnico Nacional, Mexico
Efrén Mezura-Montes	Universidad Veracruzana, Mexico
Eloísa García-Canseco	Universidad Autónoma de Baja California, Mexico
Elva Lilia Reynoso Jardon	Universidad Autónoma de Ciudad Juárez, Mexico
Eric Tellez	CICESE-INFOTEC-CONACyT, Mexico
Ernesto Moya-Albor	Universidad Panamericana, Mexico
Félix Castro Espinoza	Universidad Autónoma del Estado de Hidalgo, Mexico
Fernando Gudino	UNAM, Mexico
Garibaldi Pineda Garcia	Applied AGI, UK
Genoveva Vargas-Solar	Grenoble Alpes University, CNRS, France
Gilberto Ochoa-Ruiz	Tecnológico de Monterrey, Mexico
Giner Alor-Hernandez	Tecnológico Nacional de México - ITO, Mexico
Guillermo Santamaría-Bonfil	BBVA México, Mexico
Gustavo Arroyo	Instituto Nacional de Electricidad y Energías Limpias, Mexico
Helena Gómez Adorno	IIMAS-UNAM, Mexico
Hiram Ponce	Universidad Panamericana, Mexico
Hiram Calvo	Instituto Politécnico Nacional, Mexico
Hugo Jair Escalante	INAOE, Mexico
Humberto Sossa	Instituto Politécnico Nacional, Mexico
Iris Iddaly Méndez-Gurrola	Universidad Autónoma de Ciudad Juárez, Mexico
Iskander Akhmetov	Institute of Information and Computational Technologies, Kazakhstan
Ismael Osuna-Galán	Universidad de Quintana Roo, Mexico
Israel Tabarez	Universidad Autónoma del Estado de México, Mexico
Jaime Cerda	Universidad Michoacana de San Nicolás de Hidalgo, Mexico
Jerusa Marchi	Federal University of Santa Catarina, Brazil
Joanna Alvarado Uribe	Tecnológico de Monterrey, Mexico
Jorge Perez Gonzalez	UNAM, Mexico
José Alanis	Universidad Tecnológica de Puebla, Mexico
José Martínez-Carranza	INAOE, Mexico
José Alberto Hernández-Aguilar	Universidad Autónoma del Estado de Morelos, Mexico
José Carlos Ortiz-Bayliss	Tecnológico de Monterrey, Mexico

Juan Villegas-Cortez	UAM - Azcapotzalco, Mexico
Juan Carlos Olivares Rojas	Tecnológico Nacional de México - ITM, Mexico
Karina Perez-Daniel	Universidad Panamericana, Mexico
Karina Figueroa Mora	Universidad Michoacana de San Nicolás de Hidalgo, Mexico
Leticia Flores Pulido	Universidad Autónoma de Tlaxcala, Mexico
Lourdes Martínez-Villaseñor	Universidad Panamericana, Mexico
Luis Torres-Treviño	Universidad Autónoma de Nuevo León, Mexico
Luis Luevano	Institut National de Recherche en Informatique et en Automatique, France
Mansoor Ali Teevno	Tecnológico de Monterrey, Mexico
Masaki Murata	Tottori University, Japan
Miguel Gonzalez-Mendoza	Tecnológico de Monterrey, Mexico
Miguel Mora-Gonzalez	Universidad de Guadalajara, Mexico
Mukesh Prasad	University of Technology Sydney, Australia
Omar López-Ortega	Universidad Autónoma del Estado de Hidalgo, Mexico
Rafael Guzman-Cabrera	Universidad de Guanajuato, Mexico
Rafael Batres	Tecnológico de Monterrey, Mexico
Ramon Brena	Instituto Tecnológico de Sonora, Mexico
Ramón Zatarain Cabada	Tec Culiacán, Mexico
Ramón Iván Barraza-Castillo	Universidad Autónoma de Ciudad Juárez, Mexico
Roberto Antonio Vasquez	Universidad La Salle, Mexico
Rocio Ochoa-Montiel	Universidad Autónoma de Tlaxcala, Mexico
Ruben Carino-Escobar	Instituto Nacional de Rehabilitación - Luis Guillermo Ibarra Ibarra, Mexico
Sabino Miranda	INFOTEC-CONACyT, Mexico
Saturnino Job Morales	Universidad Autónoma del Estado de México, Mexico
Segun Aroyehun	University of Konstanz, Germany
Sofía Galicia Haro	Sistema Nacional de Investigadoras e Investigadores, Mexico
Tania Ramirez-delReal	CentroGEO-CONACyT, Mexico
Vadim Borisov	Branch of National Research University "Moscow Power Engineering Institute" in Smolensk, Russia
Valery Solovyev	Kazan Federal University, Russia
Vicenc Puig	Universitat Politècnica de Catalunya, Spain
Vicente Garcia Jimenez	Universidad Autónoma de Ciudad Juárez, Mexico
Victor Lomas-Barrie	IIMAS-UNAM, Mexico

Workshops Organization

HIS 2024 Organizing Committee

General Chairs

Martín Montes Rivera	Universidad Politécnica de Aguascalientes, Mexico
Carlos Alberto Ochoa Zezzatti	Universidad Autónoma de Ciudad Juárez, Mexico
Daniela Paola López Betancur	Universidad Autónoma de Zacatecas, Mexico
Carlos Alejandro Guerrero Méndez	Universidad Autónoma de Zacatecas, Mexico
José Alberto Hernández Aguilar	Universidad Autónoma del Estado de Morelos, Mexico

Program Committee

Edgar Gonzalo Cossio Franco	Instituto de Información Estadística y Geográfica de Jalisco, Mexico
Humberto Velasco Arellano	Universidad Politécnica de Aguascalientes, Mexico
Daniela Paola López Betancourt	Universidad Politécnica de Aguascalientes, Mexico
Carlos Alejandro Guerrero Méndez	Universidad Autónoma de Zacatecas, Mexico
Humberto Muñoz Bautista	Universidad Tecnológica Metropolitana de Aguascalientes, Mexico
Miguel Ángel Ortiz Esparza	Universidad Autónoma de Aguascalientes, Mexico
Himer Avila-George	Universidad de Guadalajara, Mexico
Alejandro Padilla Díaz	Universidad Autónoma de Aguascalientes, Mexico
Carlos Alberto Lara Alvarez	CIMAT Zacatecas, Mexico
Roberto Antonio Contreras Masse	Instituto Tecnológico Ciudad Juárez, Mexico
Irma Yazmín Hernández Báez	Universidad Politécnica el Estado de Morelos, Mexico

WILE 2024 Organizing Committee

General Chairs

María Lucía Barrón Estrada	TecNM-Instituto Tecnológico de Culiacán, Mexico
Ramón Zatarain Cabada	TecNM-Instituto Tecnológico de Culiacán, Mexico
Yasmín Hernández Pérez	TecNM-Cenidet, Mexico
Carlos A. Reyes García	INAOE, Mexico
Karina Mariela Figueroa Mora	Universidad Michoacana de San Nicolás de Hidalgo, Mexico

Program Committee

Ramón Zatarain Cabada	Instituto Tecnológico de Culiacán, Mexico
María Lucía Barrón Estrada	Instituto Tecnológico de Culiacán, Mexico
Yasmín Hernández Pérez	Cenidet, Mexico
Karina Mariela Figueroa Mora	UMSNH, Mexico
Carlos A. Reyes García	Instituto Nacional de Astrofísica, Óptica y Electrónica, Mexico
Giner Alor Hernández	Instituto Tecnológico de Orizaba, Mexico
Miguel Pérez Ramírez	Instituto Nacional de Electricidad y Energías Limpias, Mexico
Jaime Muñoz Arteaga	Universidad Autónoma de Aguascalientes, Mexico
Rafael Morales Gamboa	Universidad de Guadalajara, Mexico
Guillermo Santamaría Bonfil	BBVA, Mexico
Carlos Alberto Lara Álvarez	CIMAT Zacatecas, Mexico
Hugo Arnoldo Mitre Hernández	CIMAT Zacatecas, Mexico
María Elena Chávez Echeagaray	Arizona State University, USA
María Blanca Ibáñez Espiga	Universidad Carlos III de Madrid, Spain
Alicia Martínez Rebollar	Cenidet, Mexico
María Lucila Morales Rodríguez	Instituto Tecnológico de Cd. Madero, Mexico
Héctor Rodríguez Rangel	Instituto Tecnológico de Culiacán, Mexico
Julieta Noguez Monroy	Tecnológico de Monterrey, Mexico
Samuel González López	Instituto Tecnológico de Nogales, Mexico
Raúl Oramas Bustillos	Universidad Autónoma de Occidente, Mexico
José Mario Ríos Félix	Instituto Tecnológico de Culiacán, Mexico
Luis Alberto Morales Rosales	Universidad Michoacana de San Nicolás de Hidalgo, Mexico
Maritza Bustos López	Instituto Tecnológico de Orizaba, Mexico

CIAPP 2024 Organizing Committee

General Chairs

Héctor Gabriel Acosta Mesa	Universidad Veracruzana, Mexico
Marcela Quiroz Castellanos	Universidad Veracruzana, Mexico
Rocío Erandi Barrientos Martínez	Universidad Veracruzana, Mexico
Efrén Mezura Montes	Universidad Veracruzana, Mexico

Program Committee

Martha Lorena Avendaño	Universidad Veracruzana, Mexico
Aldo Márquez Grajales	Universidad Veracruzana, Mexico
Nancy Pérez Castro	Universidad de Papaloapan, Mexico
Rafael Rivera-López	Tecnológico Nacional de México-Instituto Tecnológico de Veracruz, Mexico
Guillermo Hoyos-Rivera	Universidad Veracruzana, Mexico
Adriana L. López Lobato	Universidad Veracruzana, Mexico
Octavio Ramos Figueroa	Universidad de Xalapa, Mexico
Jesús Adolfo Mejía de Dios	Universidad Autónoma de Coahuila, Mexico
Mario Graff Guerrero	INFOTEC Aguascalientes, Mexico
José Luis Morales Reyes	Universidad de Xalapa, Mexico

Contents – Part I

HIS 2024

Comparative Study of Dragonfly and Firefly Algorithms with Type-1 and Type-2 Fuzzy Parameter Adaptation 3
   Hector M. Guajardo, Fevrier Valdez, Patricia Melin, Oscar Castillo, and Prometeo Cortes-Antonio

Decision Support System Based on Grey Systems and Markov Chain MCGM (1,1) Applied to Improve Forecast for Demand Uncertainty 16
   Francisco Trejo, Rafael Torres Escobar, and Alberto Ochoa-Zezzatti

Integration of Alexa and Social IoT for Generation X: A Study on the Optimization Use of Smart Treadmill to Improve Physical Health 28
   Alberto Ochoa-Zezzatti, José De los Santos, Ángel Ortíz, Maylin Hernández, and Joshuar Reyes

Logarithmic Weighted Random Selector Algorithm: A Novel Approach for Biasing Selection Based on Positional Order Without Hyperparameters 43
   Iván Alejandro Ramos Herrera

Design and Implementation of a Machine Learning Model for Soccer Match Prediction Based on Player Statistics 55
   Antonio Muñoz Barrientos, Humberto Muñoz Bautista, Alejandro Padilla Díaz, and Francisco Javier Álvarez Rodríguez

Electric Scooters and Renewable Energy Integration Associated with Tourist Parks: A Dijkstra-Based Model for Smart Mobility Optimization 67
   Gilberto Espadas-Baños, César Quej-Solís, Manuel Flota-Bañuelos, and Alberto Ochoa-Zezzatti

Use of Convolutional Neural Networks for the Recognition of Bird Species in Risk Categories in the State of Chihuahua 84
   Jose Luis Acosta-Roman, Alberto Ochoa-Zezzatti, and Martin Montes-Rivera

Exploring Deep Learning Applications in Neurodegenerative Diseases: A State-of-the-Art Review 97
   Ayrton Santos, Claudia I. Gonzalez, and Mario Garcia

Automated Insights: LLMs in Neurodegenerative Disease Research
and Comparison 109
    Cesar Torres and Claudia I. Gonzalez

Convolutional Neural Network Models for Classifying of Peach (Prunus
persica L) 121
    Flossi Puma-Ttito, Carlos Guerrero-Mendez, Daniela Lopez-Betancur,
    Tonatiuh Saucedo-Anaya, Rafael Castaneda-Diaz,
    and Luis Martinez-Ytuza

Cleaning Binary Distortion on MNIST Dataset 133
    Rafael Castaneda-Diaz, Daniela Lopez-Betancur,
    Carlos Guerrero-Mendez, Efrén González Ramírez,
    Salvador Gómez-Jiménez, and Flossi Puma-Ttito

Rule-Based Expert System with Bayesian Theory and Fuzzy Inference
for Vocational Guidance: A Tool to Prevent School Dropouts 143
    Cynthia Cristina Martinez Padilla

Detection of Basic Motorcycle Faults Using a Fuzzy Bayesian Expert
System 163
    David Alonso Carranza Escobar

Intervention Model with Data Mining Techniques to Work with Dating
Violence Victims Using a Social Support Network 180
    Rogelio Rodríguez-Hernández, Nemesio Castillo-Viveros,
    and Alberto Ochoa-Ortiz

Enhanced Pest Detection Using Quaternion-Based Image Segmentation
in Yellow Sticky Trap Samples for Precision Agriculture 199
    Esquivel-Félix Ramiro, Solís-Sánchez Luis Octavio,
    Ochoa-Zezzatti Alberto, Castañeda-Miranda Celina Lizeth,
    and Guerrero-Osuna Héctor Alonso

Optimization of a Treatment Plant Through the Incorporation of New
Waste Separation Components: A TOPSIS-Focused Multi-criteria
Analysis Approach 216
    Carlos Iván Ramón Diego, José Ismael Ojeda Campaña,
    Virginia Berenice Niebla Zatarain, and Alberto Ochoa-Zezzatti

Multicriteria Analysis Applied to the Selection of Shopping Centers
for a Family-Owned Restaurant Business in Smart Cities: A Case Study
in Ciudad Juárez 233
    José Roberto Escamilla de Santiago, Alberto Ochoa-Zezzatti,
    and Aida-Yarira Reyes-Escalante

Alarm Recommendation Intelligent System for Multilayer Ceramic
Capacitor (MLCC) Electroplating Using Case-Based Reasoning
and Natural Language Processing 248
   *Juan Pablo Canizales-Martinez, Alberto Ochoa-Zezzatti,
   and Carmen Villar-Patiño*

Implementation of a Hybrid Tabu Search Algorithm for Solving
the Capacitated Vehicle Routing Problem 272
   *Carlos Condado-Huerta, José Alberto Hernández-Aguilar,
   Martín H. Cruz-Rosales, Víctor Pacheco-Valencia,
   and Julio César Ponce-Gallegos*

Intelligent System Associated with a Stochastic RoRo Shipping Problem
for a Fleet of Differentiated Vehicles and Collection of Specific Problems 284
   *Alberto Ochoa-Zezzatti, Irma Hernández-Báez, Axel Bernal,
   and Humberto García-Castellanos*

Author Index 297

Contents – Part II

WILE 2024

Expert System for Teaching Classification Systems Workflow 3
   Christian Sánchez-Sánchez

Enhancing Student Theses with Advanced Text Analysis Using NLP
and Pre-trained Models 15
   Maximiliano Ponce Marquez, Samuel González-López,
   Jesús Raúl Cruz Rentería, Gilberto Borrego Soto,
   Manuel Omar Meranza Castillon, and Guillermina Muñoz Zamora

Competence-Based Student Modelling with Dynamic Bayesian Networks 25
   Rafael Morales and L. Enrique Sucar

XploRe: XR Tool for Learning About the Solar System and Its Physical
Phenomena 37
   José Miguel Gil-Núñez, María Blanca Ibañez-Espiga,
   Ramón Zataraín-Cabada, and María Lucía Barrón-Estrada

Emotion Recognition in Virtual Reality Learning Environments:
A Multimodal Machine Learning Approach 45
   Luis Romero-Ramos, Gabriel González-Serna, Máximo López-Sánchez,
   Nimrod González-Franco, and Blanca Valenzuela-Robles

Enhancing Dropout Prediction Models Through Feature Selection
Techniques 53
   Daniel Domínguez-Gómez, Eduardo Sánchez-Jiménez,
   Yasmín Hernández, Juan de Dios González Torres,
   and Javier Ortiz-Hernandez

Assessing Cognitive Load in Programming Exercises Based on Readability
and Lexical Richness 61
   Jesús Miguel García-Gorrostieta, Samuel González-López,
   Aurelio López-López, Ulises Ponce-Mendoza,
   and José David Madrid-Monteverde

CIAPP 2024

Air Pollution, Socioeconomic Status, and Avoidable Hospitalizations
in Mexico City: A Multifaceted Analysis 73
   Carlos Minutti-Martinez, Miguel F. Mata-Rivera,
   Magali Arellano-Vazquez, Boris Escalante-Ramírez, and Jimena Olveres

Automatic Detection of Abnormal Pedestrian Flows, Using Classification
and Tracking with Pre-trained YOLOv8 87
   Adrián Núñez-Vieyra, Rogelio Ferreira-Escutia,
   Juan C. Olivares-Rojas, Arturo Méndez-Patiño,
   José A. Gutiérrez-Gnecchi, and Enrique Reyes-Archundia

Computational Time Reduction in the Induction of Convolutional Decision
Trees 99
   Adriana-Laura López-Lobato, Héctor-Gabriel Acosta-Mesa,
   and Efrén Mezura-Montes

Bean Landraces Color Identification Through Image Analysis
and Gaussian Mixture Model 112
   Adriana-Laura López-Lobato, Martha-Lorena Avendaño-Garrido,
   Héctor-Gabriel Acosta-Mesa, José-Luis Morales-Reyes,
   and Elia-Nora Aquino-Bolaños

Efficient Neural Architecture Search: Computational Cost Reduction
Mechanisms in DeepGA 125
   Jesús-Arnulfo Barradas-Palmeros, Carlos-Alberto López-Herrera,
   Héctor-Gabriel Acosta-Mesa, and Efrén Mezura-Montes

Prediction of Epileptic Seizure Using Neuroevolved Spiking Neural
Network 135
   Carlos-Alberto López-Herrera, Héctor-Gabriel Acosta-Mesa,
   Efrén Mezura-Montes, and Jesús-Arnulfo Barradas-Palmeros

Identification of Simple Geometric Figures Using Matlab and ROS 147
   Atalia-Yael Hernández-Sánchez, Jorge Aramburo-Aguilar,
   Héctor-Gabriel Acosta-Mesa, Sergio Hernandez-Mendez,
   and Antonio Marin-Hernandez

Experimental Study for Automatic Feature Construction to Segment
Images of Lungs Affected by COVID-19 Using Genetic Programming 155
   David Herrera-Sánchez, José-Antonio Fuentes-Tomás,
   Héctor-Gabriel Acosta-Mesa, and Efrén Mezura-Montes

Color Quantification in Common Bean Landraces Using a Supervised
Learning Technique 167
   José-Luis Morales-Reyes, Elia-Nora Aquino-Bolaños,
   and Héctor-Gabriel Acosta-Mesa

Explainable AI Through Decision Trees for Black-Box Models Used
to Support Bacterial Vaginosis Diagnosis 179
   Rafael Rivera-López, Juana Canul-Reich, Erick De la Cruz Hernández,
   Héctor Gibrán Ceballos-Cancino, Efrén Mezura-Montes,
   and Marco Antonio Cruz-Chávez

Improving Lactation Curve Estimation in Sheep: A Comparative Analysis
of Machine Learning Algorithms Across Milk Recording Schemes 190
   L. Guevara, F. A. Castro-Espinoza, A. M. Fernandes,
   I. Nacarati-da-Silva, T. Oliveira, E. G. Salgado Hernández,
   and J. C. Angeles-Hernadez

Author Index 201

HIS 2024

Comparative Study of Dragonfly and Firefly Algorithms with Type-1 and Type-2 Fuzzy Parameter Adaptation

Hector M. Guajardo, Fevrier Valdez[✉], Patricia Melin, Oscar Castillo, and Prometeo Cortes-Antonio

Tijuana Institute of Technology, TecNM, Tijuana, Mexico
{fevrier,pmelin,ocastillo,prometeo.cortes}@tectijuana.mx

Abstract. In this paper, we studied, optimized, and compared two algorithms: Dragonfly Algorithm (DA) and the Firefly Algorithm (FA). The Dragonfly Algorithm is an optimization method based on the swarming behavior of dragonflies, mimicking behaviors such as hunting, fleeing, and swarming to explore a search space. The Firefly Algorithm, on the other hand, is based on the flashing behavior of fireflies, which use bioluminescence to communicate and attract mates. We applied Type-1 and Type-2 Fuzzy Logic to both algorithms. For the Dragonfly Algorithm, we optimized the W and Beta parameters. For the Firefly Algorithm, the Beta parameter was optimized. We conducted experiments and compared the results using mathematical functions F1 to F10.

Keywords: Dragonfly algorithm · Firefly Algorithm · Type-1 Fuzzy Logic · Type-2 Fuzzy Logic

1 Introduction

Search and optimization algorithms such as Particle Swarm Optimization (PSO), Differential Evolution (DE) [11], Genetic Algorithm (GA), Firefly Algorithm (FA), and Dragonfly Algorithm (DA) [1] have proven to be efficient in terms of speed and convergence for certain types of problems. These algorithms are based on bio-inspired principles and have been extensively studied and tested in scientific literature. The choice of a specific algorithm depends on the optimization problem in question and its characteristics. However, combining different search and optimization algorithms in an algorithm can further improve efficiency in terms of convergence speed and the quality of the solution obtained. Algorithms that combine different search and optimization strategies can take advantage of the strengths of each of the algorithms and overcome their limitations.

Where there are insects like dragonflies, fireflies, and damselflies, there are many more in-sects with the same similarities for hunting, reproduction, or in matters of movement on the entire ecosystem. Observing the behaviors and structures of organisms in nature often suggests they perform their functions exceptionally well. The classification of many species of related insects will be reviewed in the following study [25]. Meta-heuristics encompass broad strategies that skillfully blend different methodologies to navigate through the solution space. In design optimization, the design objective can be as simple as maximizing production efficiency or minimizing production costs. An optimization algorithm is a technique that compares several solutions repeatedly until an ideal or feasible answer is identified. Currently, two types of optimization methods are frequently employed [2]. Deterministic algorithms employ a set of rules to transition from one solution to the next. Since stochastic algorithms include probabilistic translation rules and random nature, they may execute in a different sequence or produce a different outcome each time they are run with the same input. The layout of this article is outlined as follows: 2. Nature Inspiration the optimization technique we offer is based on the lecture and ideas from another investigators; 3. Study of Literature, which includes all the studies relevant to the topics of this article; 4. The application of the optimization to the Dragonfly Algorithm (DA); 5. The application of the optimization to the Firefly Algorithm, 6. Fuzzy logic explanation, 7.Results, analysis, and comparison of the experiments performed; 8. Analytical conclusions and a summary of the investigation are described in this article.

2 Nature Inspiration

In some cultures, dragonflies were called kachi-mushi (victorious insects) because they only fly forward, which gave them the character of those who never retreat and always move forward, whatever the circumstances. In the following we can review the life cycle of the dragonfly. The adult dragonfly starts to mate, after that they will lay their eggs in or around the water eggs in the water, after the eggs we will become a larva and after that the dragonfly will emerge. This can be appreciated in Fig. 1.

Fig. 1. The dragonfly cycle represents the steps the dragonfly must go through to become an adult dragonfly.

Based on these factors, it was suggested an algorithm that follows the model of insects and how they adjust to physiological changes by sharing resources and communicating to survive and grow. This algorithm was called "dragonfly". Natural optimization methods have proven to be adaptable, flexible, and efficient in handling practical problems. The fact that there is currently no optimization technique that can handle all problems is well recognized. [3]. The metaheuristic optimization techniques, Particle Swarm Optimization (PSO) [7], Artificial Colony Optimization (ACO) [6], 8, Artificial Bee Algorithm (ABC) [6], Firefly Algorithm (FA) [5], as well as other algorithms based on Hill climbing swarms [9], genetic algorithms (GA) [4], and different techniques based on trajectories Differential evolution (DE) [10], and genetic programming (GP) [12], Grey Wolf (GW) [18], are examples of evolutionary algorithms.

3 Study of Literature

There are several social behaviors used in nature to carry out various tasks. Although survival is the goal of all individuals and collective actions, organisms collaborate and interact in groups for a variety of purposes, including hunting, defending, navigating, and foraging. Wolves often follow a social hierarchy to pursue prey in various ways: chasing, circling, tormenting, and attacking [13]. Holland authored a book detailing the development of genetic algorithms (GAs). De Jong concluded his research by showcasing the considerable potential and robustness of evolutionary algorithms across various objective functions, including those that are noisy, multimodal, or discontinuous [4]. To minimize their learning and prediction errors through iterative trial and error, artificial neural networks, support vector machines, and other machine learning approaches are genetic algorithms and can be considered a heuristic optimization methodology. In 1961 Van Bergeijk, et al. Proposed artificial neurons as simple information processing units [14–16]. Particle Swarm Optimization (PSO), an optimization method inspired by the collective intelligence of fish, birds, and even human beings [17]. With time, the PSO method has demonstrated its superiority over conventional algorithms and genetic algorithms in specific problem domains, though it may not be suitable for every scenario. There isn't a universal algorithm that excels in all optimization problems; hence, current research aims to identify the most effective and efficient algorithm (s) for particular tasks. The No-Free Lunch theorems to caution the scientific community that if algorithm A outperforms algorithm B for certain optimization functions, then B is likely to outperform A for other functions [19]. Over time, researchers S. Nakrani and C. Tovey suggested the honey bee algorithm and its use as a foraging algorithm for problems including multimodal and dynamic optimization [20]. The techniques were inspired by how actual bees feed in nature. To identify spreaders utilizing a variety of targets, Amir S. and Ahman Z. developed the artificial bee colony (ABC) algorithm in 2020 [6].

Particle Swarm Optimization (PSO) and the Firefly Algorithm (FA), inspired by the flashing patterns of fireflies, were combined in a practical project undertaken by Khennak, I., et al. [5]. Inspired by grey wolves, it imitates the natural leadership structure and hunting strategy of these canines. The Coyote Optimization Algorithm, is a population-based metaheuristic for optimization that draws inspiration from the Canis latrans species [21]. Ant colony optimization (ACO) [5], additionally, new metaheuristic algorithms that are better than others at solving a particular kind of problem will continue to be developed. The general objective of this research is to improve the efficiency of bio-inspired algorithms in optimization problems, focusing on convergence speed and solution quality. To achieve this, a combination of algorithms such as FA, and DA is proposed, exploring the possibility of integrating fuzzy logic to adjust their parameters more efficiently. The application of this logic aims to make the algorithms more adaptable and robust when dealing with a wide variety of optimization problems. The main motivation of this research is to find solutions that can address the diversity of optimization problems, as no single algorithm is suitable for all cases. By combining different approaches, the goal is to leverage the advantages of each algorithm and mitigate their weaknesses. Furthermore, the inclusion of fuzzy logic seeks to improve the adaptability and tuning of the algorithms through more dynamic and personalized parameters.

4 Dragonfly Algorithm (DA)

In nature, practically all other little insects are preyed upon by dragonflies, which are thought of as small predators. Additionally, nymph dragonflies eat other maritime insects and even small fish. The intriguing characteristic of dragonflies is their uncommon and unusual swarming behavior. Only two things cause dragonflies to swarm: migration and hunting. Both are referred to as swarms—the former as a static (feeding) swarm and the latter as a dynamic (migratory) swarm. As already mentioned, the DA is an optimization technique that draws inspiration from the same-named bug. The static and dynamic characteristics of swarms serve as the primary source of inspiration for the DA algorithm. These two are extremely like the exploration and exploitation phases of metaheuristic optimization [22]. The main goal of the exploration phase is for dragonflies to organize into sub-swarms and fly in a static swarm over numerous locations. Certainly, during times of static swarming, it's observed that dragonflies tend to fly together in larger groups, all aligning their flight paths a notably advantageous behavior, particularly during the exploitation phase.

The primary goal of any swarm is survival, every member should be drawn to food sources and vigilant against external threats. As demonstrated the five essential factors that affect how individuals in swarms up as we can see in Fig. 2.

Swarm behavior follows three important principles:

1. Separation: Individual avoid static collision with neighbor.

$$S_j = \sum_{j=1}^{N} X - X_j \tag{1}$$

2. Alignment: Individual velocity matched with neighbor individuals.

$$A_i = \sum_{j=1}^{N} V_j \tag{2}$$

3. Cohesion: Individual tendency toward center of the herd.

$$C_i = \frac{\sum_{j=1}^{N} X_j}{N} - X \tag{3}$$

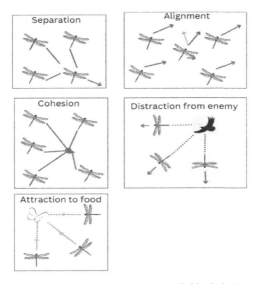

Fig. 2. The dragonfly patterns between individuals in the swarm

where:

X This represents the position of an individual, which typically denotes its current location or coordinates in each space.

V_j This represents the velocity of an individual, which typically refers to the speed and direction at which it is moving.

N This represents the number of neighborhoods or groups of individuals in your system. It's essentially a parameter that determines how individuals are grouped or organized into neighborhoods.

Attraction to food source is calculated:

$$F_i = X^+ - X \qquad (4)$$

where:

X Is the position of the current individual
X^+ Is the area of the food

Distraction from enemy is calculated:

$$E_i = X^- + X \qquad (5)$$

where:

X Is the position of the current individual
X^- Is the area of the enemy

In this research, dragonfly behavior is supposed to be a combination of these five corrective patterns. Two vectors are used to update the position of artificial dragonflies in a search space and replicate their movements: step (X) and position (X). The step vector represents the direction of the dragonfly movement and is described as follows:

$$X_{t+1} = (sS_i + aA_i + cC_i + fF_i + eE_i) + wX_i \qquad (6)$$

where (s) is the separation weight (Si) representing (i = th) individual's separation, (a) represents the alignment weight, (Ai) represents (i = th) individual's alignment, and (c) represents the cohesion weight. (C1) is the (i = th) individual's cohesiveness, (f) represents the food factor, and (Fi) is the (i = th) individual's food supply, (e) represents the enemy factor, and (Fi) is the (i = th) individual's position of the enemy, (w) is the inertia weight, and (t) is the iteration timer.

Table 1 presents the names and concise descriptions of all the parameters applied in the Dragonfly algorithm.

Table 1. Description of the algorithm parameters.

Parameter	Description	Values
Population	Population size	40
Boundaries	Number of boundaries	2
Dimensions	Dimension's size	8, 10, 16, 32 and 100
Iterations	Iteration's size	500
S	Separation weight	
A	Alignment weight	
C	Cohesion	
F	Food factor	
E	Enemy factor	
w	Inertia weight	
S_i	Separation of the i = th individual	
A_i	Alignment of the i = th individual	
C_i	Cohesion of the i = th individual	
F_i	Food source of the i = th individual	
E_i	Enemy position of the i = th individual	
t	Current iteration	

5 Firefly Algorithm

The common approach used in this algorithm is that fireflies will shine brighter when they attract other nearby fireflies. The attraction between two fireflies decreases or increases depending on their distance. If there are no nearby fireflies brighter than a particular firefly, this firefly will move randomly in the search space. All fireflies are unisex, so a firefly is attracted to other fireflies regardless of sex. From the literature, it emerges that the Firefly algorithm (FA) can outperform many other algorithms. The FA algorithm has expanded, and new.

$$X_p^t + 1 = X_p^t + \beta(r)(X_p - X_q) + \left(rand - \frac{1}{2}\right) \tag{7}$$

6 Fuzzy Logic

Fuzzy logic, alternatively referred to as fuzzy sets theory, provides a mathematical framework for addressing reasoning and decision-making in scenarios characterized by uncertainty and imprecision. Unlike traditional binary logic where statements are either true or false, fuzzy logic allows for degrees of truth between 0 and 1, representing degrees of membership or truthfulness. This allows for more nuanced modeling and analysis, particularly in areas where precise boundaries are difficult to define.[1].

Key Concepts of Fuzzy Logic:

Fuzzy Sets: In a fuzzy set, each element has a membership value that represents the degree to which the element belongs to the set. These membership values range between 0 and 1, where 0 indicates no membership, 1 indicates full membership, and values in between represent degrees of partial membership.

Membership Functions: Membership functions are mathematical functions that define the degree of membership of each element in a fuzzy set. These functions map each element from the universal set to a real number in the interval [0, 1]. There are various types of membership functions, such as triangular, trapezoidal, Gaussian, and sigmoidal, each suited for different applications and interpretations.

Fuzzy Operators: Fuzzy operators are mathematical operations defined on fuzzy sets that allow for combining and manipulating fuzzy information. These operators extend classical set operations such as union, intersection, and complement to accommodate the degrees of membership associated with fuzzy sets. The basic fuzzy operators include union (OR), intersection (AND), and complement (NOT) [23–25].

We will present Type-1 fuzzy inference systems implemented for the parameters w and beta of the Dragonfly algorithm in the following table. Here, in Fig. 5 we can observe the membership functions with the iteration parameter as input and the parameter was output, with their three fuzzy rules:

Fuzzy rules for the parameter w: If iteration is low, then w is high. If iteration is medium, then w is medium. If iteration is high, then w is low.

Here, in Fig. 6 we can observe the membership functions with the iteration parameter as input and the beta parameter as output, with their three fuzzy rules: Fuzzy rules for the parameter beta: If iteration is low, then beta is high. If iteration is medium, then beta is medium. If iteration is high, then beta is low.

In Fig. 7, we can observe the membership functions with the iteration parameter as input and the parameter was output, with its five fuzzy rules: Fuzzy rules for the parameter w: If iteration is low, then w is high. If iteration is low, then w is medium. If

Type-1 fuzzy inference systems w parameter

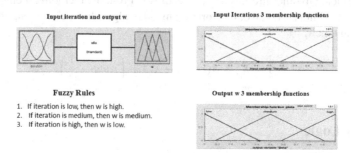

Fig. 5. Type-1 fuzzy inference systems w parameter

Type-1 fuzzy inference systems beta parameter

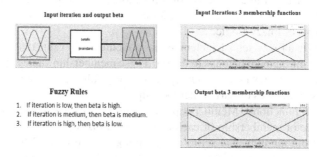

Fig. 6. Type-1 fuzzy inference systems for the beta parameter.

Type-1 fuzzy inference systems w parameter

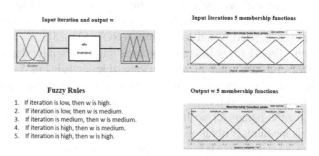

Fig. 7. Type-1 fuzzy inference systems for the w parameter

iteration is medium, then w is medium. If iteration is high, then w is medium. If iteration is high, then w is high.

Here, in Fig. 8 we can observe the membership Type-2 Gaussian functions with the iteration parameter iteration as input and the beta and w parameters as output, with their three fuzzy rules for Dragonfly Algorithm:

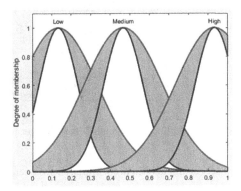

Fig. 8. Type-2 Gaussian membership functions for the beta parameter in the Firefly algorithm.

Here, in Fig. 9 we can observe the membership functions with the iteration parameter iteration as input and the beta parameter as output, with their three fuzzy rules for Firefly Algorithm: Fuzzy rules for the parameter beta: If iteration is low, then beta is high. If iteration is medium, then beta is medium. If iteration is high, then beta is low.

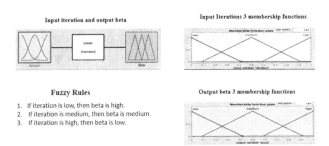

Fig. 9. Type-1 fuzzy inference systems beta parameter for Firefly algorithm.

Here, in Fig. 10 we can observe the membership Type-2 Gaussian functions with the iteration parameter iteration as input and the beta and w parameters as output, with their three fuzzy rules for Firefly Algorithm:

7 Results and Comparison

We compared each algorithm with optimization to one or two parameters to observe the functioning, with Type-1 we detect better results in comparison with the experiments with the natural algorithm, but when we compared Type-1 with Type-2 results Type-1

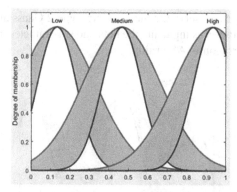

Fig. 10. Type-2 Gaussian Fuzzy inference systems beta parameter for Firefly algorithm.

was the one providing better results in both algorithms, so we decided to compared those results with a Z test.

In Table 2, we present a comparison of outcomes derived from the Dragonfly Algorithm (DA), we applied Type-1, Type-2 to the parameters w and beta. The table showcases the most favorable mean values and standard deviations obtained from each experiment across different functions.

Table 2. Comparison results for 100 dimensions of DA, DA with Type-1, DA with Type-2.

Fun	DA		DA Type-1		DA Type-2	
	Mean	Std dev	Mean	Std dev	Mean	Std dev
F1	2.85E−18	7.16E−01	2.48E+0	6.86E+01	2.61E−09	4.01E−09
F2	7.60E+00	6.79E+0	1.33E+0	1.12E+01	1.43E−04	1.32E−04
F3	1.03E−02	4.69E−03	9.17E+0	1.80E+03	1.17E−07	2.44E−07
F4	1.60E+01	9.48E+0	2.64E+0	2.17E+00	6.10E−03	8.65E−04
F5	2.31E−01	4.87E−01	8.25E+0	2.48E+02	6.06E−03	8.38E−04

In Table 3, we present a comparison of results from Firefly Algorithm (FA), we applied Type-1, Type-2 to the parameters w. The table showcases the most favorable mean values and standard deviations obtained from each experiment across different functions.

In Table 4, we present a comparison of results from FA and DA with Type-1 application to parameters. The table shows the results when we use the Z test to prove which algorithm presented better results. The FA got better results in F1, F3, F4, and F5. Meantime DA was better only in F2 results.

In Table 5, we present a comparison of results from FA and DA with Type-2 application to parameters. The table shows the results when we use the Z test to prove which algorithm presented better results. The FA got better results in every Function from F1 to F5.

Table 3. Comparison results for 100 dimensions of FA, FA with Type-1, and FA with Type-2

Fun	FA		FA Type-1		FA Type-2	
	Mean	Std dev	Mean	Std dev	Mean	Std dev
F1	2.47E+06	1.97E−08	2.61E−09	4.01E−09	7.33E−09	1.01E−08
F2	4.73E+11	4.03E+0	1.33E−04	1.20E−04	4.45E−05	2.32E−05
F3	3.22E+07	1.44E+0	9.06E−08	1.91E−07	7.43E−08	1.06E−07
F4	1.60E+01	9.48E+0	4.56E−05	4.75E−05	3.34E−05	2.66E−05
F5	2.80E+01	1.69E−01	7.65E+0	1.42E−01	4.18E−03	7.17E−04

Table 4. Comparison results for 100 dimensions of FA with Type-1, DA with Type-1

Fun	FA Type-1		DA Type-1		Z- Test
	Mean	Std dev	Mean	Std dev	Value
F1	2.61E−09	4.01E−09	2.84E+03	1.54E+03	3.88E+12
F2	1.43E−04	1.32E−04	4.45E−05	2.32E−05	1.26E+05
F3	1.17E−07	2.44E−07	2.40E+02	2.16E+02	5.38E+09
F4	6.10E−03	8.65E−04	4.35E+00	2.71E+00	2.75E+04
F5	6.06E−03	8.38E−04	5.78E+06	2.95E+07	2.90E+13

Table 5: Comparison results for 100 dimensions of FA with Type-2, and DA with Type-2

Fun	FA Type-2		DA Type-2		Z- Test
	Mean	Std dev	Mean	Std dev	Value
F1	1.17E−08	1.76E−08	6.73E+02	2.48E+03	2.09E+11
F2	2.27E−05	4.72E−06	1.17E+01	9.53E+0	1.35E+07
F3	3.07E−08	5.94E−08	1.84E+03	2.56E+03	1.69E+11
F4	4.91E−05	5.23E−05	3.06E+00	1.68E+00	3.21E+05
F5	6.93E−05	1.58E−05	2.17E+04	3.85E+04	7.56E+09

8 Conclusions

In conclusion, when we compared both algorithms in their original forms, the FA algorithm achieved better results. However, after applying Type-1 fuzzy logic, the DA algorithm outperformed the FA algorithm only in the F2 function, as evidenced by the results of the z-test. Even with the application of Type-2 fuzzy logic, the FA algorithm's performance remained superior to that of the DA algorithm. We ensured a fair comparison by using the same parameters for both algorithms: 30 experiment results, 40 agents as population, 500 iterations, and 100 dimensions for each bench mark algorithms. Even when we applied optimization to two parameters of the DA algorithm, it was not enough to defeat the FA algorithm. Future work will involve increasing the number of dimensions to observe more results.

References

1. Wikelski, M., Moskowitz, D., Adelman, J.S., Cochran, J., Wilcove, D.S., May, M.L.: Simple rules guide dragonfly migration (2006)
2. Bybee, S.M., et al.: Phylogeny and classification of Odonata using targeted genomics. Mol. Phylogenet. Evol. (2021)
3. Carreres-Prieto, D., Ybarra-Moreno, J., García, J.T., Cerdán-Cartagena, J.F.: A comparative analysis of neural networks and genetic algorithms to characterize wastewater from LED spectrophotometry. J. Env. Chem. Eng. **11**(3), 110219 (2023). https://doi.org/10.1016/j.jece.2023.110219
4. Sircar, A., Yadav, K., Rayavarapu, K., Bist, N., Oza, H.: Application of machine learning and artificial intelligence in oil and gas industry. Petrol. Res. **6**(4), 379–391 (2021). https://doi.org/10.1016/j.ptlrs.2021.05.009
5. Dorigo, M., Mondada, F., Stützle, T., et al. (eds.): Ant Colony Optimization and Swarm Intelligence: 4th International Workshop, ANTS 2004, Brussels, Belgium, 5–8 Sep 2004. Proceedings. Springer Berlin Heidelberg, Berlin, Heidelberg (2004)
6. Gutjahr, W.J.: ACO algorithms with guaranteed convergence to the optimal solution. Inform. Process. Lett. **82**(3), 145–153 (2002). https://doi.org/10.1016/S0020-0190(01)00258-7
7. Amir S., Ahmad Z.: Identifying influential spreaders using multi-objective artificial bee colony optimization (2020)
8. Khennak, I., Drias, H., Drias, Y., et al.: I/F-Race tuned firefly algorithm and particle swarm optimization for K-medoids-based clustering. Evol. Intel. **16**, 351–373 (2023)
9. Yang, X.-S.: Cuckoo search and firefly algorithm: overview and analysis. In: Yang, X.S. (eds.) Cuckoo Search and Firefly Algorithm. Studies in Computational Intelligence, vol. 516. Springer, Cham (2014)
10. Jacobson, S.H., Yücesan, E.: Analyzing the performance of generalized hill climbing algorithms. J. Heuristics **10**(4), 387–405 (2004). https://doi.org/10.1023/B:HEUR.0000034712.48917.a9
11. Storn, R., Price, K.: Differential evolution - a simple and efficient heuristic for global optimization over continuous spaces. J. Global Optim. **11**, 341–359 (1997)
12. Cordes, K., Rosenhahn, B., Ostermann, J.: Increasing the accuracy of feature evaluation benchmarks using differential evolution. IEEE (2011)
13. Nicolau, M.: Evostar. Genetic Programming. In Lecture Notes in Computer Science (2014)
14. Saophan, P., Pannakkong, W., Singhaphandu, R., Huynh, V.: Rapid production re-scheduling for flow shop under machine failure disturbance using hybrid perturbation population genetic algorithm-artificial neural networks (PPGAANNs). IEEE Access **11**, 75794–75817 (2023)
15. Van Bergeijk, W.A.: Studies with artificial neurons, II: analog of the external spiral innervation of the cochlea. Biol. Cybern. **1**(3), 102–107 (1961). 21. Harmon, L.D.: Studies with artificial neurons, I: properties and functions of an artificial neuron. Biol. Cybern. **1**(3), 89–101. (1961)
16. Levinson, J.Z., Harmon, L.D.: Studies with artificial neurons, III: mechanisms of flicker-fusion. Biol. Cybern. **1**(3), 107–117 (1961)

17. Olivas, F., Valdez, F., Castillo, O.: Particle swarm optimization with dynamic parameter adaptation using interval type-2 fuzzy logic for benchmark mathematical functions. In: 2013 World Congress on Nature and Biologically Inspired Computing (2013). https://doi.org/10.1109/nabic.2013.6617875
18. Barraza, J., Rodríguez, L., Castillo, O., Melin, P., Valdez, F.: A new hybridization approach between the fireworks algorithm and grey wolf optimizer algorithm. J. Optim. Res. **2018**, 6495362 (2018)
19. Saptarshi, S., Sanchita, B,. Richard, A.P.: II. Particle Swarm Optimization: A Survey of Historical and Recent Developments with Hybridization Perspectives (2018)
20. Wolpert, D.H., Macready, W.G.: No free lunch theorems for optimization. IEEE Trans. Evol. Computation **1**, 67–82 (1997)
21. Baig, A.R., Rashid, M.: Honeybee foraging algorithm for multimodal and dynamic optimization problems. In: Genetic and Evolutionary Computation Conference, Proceedings, London, England, UK, pp. 7–11 (2007)
22. Abdolkarim, M.-B., Mahmoud, D.N., Adel, A., Mohammadreza, T.: Golden eagle optimizer: A nature-inspired metaheuristic algorithm. Comput. Industrial Eng. **152** (2021)
23. Klir, G.J., Yuan, B.: Fuzzy Sets and Fuzzy Logic: Theory and Applications. Prentice Hall (1995)
24. Jang, J.S.R., Sun, C.T., Mizutani, E.: Neuro-fuzzy and soft computing-a computational approach to learning and machine intelligence [Book Review]. IEEE Trans. Automat. Contr. **42**(10), 1482–1484 (1997)
25. Yager, R. R., Filev, D.P.: Essentials of Fuzzy Modeling and Control. Wiley Interscience (1994)

Decision Support System Based on Grey Systems and Markov Chain MCGM (1,1) Applied to Improve Forecast for Demand Uncertainty

Francisco Trejo[1](✉), Rafael Torres Escobar[1], and Alberto Ochoa-Zezzatti[2]

[1] Facultad de Ingeniería, Universidad Anáhuac México, Lomas Anáhuac, Mexico
{francisco_trejo,rafael.torrese}@anahuac.mx
[2] Department of Computer Science, ITCJ, Universidad Autónoma de Ciudad Juárez (UACJ), 32310 Ciudad Juárez, Mexico
alberto.ochoa@uacj.mx

Abstract. Real-world information often comes with a degree of vagueness and imprecision, leading to potential unreliability. To address such uncertainties, various models have been developed, including grey systems, fuzzy logic systems, and rough sets. Each of these models offers different ways to handle the inherent uncertainty in data. This paper introduces a novel forecasting model that synergistically combines Markov Chains with the GM (1,1) model. The integration of these two methodologies aims to enhance forecasting accuracy beyond what either model can achieve individually. The proposed hybrid approach utilizes a Markov Chain to model the transition probabilities between states and the GM (1,1) model to forecast future demand based on accumulated data. By merging these two techniques, the model produces an interval forecast that is more robust and reliable compared to traditional forecasting methods like ARIMA. The interval forecast derived from this combined approach provides a range of potential future values, offering a clearer picture of the uncertainty and variability in demand predictions. The application of this new forecasting model was demonstrated in the context of warehousing services (3PL). Comparative analysis with previous forecasting methods has shown that the hybrid model outperforms its predecessors. The results confirm the effectiveness of the new scheme, highlighting its superiority in providing accurate and reliable forecasts. This paper contributes to the field by presenting a practical solution for handling uncertainty in demand forecasting and offering a validated method that surpasses traditional approaches.

Keywords: Markov Chain · Forecasting · Grey Systems · Supply Chain · Decision support systems

1 Introduction

Humans are not designed to be perfect decision-makers, and life is full of uncertainties. When our cognitive processes for dealing with uncertainty introduce error in our judgments, we are not always prepared to determine whether should follow these facts, correct the way or just simply follow our gut feelings. Of course, uncertainty is not limited to our private lives, it is present on all the decision-making processes.

This paper is aimed to provide some tools to reduce the uncertainty and to use at the same time the bast knowledge accumulated with our experience. As Francis Galton [1] already explained as: "the wisdom of the crowd", it is not about finding the right person who will have that answer, but how as a whole certain people or tools, grouped or coordinated in the right way, will have it; looking for the expert could be a mistake. A naïve assumption is to deal with the uncertainty assuming the data we are receive is correct, but because the imperfect measurement devices, the incomplete information about a system or simple human errors we can take a disastrous decision. We can find random uncertainties that are intrinsic to the environment, for example, partial observations due to noise measurement, while epistemic uncertainty is found when there a lack of knowledge and it can be reduced by making more measurements [2], however if you are not able to take more measurements and are not able to know more about the process or the system, and yet you need to take a decision. "Combining multiple forecasting methods for a given time series improves accuracy through integration obtained from different sources, thus avoiding the need to identify a best method" [3].

In this paper, in a decision support system based on Grey Systems and Markov Chain MCGM (1,1) is applied to improve the forecast for demand under high level of uncertainty, and its application in a third-party warehouse (3PL), to store finished product or raw materials. The demand volatility has forced organizations to question the validity and application of traditional forecasting models, and in this paper both the experience of the organization and the ambiguity of the data are considered when making decision is required nevertheless them are or not reliable.

The proposed model combines, one-variable model of GM (1,1) grey systems [4, 5] and Markov chains, that combined provides a better approach to demand forecasting within the supply chain, by establishing range parameters more precise than ARIMA. This combination is a novel alternative to the various current and most popularly used methods, with the difference that the proposed model, by making use Markov Chaon and the GM (1,1) model, integrates the user's experience precisely, compared other hybrid grey systems and Markov proposals, it considers the limited available information (>4 data), to perform demand forecasting with a higher level of accuracy. It also proposes to establish training data for further modelling development.

The proposed model makes use the exponential function of GM (1,1) grey systems and Markov chains, by developing a stationary transition matrix v, defined by the time series data set x. This develops: 1) a controller function of the type vector, where $\hat{x}_n$, is the forecast data at period n, $\hat{x}_{n-1}$ is the forecast at period n-1, $\hat{x}_{\mu n}$ is the mean forecast for the period n and $n-1$, defined by the lower $\hat{x}_{LO}$ and upper limits $\hat{x}_{HI}$ of the forecast for periods n and n-1, $v(\hat{x}_n, \hat{x}_{n-1}) = \hat{x}_{\mu n} = \frac{(\hat{x}_{LO} + \hat{x}_{HI})}{2}$, , 2) develops a forecast, 3) and measure the performance based on MAPE and MPE comparison, this function simplifies the result, by only considering the most recent input information in the time series x. The stationary transition matrix v can also be an impulse function provided by the user based on a) specify data provided by the organization, b) previous experience o c) using historical reference from similar models or customers.

These outcoming data are evaluated by the critical values defined based on MAPE and MPE. This process: i) simplifies the calculations, ii) allow to use part of the time series as training data, iii) the includes the user experience or any other relevant data for the organization, and iv) allows to apply sensitivity analysis to the system.

2 Grey Systems

The theory of grey systems is a relatively young discipline of knowledge, which has demonstrated its usefulness in solving many real problems. In recent years it has been used to explain and forecast several phenomena in many aspects of reality, such as in the analysis of multiple social phenomena, economic, technical models and natural disasters, accidents or even tourism, [4, 6–13].

However, it appears that there is little or no application of grey systems as a forecasting method in the supply chain, from what has been found in the literature.

Using a first order differential equation, with one variable, characterized by an unknown system, i.e., one in which there is no additional information that defines it and for which there is little information, including few records (>4). The GM (1,1) model is suitable for competitive forecasting in an environment where decision makers can refer to a limited amount of historical data [7], obtaining, however, highly accurate results, above the traditional models of: simple moving averages, weighted, exponential, simple linear regression, etc. The Markov chain forecasting model can be used to forecast a system with time series that vary randomly, with no causal reasons identified to justify this behavior. It is a dynamic system that forecasts the development of the system according to the transition probabilities between states reflecting the influence of all random factors [7].

$$x^{(0)} = \left\{ x^{(0)}(1), x^{(0)}(2), x^{(0)}(3), \ldots x^{(0)}(n) \right\} \tag{1}$$

where x (0) is assumed to be the original data sequence. To subsequently obtain the one-time accumulated generating operator 1-AGO (one-time accumulated generating operator 1-AGO), as shown in Eq. (2) and (3):

$$x^{(1)}(k) = \left\{ x^{(1)}(1), x^{(1)}(2), x^{(1)}(3), \ldots x^{(1)}(n) \right\} \tag{2}$$

where:

$$x^{(1)}(k) = \sum_{i=1}^{k} x^{(0)} i \tag{3}$$

The GM differential equation (1,1) and its "whitening" equation are obtained respectively in Eqs. (4) and (5):

$$x^{(0)}(k) + ax^{(1)}(1) = b, k = 1, 2, 3, ...n \qquad (4)$$

$$\frac{dx^{(1)}}{dt} + ax^{(1)}(k) = b \qquad (5)$$

where *a* represents the development coefficient and *b*, denotes the uncertainty factor or input error, or "grey factor". Let $\hat{u}$ now be the parameters of the vector defined in Eq. (6):

$$\hat{u} = (\hat{a}, \hat{b})^T = (B^T B)^{-1} B^T Y_N \qquad (6)$$

where *B* denotes the cumulative matrix and, *Y* is the constant vector, such that *a* and *b* can be obtained using the solution of the equation by the least squares method, where $z^{(1)}(k)$ is defined by Eq. (7):

$$z^{(1)}(k) = \frac{x^{(1)}(k) + x^{(1)}(k-1)}{2}, (k = 2, 3, 4, ..., n) \qquad (7)$$

And the matrix **B** and **Y_N** are defined by Eq. (8):

$$= \begin{bmatrix} -z^{(1)}(2) & 1 \\ -z^{(1)}(3) & 1 \\ \vdots & \vdots \\ -z^{(1)}(n) & 1 \end{bmatrix}, Y_N = \begin{bmatrix} x^{(0)}(2) \\ x^{(0)}(3) \\ \vdots \\ x^{(0)}(n) \end{bmatrix} \qquad (8)$$

The solution of the differential equation is obtained by Eq. (9):

$$\hat{x}^{(1)}(k+1) = \left(x^{(0)}(1) - \frac{b}{a}\right)e^{-ak} + \frac{b}{a} \qquad (9)$$

Re-expressing by Eq. (10):

$$\hat{x}^{(1)}(k) = \left(x^{(0)}(1) - \frac{b}{a}\right)e^{-a(k-1)} \qquad (10)$$

To obtain the forecast data, the equation obtained in the previous step is applied and the cumulative inverse operation is performed represented by Eq. (11):

$$\hat{x}^{(0)}(k) = (1 - e^a)\left[x^{(0)}(1) - \frac{b}{a}\right]e^{-a(k-1)} \qquad (11)$$

3 Markov Chains

The critical aspect of a Markov model, unlike any other set of random variables, is that it forgets everything but its immediate past. A Markov model through time is characterized by the fact that the future of the process is independent of the past given only its present value. Likewise, for a process Φ, which evolves in a space *X*, is governed by a general probability law *P*, be a Markov chain homogeneous in time, there must be a set of "transition probabilities" {P n (x, A), x ∈ X, A ⊂ X} for sets A such that for times n, m in Z+, defined in Eq. (12):

$$P(\phi_{n+m} \in A|\phi_j, j \leq m; \phi_m = x) = P^n(x, A) \tag{12}$$

The elements of the transition matrix P are called the transition probabilities. The transition probability $P_{i,j}$ is the conditional probability of being in state s_j "tomorrow" given that in the current state "today" is s_i [14]. It is very important to note that this work considers the fundamental premise of long-term "asymptotic" behavior: with a stationary distribution, which satisfies the steady vector $v(k)$ of a transition matrix P, is the unique probability that satisfies the Eq. (13):

$$v^k x P = v^k \tag{13}$$

The steady state vector v is independent of the initial state vector and unchanged when transformed by P. This means that v is an eigenvector with eigenvalue 1, and can be derived from P. It continuous to the point, where the components of v (the steady state vector) consist of two components, namely x and y, as shown in Eq. (14):

$$v = [v_x, v_y] \tag{14}$$

This means that the steady transition matrix v can predict the future state of the system. The proposed transition matrix is based on identifying the behavior that the system records have presented over time. Thus, if within the records the change is determined as follows in Table 1:

Table 1. Transition probability matrix v

State	Increase	Reduction
Increase	v_{ii}	v_{ir}
Reduction	v_{ri}	v_{rr}

In such a way that, if between two consecutive periods or records where one comes from a state s_i and a transaction is made to another s_j, where this s_i was an increment and the new s_j is also one, there will be a frequency of occurrence v_{ii}, if on the contrary, the new state s_j presents a decrease, it will correspond to a v_{ir} occurrence.

Once each one of the records are computed, the different transition probabilities between the different states are obtained, and the complete stationary transition matrix v is obtained.

4 Combined Forecasting Model GM (1,1) and Markov Chains

Markov chain can modify the prediction results of grey model, to improve the prediction accuracy [15]. The MCGM (1,1) correction process is described as follows:

Step 1. An important prerequisite for the model implementation, is necessary to perform an exploratory analysis to determine the main data series indicators, such as: mean, variance, trend, seasonality, and stationarity. Among other tools, it is recommended to use R. Thus, before applying the GM model (1,1) or any other model, it must first be determined whether it is the appropriate model from a range of models available in the bibliography. If the data series from Eq. (1) does not have trend, there is no stationarity, there are no cyclical or temporality patterns and growth trend, it is suitable for the GM (1,1) model.

This can be confirmed by Dickey-Fuller test, by comparing its critical point at 1% and the corresponding test-statistics if it is greater than, H_0 is not rejected and then the time series is not stationary [16]. This can be easily obtained by applying R Dickey-Fuller test.

Step 2. Once the data series is confirmed not to be stationary; the forecast is calculated based on Eq. (11) and $\hat{x}^{(1)}$, is obtained.

Step 3. The transition probability matrix v is defined based on the criteria expressed in Table 1, so it predicts the future state of the system when it is steady, based on Eq. (15) and provides the pair of data x and y defined in Eqs. (16) and (17).

Step 4. The estimated forecast range then is obtained based on:

$$(\hat{x}_{n-1}, \hat{x}_n)v = (\hat{x}_{n-1}, \hat{x}_n)(v_x, v_y) \qquad (15)$$

where:

$$(\hat{x}_{n-1})(v_x) = \hat{x}_{nLO} \qquad (16)$$

$$(\hat{x}_n)(v_y) = \hat{x}_{nHI} \qquad (17)$$

And $\hat{x}_{nLO}$ is the lower forecast of limit $\hat{x}_n$ and $\hat{x}_{nHI}$ is the upper limit of $\hat{x}_n$.

Step 5. The center value of $\hat{x}_n$ is calculated based on Eq. (18):

$$\hat{x}_{\mu n} = \frac{\hat{x}_{nHI} + \hat{x}_{nLO}}{2} \qquad (18)$$

This $\hat{x}_{\mu n}$ represents the best estimate value of the $\hat{x}_n$, and will be compared to other traditional forecast methods such as: moving average, linear regression, exponential smoothing or ARIMA.

The modeling process for MCGM (1,1) flow diagram is shown in Fig. 1.

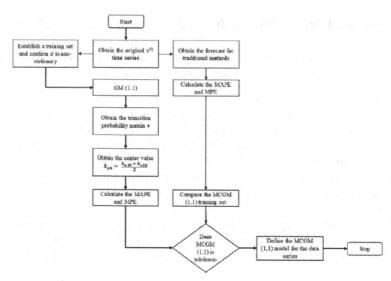

Fig. 1. Flow chart of the proposed hybrid GM (1,1) and Markov Chain model MCGM (1,1).

5 Example of the Application of the MCGM (1,1) Model for Demand Forecasting in a 3PL Warehousing Company

The combination of the GM (1,1) and Markov Chain will be exemplified with real data on the storage demand level of a logistics service provider (3PL) in the city of Tijuana BC, Mexico. The data available are 24 records, corresponding to two full years.

Step 1. We start with the known demand data and its behavior over time. The monthly demand data for the last 24 months are shown in Table 2 and the resulting graph of the same data in Fig. 2.

Table 2. Demand data for a customer of a 3PL warehousing company.

MM-YY	Period	Observed Data	MM-YY	Period	Observed Data
jan-2021	1	143	jan-2022	13	206
feb-2021	2	152	feb-2022	14	193
mar-2021	3	161	mar-2022	15	207
apr-2021	4	139	apr-2022	16	218
may-2021	5	137	may-2022	17	229
jun-2021	6	174	jun-2022	18	225
jul-2021	7	142	jul-2022	19	204

(*continued*)

Table 2. (*continued*)

MM-YY	Period	Observed Data	MM-YY	Period	Observed Data
aug-2021	8	141	aug-2022	20	227
sep-2021	9	162	sep-2022	21	223
oct-2021	10	180	oct-2022	22	242
nov-2021	11	164	nov-2022	23	239
dec-2021	12	171	dec-2022	24	266

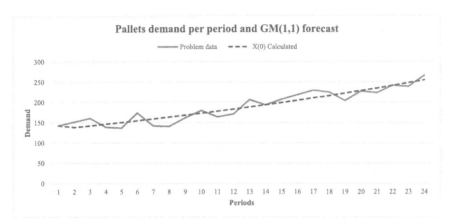

Fig. 2. $x^{(0)}$ time series data and the resulting $\hat{x}^{(0)}$.

Step 2. The forecast values are determined based on Eq. (11). We obtain Eq. (19) to calculate the corresponding period forecast.

$$\hat{x}^{(1)}(k) = 4941.49324 e^{0.02768983(k-1)} \quad (19)$$

Step 3. The transition probability matrix v is obtained from the series on Table 3.

Table 3. Transition probability matrix v from series $x^{(0)}$

State	Increase	Reduction
Increase	0.50	0.50
Reduction	0.78	0.22

Then the steady probability matrix would be as shown in Table 4:

Table 4. Steady transition probability matrix v from series $x^{(0)}$

State	Increase	Reduction
Increase	0.6086956	0.3913043
Reduction	0.6086956	0.3913043

Step 4. The estimated forecast range then is obtained based on Eqs. (16), (17) and (18), and its results are shown in Table 5 and in Fig. 3.

Table 5. Estimated $\hat{x}^{(0)}$ and forecast range.

Period	HI	LO	Real	$\hat{x}_{\mu n}$	MPE	MAPE
11	198.477	152.674	164.000	175.575	(0.071)	0.071
12	204.049	156.961	171.000	180.505	(0.056)	0.056
13	209.778	161.368	206.000	185.573	0.099	0.099
14	215.668	165.899	193.000	190.783	0.011	0.011
15	221.723	170.556	207.000	196.140	0.052	0.052
16	227.949	175.345	218.000	201.647	0.075	0.075
17	234.349	180.268	229.000	207.308	0.095	0.095
18	240.928	185.330	225.000	213.129	0.053	0.053
19	247.693	190.533	204.000	219.113	(0.074)	0.074
20	254.647	195.883	227.000	225.265	0.008	0.008
21	261.797	201.382	223.000	231.590	(0.039)	0.039
22	269.147	207.037	242.000	238.092	0.016	0.016
23	276.704	212.849	239.000	244.777	(0.024)	0.024
24	284.473	218.826	266.000	251.649	0.054	0.054
				$\mu=$	1.43%	5.19%

6 Model Performance Evaluation

The model performance is evaluated by the Mean percentage of error Equation and mean absolute percentage of error Equation. The smaller the values, the greater the accuracy.

In Table 6, a comparison among the different methods can be found where the MCGM (1,1) show that the MCGM (1,1) combined performance (MPE and MAPE) exceed the other models.

The application of the combined the GM model (1,1) and Markov Chain for demand forecasting in a 3PL warehousing company has the enormous advantage, the GM model

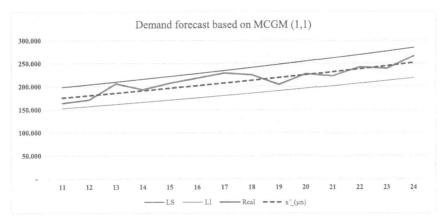

Fig. 3. Demand forecast range based on MCGM (1,1) model

Table 6. Forecast performance metrics comparison.

	MCGM (1,1)	MA (2)	MA (3)	Linear regression
MPE	1.43%	−2.74%	1.69%	−0.33%
MAPE	5.19%	5.49%	3.37%	4.88%

(1,1) alone provide a high level of accuracy and together with the Markov chain, it provides a forecast range application, with different levels of greyness for the decision-making process. This model it is easy to apply, it can include the user experience by replacing the transition matrix function, to provide a forecast range that include relevant to the user information from the organization, It has the advantage that can be compared with other forecast methods, such ARIMA but with less level of complexity. The mathematical requirement of this model is minimal.

The quality of the results can be further improved, depending on a) the user experience and b) the among of training data. The advantage of a highly accurate forecasting model GM (1,1), which is characterized by the small amount of data and the level of user experience, expressed through the transition matrix through the Markov Chains, which allow them to integrate their experience, all this without having to perform complex mathematical calculations.

Finally, given the nature of the data generated over time it is broadly categorized as stationary, seasonal, trending, and random. The type of data considered in this paper proposes a forecasting method considering that these cannot be stationary.

7 Conclusions and Future Research

The application of the combined GM model (1,1) and Markov Chain for demand forecasting in a 3PL warehousing company has the enormous advantage that, it is easy to apply, it includes the user experience of the organization, it has the advantage vs. ARIMA since it provides a lower range of greyness or uncertainty, it saves time in calculations and allows transition matrix functions or impulse functions to assess different inference systems. The mathematical requirement of these is minimal.

The results obtained in this paper could be comparable to those obtained when GM (1,1) is applied on kernel and degree of greyness [17], a such: "applicable to predictions for known possibility functions, and the latter mainly to predictions for interval grey numbers with unknown possibility functions".

The proposed model can provide simulation and predictions where the user experience is significant to the of problems and where the possibility function is known, in this case represented as the transition probability matrix.

Finally, the grade of greyness can be measured, compared to other forecasting methods, such ARIMA, where the results accuracy is greater.

There are studies on uncertain linguistic group decision making methods, for instance focus on group consensus [18], aggregation operators [19], where the influence of uncertain information is ignored, or information to expert weights, and calculations are complex, other where the degree of greyness is applied [20]. This is where expressing uncertain information as a degree of greyness is proposed to express this specific user experience in the process or system. Thus, generation of expert weights based on the degree of greyness, where the larger degree of greyness means more uncertainty, could be used to generate further simulations. A translation of the MCGM (1,1) to express the degree of uncertainty or greyness would represent the next step in this research.

References

1. Surowiecki, J.: The wisdowm of crowds, Anchor books (2005)
2. Badings, T., Simão, T.D., Suilen, M., Jansen, N.: Decision-making under uncertainty: beyond probabilities: challenges and perspectives. Int. J. Softw. Tools Technol. Transfer **25**(3), 375–391 (2023)
3. Wang, X., Hyndman, R.J., Li, F., Kang, Y.: Forecast combinations: an over 50-year review. Int. J. Forecast. **39**(4), 1518–1547 (2023)
4. Mierzwiak, R., Xie, N., Dong, W.: Classification of research problems in grey system theory based on grey space concept. J. Grey Syst. **31**(1), 100–111 (2019)
5. Lin, Y., Chen, M.-Y., Liu, S.: Theory of grey systems: capturing uncertainties of grey information. Kybernetes **33**(2), 196–218 (2004)
6. Lin, L.-C., Wu, S.-Y.: Analyzing Taiwan IC assembly industry by grey-markov forecasting model. Math. Probl. Eng, **2013**, 1–6 (2013). https://doi.org/10.1155/2013/658630
7. He, Y., Huang, M.: A grey-markov forecasting model for the electric power requirement in China. In: Gelbukh, A., de Albornoz, Á., Terashima-Marín, H. (eds.) MICAI 2005: Advances in Artificial Intelligence, pp. 574–582. Springer Berlin Heidelberg, Berlin, Heidelberg (2005). https://doi.org/10.1007/11579427_58
8. Mi, J., Fan, L., Duan, X., Qiu, Y.: Short-term power load forecasting method based on improved exponential smoothing grey model. Math. Probl. Eng. **2018**, 1–11 (2018). https://doi.org/10.1155/2018/3894723
9. Zhan-li, M., Jin-hua, S.: Application of grey-markov model in forecasting fire accidents. Procedia Eng. **11**, 314–318 (2011)
10. Li, X., Wang, X., Shao, W., et al.: Forecast of flood in Chaohu lake basin of China based on grey-Markov theory. Chin. Geograph. Sci. **17**, 64–68 (2007)

11. Li, H., et al.: Water demand prediction of grey markov model based on GM(1, 1). In: 3rd International Conference on Mechatronics and Information Technology, p. 6 (2016)
12. Chen, L.H., Guo, T.Y.: Forecasting financial crises for an enterprise by using the Grey Markov forecasting model. Qual. Quant. **45**, 911–922 (2011)
13. Hu, Y.-C.: Predicting foreign tourists for the tourism industry using soft computing-based Grey–Markov models. Sustainability **9**(7), 1228 (2017)
14. Haggstrom, O.: Finite Markov Chains and Algorithmic Applications. Cambridge University Press, Melbourne (2002)
15. Xu, Y., Lin, T., Du, P.: A hybrid coal prediction model based on grey Markov optimized by GWO – A case study of Hebei province in China. Expert Syst. Appl. **235**, 1–16 (2024)
16. Dickey, D.A., Fuller, W.A.: Distributions of the estimators for autoregressive time series with a unit root. J. Am. Stat. Assoc. **74**(366), 427–431 (1979)
17. Shu, H., Xiong, P., Chen, S.: A Novel EGM(1,1) model based on kernel and degree of greyness and its application on smog prediction. J. Grey Syst. **32**(4), 1–15 (2020)
18. Zhang, Z., Guo, C.: A method for multi-granularity uncertain linguistic group decision making with incomplete weight information. Knowl.-Based Syst. **26**, 111–119 (2012)
19. Xian, S.D., Sun, W.J., Xu, S.H., Gao, Y.Y.: Fuzzy linguistic induced OWA Minkowski distance operator and its application in group decision making. Pattern Anal. Appl. **19**(2), 325–335 (2016)
20. Ma, Z.Z., Zhu, J.: An uncertain pure linguistic approach on evaluation of enterprise integrity based on grey information. Grey Syst.: Theory Appl. **6**(3), 353–364 (2016). https://doi.org/10.1108/GS-09-2015-0050

Integration of Alexa and Social IoT for Generation X: A Study on the Optimization Use of Smart Treadmill to Improve Physical Health

Alberto Ochoa-Zezzatti, José De los Santos(✉), Ángel Ortíz, Maylin Hernández, and Joshuar Reyes

Universidad Autónoma de Ciudad Juárez (UACJ), Juárez, Mexico
AI212177@alumnos.uacj.mx

Abstract. The Internet of Things (IoT) is an emerging technology with a significant impact on social and technological domains. Its focus is on the interconnectivity between electronic devices for data transfer and retention through wireless networks, with minimal human intervention, under the premise that any device with these characteristics is intelligent or autonomous. People's interest in acquiring technological advancements that make life more efficient is increasing daily. Among these advancements are virtual assistants like Alexa, Siri, and ChatGPT, which have gained relevance in contemporary times by offering intuitive and wireless control over a wide range of connected devices. Although these advancements have a considerable market among current generations, there is also a promising field of exploration in Generation X. Due to their lifestyle, the help provided by IoT and virtual assistants can be crucial to improving their well-being. In this article, we will explore how the integration of voice commands and IoT capabilities can maximize the use of a treadmill. Sensors will be added to initiate a prototype capable of capturing variables such as heart rate and blood oxygenation, with the future implementation of a virtual assistant backed by artificial intelligence to achieve optimal personalization aimed at promoting a healthy lifestyle. The creation of this device not only focuses on facilitating the manipulation and personalization of workout routines but also seeks to generate economic benefits through monetary comparisons with a conventional gym. Furthermore, it will focus on a smooth, feedback-driven adaptation for optimal integration into the user's daily life, thus improving habits and quality of life.

Keywords: Internet of Things (IoT) · Voice Assistants · Alexa · Smart Home · Generation X · Home Automation · Energy Efficiency

1 Introduction

As contemporary technology has become essential in the daily routine of the average person, it has brought with it a multitude of advantages. Among these are easier access to specialized information, the promotion of globalization among communities, and the simplification of daily tasks, which has led to a notable increase in productivity and efficiency. However, along with these positive aspects, there are distinctive characteristics that have drawn attention for their potential negative impact, such as technological dependency and the promotion of sedentary lifestyles [1]. It is widely known that most Mexicans lead sedentary lives. Only 39.8% of the population over 18 years of age engages in any type of physical activity, placing this percentage among the lowest recorded in Mexico since the start of censuses measuring this social variable. As seen in Fig. 1, obtained from the National Institute of Statistics and Geography (INEGI), the results tend to decline over the years [2, 3].

Fig. 1. Active population in Mexico 2023.

When evaluating physical activity more precisely, according to the level of sufficiency, it was found that 23.6% of the population aged 18 and older engages in physical activity at a sufficient level to obtain health benefits, while 15.3% does so insufficiently. The population that does not engage in any physical activity reaches a worrying 60.2%, as indicated in Fig. 2 [3].

This data is a major cause for concern, as a sedentary lifestyle is one of the leading risk factors for mortality, being associated with non-communicable chronic diseases such as diabetes, hypertension, osteoporosis, and cancer. Sedentary behavior is not only due to technological dependency but also to factors such as the lack of green spaces, pollution, economic issues in specific cases, and lack of time [4]. Generation X especially faces problems like obesity and sedentary lifestyles due to the nature of their daily activities, predominantly in offices or jobs that limit mobility. Additionally, the time available for physical activity is often used to rest after a long workday [5]. Therefore, in this document, a treadmill powered by the Internet of Things (IoT) will be developed to promote the physical well-being of people from these generations. This device will incorporate components such as AD8232 sensors for heart rate monitoring and the MAX30102 as an oximeter. All this data will be crucial, as thanks to a voice assistant powered by artificial intelligence, an exercise routine can be personalized according to the user's individual needs. Additionally, it will be justified that this treadmill will be more economical than a traditional gym subscription, and the possibility of integrating the detection of cardiac irregularities, such as cardiac arrest, will be explored, thus increasing user safety during exercise.

Fig. 2. Effectively active population in Mexico 2023.

2 Objective of Our Research

The relevance of addressing this issue lies in the need to develop an intelligent system that integrates the collaboration of a virtual assistant (such as Alexa, Chat-GPT, among others) along with Internet of Things (IoT) capabilities in a treadmill, in order to optimize its use and promote healthy habits in Generation X. This approach is essential for personalizing exercise routines through artificial intelligence, tailoring them to the individual needs of users. This research also aims to justify the economic viability of this system compared to gym subscriptions. Ultimately, the goal is for this system to contribute to reducing sedentary lifestyles in Mexico, thereby improving the overall quality of life for the population.

3 Benefits to Generation X

The implementation of a smart treadmill supported by the Internet of Things (IoT) and virtual assistants offers several key benefits for Generation X, who often face specific challenges related to time, health, and the adoption of new technologies.

- **Adaptation to Busy Schedules:** Generation X often manages strict schedules due to work and family commitments. This device allows them to exercise at home at the time that suits them best, eliminating the need to commute to a gym. Additionally, virtual assistants can schedule personalized workouts and send reminders, helping to smoothly integrate physical activity into their daily routine.
- **Personalized Health Monitoring:** Since people in this generation are at an age where health issues such as heart disease, hypertension, and diabetes are common, the smart treadmill becomes a crucial tool. The integrated sensors enable real-time

monitoring of key indicators like heart rate and blood oxygen levels. These data allow for personalized exercise routines to improve the user's fitness without risking their health.
- **Prevention of Chronic Diseases:** Sedentary behavior is a major risk factor for developing chronic diseases. Continuous use of the treadmill, along with automatic adjustments based on health data, not only promotes regular physical activity but also contributes to the prevention and control of conditions such as hypertension, high cholesterol, and obesity, which are prevalent in Generation X.
- **Technological Ease:** Although this generation did not grow up with the same exposure to technology as younger generations, the treadmill offers a user-friendly interface through voice commands with virtual assistants like Alexa. This removes technological barriers, allowing any user to interact with the device intuitively and personally, without needing advanced knowledge.
- **Economic Savings:** By opting for a smart treadmill instead of a gym membership, users can achieve long-term economic benefits. While the initial investment may be significant, the savings on gym fees, transportation, and time justify this approach. Additionally, constant health monitoring can help avoid costly medical issues in the future, creating a positive impact on both the user's health and finances.

3.1 Economic Justification

The cost of a gym membership in Mexico varies depending on the type of establishment. For example, an annual subscription to premium gyms like Sport City can range from $8,000 to $15,000 MXN per year [6]. These gyms offer additional services such as group classes, access to multiple locations, and personalized coaching, but they also represent a high cost for those who do not fully utilize these services. On the other hand, low-cost gyms like Smart Fit charge approximately $4,788 to $7,200 MXN annually [7]. Although more affordable, users must travel to the physical location of the gym, which may not be practical, especially for Generation X individuals with long workdays. In contrast, while the initial investment in a smart treadmill may seem high, it is amortized over time. For example, a treadmill equipped with sensors, IoT, and a virtual assistant could cost between $6,000 and $8,000 MXN, depending on the features included. However, this is a one-time investment. Unlike an annual gym membership, there would be no need to pay recurring fees, and the user can exercise at home without the additional costs of transportation or extra charges for specific services. Moreover, connected treadmills allow for personalized workout routines based on the user's needs, improving training efficiency without requiring constant personal trainer assistance. In the long term, this device can be more economical and convenient, especially for those who value the flexibility of exercising at home and avoiding gym commutes. This proposal promotes the efficient use of time and economic resources, making the smart treadmill a more cost-effective solution in the medium and long term compared to an annual membership to a traditional gym.

3.2 Long-Term Cost Comparison

The following is a comparison of the accumulated cost of a smart treadmill versus a membership to a low-cost and premium gym over five years. The prototype of the treadmill represents an initial investment of $6,000 MXN, while gym memberships involve recurring annual payments. As shown in Fig. 3, the treadmill, being a one-time investment, becomes significantly more economical after the first few years, making it a cost-effective option for those seeking a healthy lifestyle without the long-term financial commitments of a gym subscription.

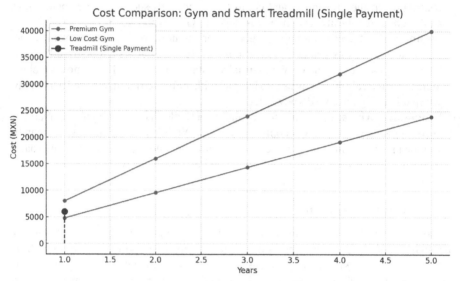

Fig. 3. Economic comparison between gym and prototype IoT treadmill.

4 Development

The development of the integration between Alexa and Social IoT for optimizing treadmill use among Generation X focuses on the seamless interaction between users and smart devices. By utilizing voice commands, this research aims to enhance the experience of treadmill users, specifically targeting the technological engagement of Generation X, a demographic known for being less familiar with cutting-edge technology than younger generations. Alexa, when integrated with IoT, allows for effortless control of the treadmill and other smart devices in the user's environment, fostering a more intuitive exercise routine. Sensors embedded in the treadmill monitor users' performance and health metrics in real-time, while Alexa's AI-driven capabilities offer personalized workout routines, track progress, and provide motivation based on the individual's preferences. This interconnection exemplifies the future of fitness, where the treadmill becomes an intelligent and interactive part of the smart home ecosystem. In addition to improving the user experience, the integration of IoT with voice assistants like Alexa highlights the potential economic and health benefits for Generation X. With customized workout plans and real-time feedback, users can avoid the costs associated with gym memberships, while still maintaining a personalized fitness routine in the comfort of their homes. This approach not only enhances accessibility to fitness for those with busier lifestyles but also ensures continuous improvement in health monitoring. The treadmill's adaptive features, coupled with Alexa's integration, support a more sustainable fitness journey, ensuring that users can remain engaged and motivated in their routines. As the research progresses, the primary focus is to refine this technology to be user-friendly, cost-effective, and sustainable for long-term health benefits.

4.1 Conceptual Diagram

Figure 4 provides an overview of the system designed to optimize the user experience of the smart treadmill, specifically targeting Generation X. This system integrates a series of advanced sensors that allow continuous monitoring of key health variables, such as heart rate and blood oxygen levels. These data are stored and processed in real time through a central monitoring system, enabling the immediate personalization of exercise routines. One of the key features of the system is the inclusion of a virtual assistant, in this case, Amazon Alexa, which acts as the brain behind personalization. Using artificial intelligence algorithms, Alexa analyzes the user's health data to dynamically and automatically adjust the intensity and duration of exercise sessions. This continuous interaction ensures that routines are both safe and effective, adapting not only to the user's immediate needs but also to their long-term health goals. Finally, the economic and long-term benefits offered by the smart treadmill are highlighted when compared to traditional gym memberships. As illustrated, the user-centered design and the use of cloud-based data provide a comprehensive and connected approach to personal well-being, ensuring tangible health improvements over time.

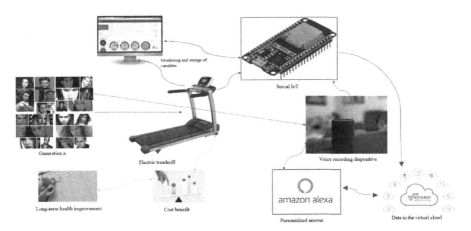

Fig. 4. Conceptual diagram of the treadmill.

4.2 Main Components of the System

I. **Health Monitoring Sensors:**

- *MAX30102:* This optical sensor employs photoplethysmography (PPG) technology to measure heart rate (BPM) and blood oxygenation (SpO2). Through LEDs and a photodetector, it captures variations in light reflected by blood flow, providing accurate indicators of cardiovascular and respiratory health [8, 9].
- *AD8232:* Electrocardiogram (ECG) sensor designed to monitor the electrical activity of the heart. It is essential for detailed analysis of heart rhythm and the identification of possible abnormalities in the heart's electrical signals [10].

II. **Processor and Communication:**

- *ESP32:* Microcontroller that manages the input from sensors. It functions as a node in an IoT network, responsible for data collection, control logic execution, and communication with other devices in the system.

III. **User Interface:**

- *Oled Display*: Provides real-time visualization of vital metrics such as heart rate and blood oxygenation. It offers a clear and accessible interface for user interaction.
- *Alexa:* Integrated via Amazon Web Services (AWS), this voice assistant facilitates interaction by providing personalized feedback and alerts about anomalies based on collected health data.

IV. **Connectivity and Remote Monitoring:**

- *AWS (Amazon Web Services):* This cloud platform receives, processes, and stores the data collected by the system. Integration with AWS not only enables long-term data storage but also facilitates advanced analysis and the personalization of exercise routines through Alexa.

4.3 Connecting the Main Components of the System

The following table details the distribution and specific allocation of the main components in our smart treadmill system. This table organizes each component according to its location and function within the system, providing a clear overview of how each element contributes to the overall integration and efficiency of the project (Table 1).

4.4 Signal Discretization

Sampling, also known as "Signal Discretization," is the first step in the process of converting an analog signal (continuous time and amplitude) into a digital signal (discrete time and amplitude) [11]. The need to convert an analog signal to digital has an essential justification, especially for its role in the integration and functionality of modern electronic systems that we use in our daily lives and various industries. Devices such as cell phones and computers are capable of capturing analog signals, but processing

Table 1. Component connections.

COMPONENT	COMPONENT PIN	PIN DESCRIPTION	PIN ON ESP32
MAX30102	SDA	I2C Serial Data	GPIO 21
	SCL	I2C Serial Clock	GPIO 22
	INT	Interrupt (not used in current code)	Not connected
AD8232	OUT	ECG Signal Output	GPIO 34
	LO+	Electrode disconnection detection (positive)	Not used in current code
	LO-	Electrode disconnection detection (negative)	Not used in current code
	GND	Ground	GND
	VCC	Power supply	3.3 V
DISPLAY OLED	SDA	I2CSerialData	GPIO 21
	SCL	I2C Serial Clock	GPIO 22
	VCC	Power supply	3.3 V
	GND	Ground	GND

them directly is a complex task due to the continuous nature of these signals [12]. Analog signals have an infinite number of possible values within a given range, which can make processing computationally intensive for digital systems [13]. Signal discretization emerges as a solution to this problem. This process involves taking samples at defined intervals from the continuous signal, converting it into a digital version that can be interpreted by electronic systems without losing critical information. This facilitates signal

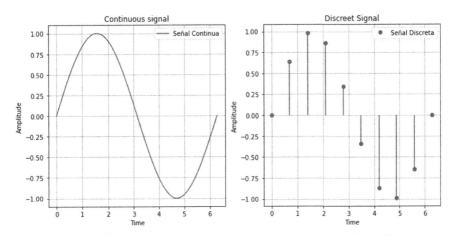

Fig. 5. Comparison between a continuous signal and a discrete signal.

processing, allowing devices to perform their functions more efficiently and accurately [14] (Fig. 5).

Once we have a discretized signal, the next step is quantization, which is the process by which the continuous values of a sinusoidal wave are converted into a series of discrete decimal numeric values corresponding to the different levels or voltage variations contained in the original analog signal [15, 16]. This allows us to assign corresponding values in the decimal number system, paving the way for conversion to binary (Fig. 6).

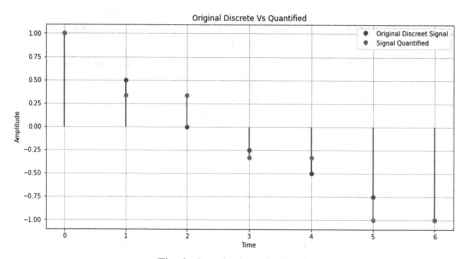

Fig. 6. Quantization of a signal.

A clear example of this would be an electrocardiogram (ECG). Sensors like the AD8232 capture the mechanical signals produced by cardiac activity and convert them into electrical signals. These electrical signals, which are still analog, represent the continuous behavior of the heart over time. However, for them to be processed by electronic systems, these signals must be digitized. The process of converting an analog signal to digital involves several steps: first, the sensor detects the analog signal, and then an analog-to-digital converter (ADC) samples this continuous signal at predefined time intervals. These samples are quantified and converted into a sequence of digital data that can be stored, analyzed, and used for decision-making, such as customizing exercise routines or detecting cardiac irregularities in real-time [17, 18] (Fig. 7).

This is highly relevant to understand for our system, as we are working with a very small voltage range, and disturbances within our signals are constant. Understanding these processes helps us as developers implement filters, reducing reading errors and opening up new possibilities.

4.5 Communication Protocols

The prototype integrates a real-time communication flow designed to monitor the user's health parameters and offer personalized responses through interaction with Alexa. This

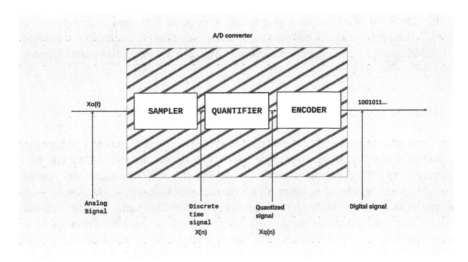

Fig. 7. Analog to digital converter.

process begins with the collection of health data via sensors connected to an ESP32 microcontroller. The monitored parameters include heart rate and oxygen levels, which are sent in real-time to the Amazon Web Services (AWS) platform using the MQTT protocol [19, 20]. Once in the cloud, the data is published to a specific AWS IoT topic, making the information accessible to other cloud services. The user can interact with the system through voice commands to Alexa, such as "Alexa, how are my heart parameters?" This request triggers an Alexa skill, which in turn triggers an AWS Lambda function designed to retrieve the user's most recent data from AWS IoT. The Lambda function processes the received information and generates a personalized response based on the user's current health status. This response is then transmitted back to the Alexa skill, which communicates the information to the user through an Alexa device, providing

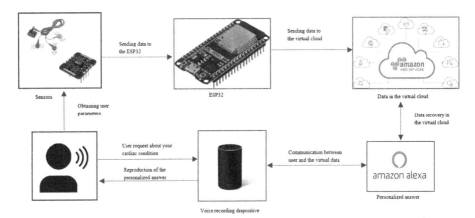

Fig. 8. Communication between devices.

details such as "Your current heart rate is 150 beats per minute, and your oxygen level is 90%." This system enables constant monitoring and delivers direct, tailored feedback to the user, optimizing the experience during physical activities and promoting greater health awareness (Fig. 8).

5 Results

In this first prototype, significant progress was made, such as obtaining key biometric parameters, including oxygen saturation (SpO2), beats per minute (BPM), and basic sampling of an electrocardiogram (ECG). Additionally, the interconnectivity between these sensors and the virtual assistant Alexa was successfully established, allowing Alexa not only to monitor but also to report in real-time on the user's various health variables. This advancement represents an important step toward the personalization of health monitoring through the integration of IoT and intelligent voice assistants (Figs. 9 and 10).

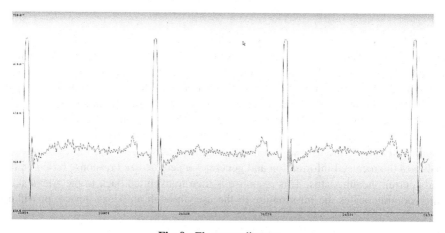

Fig. 9. Electrocardiogram.

5.1 Prototype Reaction Levels

A reaction level system is also integrated, which serves as a prelude to the final functionality of the prototype, aimed at managing the user's physical conditions in an automated and progressive manner. As shown in Fig. 11, this system is broken down into four steps.

This structure allows the system to be proactive, ensuring the user's safety during exercise and optimizing their experience based on their real-time physical conditions.

Fig. 10. Oximeter functioning

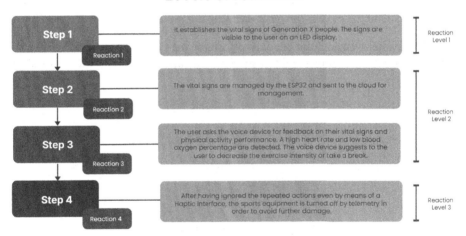

Fig. 11. Prototype reaction levels.

6 Conclusions

In this work, we have developed an innovative system specifically designed to improve the quality of life of Generation X. This system facilitates physical activity from the comfort of home, directly addressing the issue of sedentary behavior prevalent in this demographic group due to their demanding daily routines. Our prototype promotes a more active lifestyle and aims to mitigate the risk of chronic diseases associated with sedentary habits. The implementation of this system represents a significant step toward proactive health support, offering a practical solution that adapts to the time and space limitations often faced by individuals of this generation. Additionally, in cases where chronic diseases are already present, the system is designed to act preventively and therapeutically in the future, adjusting the recommended activities to the specific needs of each user. Looking to the future, this prototype has the potential to serve as a catalyst for positive changes, not only for Generation X but also as a model for future interventions aimed at other demographic groups. By integrating technology into people's daily routines, we are paving the way for healthier and more active aging.

7 Future Research

For future iterations, the complete implementation of the automation proposed in this document is anticipated. The goal is to develop Alexa's ability to adjust users' routines based on their current states. Additionally, users will be able to modify their own routines through direct voice commands. On the technical side, advanced mathematical models are planned for the development of digital filters. This implementation will enable the system to detect cardiac anomalies, thus contributing to the prevention of critical incidents such as cardiac arrest (the system would call emergency services if necessary). The analysis of discretized signals mentioned earlier will serve as the foundation for these developments. Lastly, the sensing of various physiological and environmental parameters will be explored to enhance the user experience. This will include monitoring hydration levels, sleep hours, and other key aspects related to the user's safety and well-being. These improvements aim not only to increase the system's functionality but also to enhance its ability to provide personalized and proactive support to users, as shown in Fig. 12.

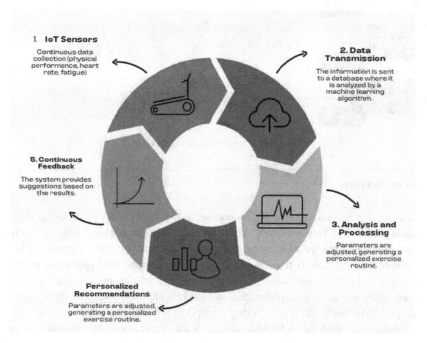

Fig. 12. Demonstrative diagram of functioning.

References

1. Universidad Anáhuac: La influencia de la tecnología en nuestra vida cotidiana (2019). Recuperado de https://www.anahuac.mx/generacion-anahuac/la-influencia-de-la-tecnologia-en-nuestra-vida-cotidiana
2. Darinka, R.: México sedentario: solo 39% de la población hace ejercicio con regularidad (2024). Recuperado de https://elpais.com/mexico/2024-01-26/mexico-sedentario-solo-39-de-la-poblacion-hace-ejercicio-con-regularidad.html?event_log=regonetap
3. Instituto Nacional de Estadística y Geografía (INEGI): Módulo de práctica deportiva y ejercicio físico (MOPRADEF) 2023. Comunicado de prensa número 31/24 (2024). Recuperado de https://www.inegi.org.mx/contenidos/saladeprensa/boletines/2024/MOPRADEF/MOPRADEF2023.pdf
4. Secretaría de Salud: Qué es el sedentarismo? (2015). Recuperado de https://www.gob.mx/salud/es/articulos/que-es-sedentarismo#:~:text=El%20sedentarismo%20ocurre%20por%20diversos,al%20televisor%20o%20a%20la%20computadora
5. Genomawork: Cómo es la Generación X en el trabajo ¿Los tienes identificados? (2023). Recuperado de https://www.genoma.work/post/como-es-la-generacion-x-en-el-trabajo-habilidades-caracteristicas
6. Javier, T.: Cuánto cuesta la anualidad en Sport City: Precios actualizados (2024). Recuperado de https://unives.com.mx/cuanto-cuesta-la-anualidad-en-sport-city-precios-actualizados/?expand_article=1
7. Cristina, O.: Smart Fit se lleva más de 20% de las membresías en gimnasios en México (2024). Recuperado de https://www.milenio.com/negocios/smart-fit-tiene-mas-de-20-por-ciento-de-las-membresias-gimnasios
8. Luis, L.l.: Pulsímetro y oxímetro con Arduino y MAX30102 (2020). Recuperado de https://www.luisllamas.es/pulsimetro-y-oximetro-con-arduino-y-max30102/
9. Castellano, C., Perez de Juan, M.A., Attie, F.: Electrocardiografía clínica, segunda edición. Elsevier, España (2004)
10. Nilda, F., Rodrigo, V., Juan, H.: Registro de la actividad electromiografía con AD8232 (2018). Recuperado de https://www.boletin.upiita.ipn.mx/index.php/ciencia/791-cyt-numero-69/1601-registro-de-la-actividad-electromiografica-con-ad8232
11. Barco, R.: Discretización de señales (2013). Recuperado de http://lcr.uns.edu.ar/fvc/NotasDeAplicacion/FVC-RodrigoBarco.pdf
12. Ogata, K.: Sistemas de control en tiempo discreto, 2a ed., pp. 1–7 (1996)
13. Alan, V., Alan, S.: Señales y sistemas, 2a ed., pp. 534–543 (1997)
14. Antony, G.: Discretización de señales analógicas (2023). Recuperado de https://panamahitek.com/discretizacion-de-senales-analogicas/
15. Jimmy, C., Hugo, C., José, C.: Fundamentos y aplicaciones del muestreo en señales ubicadas en las bandas altas del espectro. Scientia et Technica Año XIV, No **39**, 37–38 (2008)
16. Ingrid, D.: Diferentes tipos de señales y ejemplos (2013). Recuperado de https://apuntes-de-analisis-de-sistemas.webnode.es/news/diferentes-tipos-de-senales-y-ejemplos/
17. Universidad Complutense Madrid. (s.f.). Cuantificación. Recuperado de https://www.ucm.es/innovasonora/cuantificacion
18. Vicente, G.: Muestreo de señales (2015). Recuperado de https://w3.ual.es/~vruiz/Docencia/Apuntes/Signals/Sampling/index.html

19. Alejandro, L.: Usando Amazon Lex y Alexa para controlar un coche de carreras en el AWS Summit Madrid 2022 (2022). Recuperado de https://ifgeekthen.nttdata.com/s/post/usando-amazon-lex-y-alexa-para-controlar-un-coche-de-carreras-en-el-aws-summit-m-MCJSD5IYZR4NBYNI536QPHNKRQH4?language=es
20. Solectro: ¿Qué es MQTT? El protocolo de comunicación IoT (2022). Recuperado de https://solectroshop.com/es/blog/que-es-mqtt-el-protocolo-de-comunicacion-para-iot-n117?srsltid=AfmBOooLADne13jnzGbUuZaM_cqgGepslf9t1ohUplrMpNO2tHTgzl8K

Logarithmic Weighted Random Selector Algorithm: A Novel Approach for Biasing Selection Based on Positional Order Without Hyperparameters

Iván Alejandro Ramos Herrera(✉)

Autonomous University of Aguascalientes, Aguascalientes, Mexico
arhcoder@gmail.com

Abstract. The *"Logarithmic Weighted Random Selector"* (*LWRS*) introduces a novel algorithm designed for selecting randomly one item from a list, with the bias that the further to the beginning of the list each one is, the more likely it is to be selected. Unlike traditional selection and sampling methods that rely on numerical fitness scores or require hyperparameter tuning, *LWRS* employs a logarithmic weighting mechanism to naturally favor items based on their position. It arose from the problem of requiring a method that allows weighting a selection based on the order of preference on a list of elements, for phenomena for which numerical weighting values are unknown but only their order, which unlike algorithms such as *Rank-based Selection*, does not require configuration of hyperparameters. Different mechanisms were explored, such as a *Linear Selection* method and *Exponential Selection*. Discovering that utilizing a logarithmic scale, *LWRS* achieves a non-numerical preference bias, distinguishing itself from classic methods of weighted random sampling as *Fitness-proportionate selection* and *Tournament Selection*, which do require defined numerical weights, contributing to the field with a straightforward, parameter-free approach to weighted random selection.

Keywords: Algorithm Design · Computer Science · Random Sampling · Randomized Algorithm · Selection Method · Evolutionary Computing

1 Introduction

When dealing with computer models that require selection mechanisms, probability-based algorithms are often employed [1]. These chance-based methods work like a roulette wheel that selects elements from a set, either with the same probability of selection for each element, or with a weight based on a fitness score; for example, the *Fitness-proportionate selection algorithm* used commonly in *Genetic algorithms* [2], which, following the allegory of a roulette, give more size of decision -higher probability-, based on a score in the population of elements. However, it could be necessary to give an importance bias to each element not based on a number, but on its preference or favoritism. *LWRS* addresses this case by introducing a logarithmic scale to provide a growth ratio for bias in the selection process.

This proposal aims to address the need of models that make decisions based on importance biases attributed to available options, for phenomena for which no numerical value is known that allows defining said importance but the order of favoritism between the options, considering that the decision mechanism should allow variety, so any element of the options can be selected, but exponentially favoring the elements that are higher up, without the need for extra hyperparameters for the selection configuration.

2 Motivation

The inspiration of this proposal arises from an examination of phenomena for which there are events with a greater frequency of being present, but for which the factor that determines this magnitude is not known, when the need arises to sample this population of options using selection methods. More specifically, *LWRS* originates an attempt to emulate music composition through the application of computational-generative algorithm with inherent randomness, which in its generation process requires making decisions about chords, pitch, tempo and harmony. In some cases, depending on musical style, genre, and era, distinct composition patterns emerge as more prevalent. For example, popular music exhibits a proclivity towards the *4/4-time* signature, simple major-minor chords, short chord progressions or repetitive rhythmic patterns [3]. Consequently, in attempting to replicate popular music composition with generative algorithms without classic *Machine learning* techniques, it is imperative to consider these characteristics as the most common, but without neglecting the fact that it is possible to have more variety of decisions while generating, emphasizing that there is no numerical value to measuring said distribution of selection favoritism.

The present proposal aims to create a decision mechanism that achieves weighted random selection with this top-favoritism bias, without the need for extra hyperparameters and variables to configure it.

3 Related Techniques

In the context of needing a selection mechanism based on randomness, there are different classical methods in state of art, of which those that offer a numerical bias and those that do not are distinguished.

3.1 Random Weighted Sampling Without Numerical Bias

When referring to the need for probability-based selection mechanisms, the concept aligns with *random sampling* techniques. In the literature, four main types are distinguished: *Systematic*, *Stratified*, *Cluster*, and *Simple Random Sampling* [4].

These methods seek to find a representative sample of a population but attempting to have an equitable selection for all subdivisions of the population. For example, the *Simple* random sampling technique takes elements from the population list in a uniformly random manner, such that any element has the same probability of being selected. On the other hand, the *Stratified* and *Cluster* random sampling methods seek to generate groups and biases from the population based on their characteristics, whether these are heterogeneous or homogeneous. In *Systematic* sampling, a starting point is randomly selected in the population list and subsequent elements are selected following a spacing rule given by a hyperparameter formed by the population size divided by the size of the sample to be selected [5].

Although it could be inferred that there is a bias in *Systematic Sampling* technique, since in an ordered population list the elements that tend to be in certain positions would be taken with higher probability, the truth is that this bias continues to be random to the extent that the starting point is.

It is important to clarify that *Simple Random Sampling*, *Stratified Random Sampling*, and other techniques such as *Multistage Random Sampling* and *Independent Sampling* are based on randomness and generally designed to minimize bias. However, the inherent design in *Multistage Random Sampling* can sometimes increase sampling error, but it is not directly due to bias in the selection process. The *Quota Sampling* technique in the other case involves bias in its selection because the second step of selecting units is non-random, but it is not based on randomness. All these techniques can be found defined and discussed in [5] and [6].

3.2 Random Weighted Sampling with Numerical Bias

When biases that favor elements are needed, *Genetic Algorithms* use various types of *Selection Algorithms*, since an intrinsic part of building these involves selecting elements from a population considering a *fitness* given to each element with respect to its performance in solving an optimization problem.

A common selection method used on *Genetic Algorithms* is *Fitness-proportionate selection*, which assigns a probability of selection based on the *fitness* of each element in the population, using an objective function to measure it. It achieves it computing the total fitness sum of a population and assigning to each element the probability using its fitness value divided by the computed total, thus, the greater the fitness of an element, the greater its probability of being selected.

Another common algorithm is the *Tournament Selection* algorithm, which selects individuals by running "tournaments" among a few randomly chosen elements from the population. The element with the highest fitness in each tournament is chosen for the next generation. This method introduces a controlled amount of selection pressure by adjusting the tournament size: larger tournaments increase the likelihood of selecting higher-fitness individuals, whereas smaller tournaments introduce more diversity by allowing lower-fitness elements a chance to be selected.

Additionally, the *Rank-based selection* method [7] sorts the entire population by fitness and assigns selection probabilities based on this rank rather than the absolute fitness values. This algorithm handles the posed question of selecting elements based on a ranking order, just like the proposal in this work, however, it requires the definition of the hyperparameter $\beta rank$ to control the selection pressure by determining how much more likely the best individual is to be selected compared to the worst. Moreover, $\alpha rank$ parameter is needed to ensure the selection probabilities sum to 1 and balance the selection process. Other key hyperparameters include α and β for adjusting the probability distribution, and μ and λ for balancing parent and offspring populations. Finally, the threshold parameter T determines the fraction of individuals selected for reproduction, impacting selection dynamics and algorithm performance.

All these analyzed selection methods [8] offer selection bias, but they require a numerical value that defines said bias, in addition to requiring the extra configuration of hyperparameters. In this context is where the contribution of the *LRWS* algorithm stands out, since it does not require hyperparameters for its configuration and the bias is given by an exponential form assigned to the elements according just to their order of preference.

4 Methodology

To devise a solution to the problem posed, the following steps were followed, as an adaptation of the steps of *"How to solve a problem"* from the book in reference [9].

1. Understanding and establishment of the problem conditions: where it was delimited to devise an algorithm that, without defined numerical weight values, manages to weigh a random selection based on the order of preference.
2. Analysis of similar algorithms: where the classical methods for performing weightings with numerical and non-numerical biases were analyzed.
3. Discovery of weighting mechanisms: finding linear, exponential and logarithmic mechanisms.
4. Implementation: mechanisms implemented into random selection algorithms.
5. Experimentation and results: where having the algorithms, they were subjected to selection repetitions with a defined list of elements, graphing the selection preference of each algorithm.
6. Conclusions and discussion: where it was found that the logarithmic mechanism best solves the problem posed, offering a selection with bias based on the order of importance and without the need for hyperparameter configuration.

5 Perform

LWRS "constructs" a logarithmic scale based on the number of objects to be selected and calculate the distances. As example, consider a list of elements: $S = [A, B, C, D, E]$. S contains the elements e_i that represent objects to select, and in turn, the position i in the list represents the favoritism/preference index for each object e_i, being that A is the object with the greatest importance of being chosen, then B, then C, ..., etc. With $n = 5$ being the number of objects in S. *LWRS* divides the space of the scale into n sectors, following the growth ratio based on exponential mathematical nature [10].

A representation of what is the difference between linear and logarithmic scales looks like is shown in the Fig. 1, and taking it as an example, the difference between *LWRS* decision algorithm and a *Simple Random Sampling* selection step can be understood as placing a point randomly in any of the two scales, where the space between two divisions represents an object to select. For the case of the linear scale, the probability of finding the object between 0 and 1 is the same as that of the object between 1 and 2, or 2 and 3 -like a normal roulette wheel-. On the other hand, with the logarithmic scale,

Linear scale

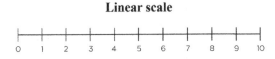

Logarithmic scale

Fig. 1. Comparison between linear and logarithmic scale.

the probability of falling into the object between 0 and 1 is noticeably different from that of the object between 5 and 6.

Figure 2 shows how the selection with *LWRS* algorithm's scale looks compared to *Simple Random Sampling* selection step (not-weighted roulette wheel selection). Throwing a random point in this space results in obtaining decisions for both of methods:

– C element for a *Random Sampling* selection step.
– B element for the *LWRS* algorithm.

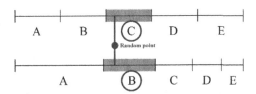

Fig. 2. Selection example for five elements on linear and logarithmic scales.

6 Formulation

The way to build a logarithmic scale adapted to the n quantity of objects to be selected is possible through simulating a $\mathbb{P}$ set of points on a Cartesian axis in which each point delimits the selection space of each of the n elements on the list, using the formula (1):

$$\mathbb{P} = \{p_i : p_i = \left(n \log_{n+1}(i+1), 0\right), n, i \in \mathbb{N}, 1 \leq i \leq n\} \tag{1}$$

where:

– p_i is (x, y) coordinate.
– n is the number of items.
– i is the index of point.

That is, given the same case of building the scale on a Cartesian axis, with n number of spaces (objects) and therefore n number of division points, the distance between each point i and the origin (coordinate (0, 0)) is given by the formula (2):

$$d_i = n \log_{n+1}(i+1) \tag{2}$$

To build a logarithmic scale of five objects ($n = 5$):

$$p_1 = \left(5\log_{5+1}(1+1), 0\right) = (\mathbf{1.934}, \mathbf{0}) \tag{3}$$

$$p_2 = \left(5\log_{5+1}(2+1), 0\right) = (\mathbf{3.066}, \mathbf{0}) \tag{4}$$

$$p_3 = \left(5\log_{5+1}(3+1), 0\right) = (\mathbf{3.869}, \mathbf{0}) \tag{5}$$

$$p_4 = \left(5\log_{5+1}(4+1), 0\right) = (\mathbf{4.491}, \mathbf{0}) \tag{6}$$

$$p_5 = \left(5\log_{5+1}(5+1), 0\right) = (\mathbf{5}, \mathbf{0}) \tag{7}$$

Logarithmic scale for n = 5

Fig. 3. Five spaces logarithmic scale graphing.

The use of the Cartesian axis is unnecessary for the selection algorithm, since it is only necessary to know the distance between the origin and any other point on the scale; so only the distance formula (2) will be necessary. Figure 3 shows the distances of each point on the scale.

7 Algorithm

```
Algorithm 1 lwrs(list objects)
1.  n ← objects.length()
2.  if n = 0 then
3.      return "No objects"
4.  end if
5.  if n = 1 then
6.      return objects[0]
7.  end if
8.  randompoint ← random(0, n)
9.  for i ← 1 to n do
10.     if randompoint ≤ n·log_{n+1}(i+1) then
11.         return objects[i-1]
12.     end if
13. end for
```

Within the loop on the pseudocode of the Algorithm 1, the distance from the *"origin"* to the points of the logarithmic scale p_i are calculated using the same analogy of a Cartesian plane, with the formula (2) described above. If a throwed random point is between 0 and the distance of the first point of the scale; that is, if the random point was before p_1 (random point $\leq$ distance to p_1), then it selects the first object. In case the random point is next to this first scale point, it continues with the loop until finding the space in which the random point is.

This algorithm is shown to be an optimal alternative in terms of execution time complexity. Its complexity in the best case is $\Omega(1)$ (using *Big-O* notation), since it could finish its execution only in the first iteration, in which it selects the first element of the list. In the worst case its complexity is $O(n)$, a case in which the last element of the list was selected (n). Also due to the nature of the probability described above and given as already stated by the form of an *exponential curve*, it is known that as much as the algorithm will tend to select the elements at the top of the list, it will also tend to finish in fewer iterations, so that the same form of the exponential selection curve gives us an indication that the average case of complexity is close to $\Theta(\log(n))$.

8 Experiments

Doing count tests for the decisions made by the algorithm to check the bias in the choices, repeating the selection making a certain number of times and counting how many times each of the objects was chosen, was graphed. In the following Figs. 4, 5 and 6 results of the experiments are shown.

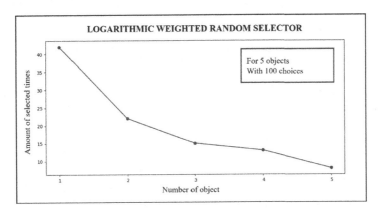

Fig. 4. Experiment with 5 objects and 100 repetitions.

It can be verified in Figs. 4, 5 and 6 that the shape of a logarithm curve is obtained by counting the selections of each element, and it is noted that the *LWRS* algorithm achieves to weight the elements with a bias based on their positional order, giving less weight with respect to the closer each element is to the end in the list and greater weight the closer it is to the top, without depending on hyperparameters.

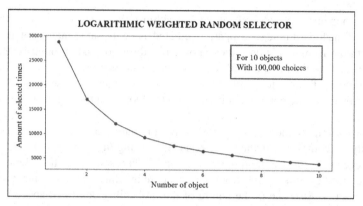

Fig. 5. Experiment with 10 objects and 100,000 repetitions.

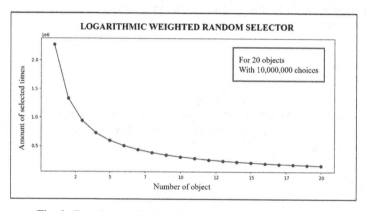

Fig. 6. Experiment with 20 objects and 10,000,000 repetitions.

In theoretical terms, no element has zero probability of being selected, just in possible practical cases in which a calculation engine cannot compute distances with sufficient decimal precision to be differentiated from the distance of other elements, in lists of sufficient size for this, and for the random numbers' generation on this experiments, *"random"* library of the *64-bit Python v.3.11* core was used.

For a particular case requiring equal selection probabilities among two or more objects, but not for all in the list, it is feasible to treat these objects collectively as a single entity element. When selected by the *LWRS* algorithm, the grouped objects can then be subjected to a simple roulette-type decision, such as the linear scale, to determine the specific chosen one.

9 Alternatives

There are two additional proposals for selection based on the order of importance of the elements according to their position in a list.

9.1 Linear Weighted Selection

The importance weight of each element is given as the result of the inverse index of the element in the list divided by the sum of indexes of the list:

$$p = \frac{w_i}{\sum_{j=1}^{n} w_j} \qquad (8)$$

where:

- n is the total number of elements in the list.
- w_i is the weight assigned to the element i of the list (given by $n - i + 1$).
- $\sum_{j=1}^{n} w_j$ is the total sum of all the weights.

```
Algorithm 2 linear(list objects)
1.   n ← objects.length()
2.   weights ← [n, n-1, …, 1]
3.   weights_sum ← sum(weights)
4.   probabilities ← [wᵢ ÷ weights_sum for wᵢ in weights]
5.   randompoint ← random(0, n)
6.   cumulative ← 0
7.   for i ← 1 to n do
8.       cumulative ← cumulative + probabilities[i]
9.       if cumulative ≥ randompoint then
10.          return objects[i]
11.      end if
12.  end for
```

Pseudocode of Linear Weighted Selection Algorithm is shown on Algorithm 2.

9.2 Exponential Weighted Selection

The same as *Linear Weighted Selection*, with the difference that the weight of each object is given by the result of the inverse index of the element in the list raised to a power x divided by the sum of indexes of the list:

$$p_i = \frac{(w_i)^x}{\sum_{j=1}^{n} w_j} \qquad (9)$$

where:

- x is a definite value of power.

Pseudocode of Exponential Weighted Selection is shown on Algorithm 3.

Algorithm 3 exponential(list objects, int x)
1. n ← objects.length()
2. weights ← [n, n-1, ..., a 1]
3. weights ← power(weights, x)
4. weights_sum ← sum(weights)
5. probabilities ← [w_i ÷ weights_sum **for** w_i **in** weights]
6. randompoint ← random(0, n)
7. cumulative ← 0
8. **for** i ← 1 **to** n **do**
9.   cumulative ← cumulative + probabilities[i]
10.   **if** cumulative ≥ randompoint **then**
11.   **return** objects[i]
12.   **end if**
13. **end for**

Figure 7 shows the result of the different experiments with the exponential and linear weighted selector, where each different exponent value shows an exponential nature in the selection preference, so that this alternative solves the problem posed, although with the need for hyperparameter adjustment.

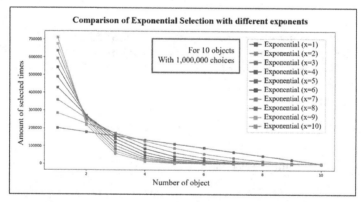

Fig. 7. Comparison of different exponent values on Exponential Weighted Selection, including Linear Weighted Selection being x value equals 1.

10 Comparison

Having proposed two alternatives to *LWRS* algorithm, a comparison between their performance is shown in Fig. 8, considering also a *Simple Random Sampling*.

As can be seen on Fig. 8, the proposed alternatives solve the problem raised, however; *Linear* method does not offer any relevant bias regarding the position of elements in the list, and *Exponential* suffers from needing a hyperparameter that adjusts the decision distribution. In this context, *LWRS* does not need hyperparameters, as it adjusts the scale to the size of the list to be used, and at the same time offers a notable preference bias with respect to the top elements in the list.

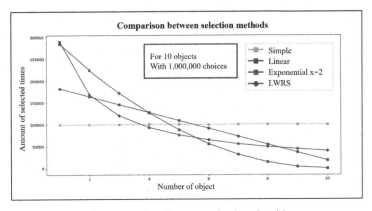

Fig. 8. Comparison between selection algorithms.

11 Limitations

It is beyond the scope of this work to provide a quantitative metric that assesses the usefulness of *LWRS* algorithm over others, as its focus is on offering a selection mechanism based on positional order without requiring hyperparameters. However, this approach has limitations as the potential limitation of its effectiveness in complex environments where factors such as explicit numerical weights, may be critical. Additionally, in contexts where metrics of effectiveness are unclear, the bias towards earlier elements can make decision-making difficult to control, requiring caution to avoid unintended outcomes in critical and sensitive applications.

12 Conclusions

The *Logarithmic Weighted Random Selector* (*LWRS*) offers a novel approach to selection algorithms by incorporating a logarithmic scale to bias selections based on the positional order of elements within a list. Unlike traditional methods such *as Fitness-proportionate selection* or linear weighted approaches, *LWRS* requires no additional hyperparameters and provides a natural, exponential favoritism towards elements earlier in the list. This makes it particularly suitable for scenarios where selection importance is determined by order rather than explicit numerical values.

The experiments validate *LWRS*' ability to consistently prioritize top elements while maintaining variability, demonstrating its potential for applications in areas like generative models and decision-making systems as *Reinforcement-Learning* agents, which work with encoding states based on preference. Overall, *LWRS* presents a balanced, efficient alternative for weighted random selection, particularly when the selection bias must be inherently linked to element order without the complexity of additional tuning parameters.

References

1. Molina, L.C., Belanche, L., Nebot, A.: Feature selection algorithms: a survey and experimental evaluation. 2002 IEEE International Conference on Data Mining, 2002. Proceedings, pp. 306–313. Maebashi City, Japan (2002). https://doi.org/10.1109/ICDM.2002.1183917
2. Dang, D.-C., Eremeev, A., Lehre, P.K.: Runtime analysis of fitness-proportionate selection on linear functions. arXiv preprint arXiv:1908.08686 (2019)
3. Burgoyne, J.A., Wild, J., Fujinaga, I.: Compositional data analysis of harmonic structures in popular music. In: Yust, J., Wild, J., Burgoyne, J.A. (eds.) Mathematics and Computation in Music. MCM 2013. Lecture Notes in Computer Science (), vol 7937. Springer, Berlin, Heidelberg (2013). https://doi.org/10.1007/978-3-642-39357-0_4
4. Novosel, L.M.: Understanding the evidence: sampling and probability (random) sampling designs. Urologic Nursing **43**(4) (2023)
5. Singh, A., Masuku, M.: Sampling techniques and determination of sample size in applied statistics research: an overview. Int. J. Commer. Manag. **2**, 1–22 (2014)
6. Cochran, W.G.: Sampling Techniques, 2nd edn. John Wiley and Sons Inc, New York (1963)
7. Grefenstette, J.: Rank-based selection. Evolutionary computation **1**, 187–194 (2000)
8. Shukla, A., Pandey, H.M., Mehrotra, D.: Comparative review of selection techniques in genetic algorithm. 2015 International Conference on Futuristic Trends on Computational Analysis and Knowledge Management (ABLAZE), pp. 515–519. Greater Noida, India (2015). https://doi.org/10.1109/ABLAZE.2015.7154916
9. Polya, G., Pólya, G.: How to solve it: A new aspect of mathematical method. Vol. 34. Princeton University Press (2014)
10. Hobbie, R.K., Roth, B.J., Hobbie, R.K., Roth, B.J.: Exponential growth and decay. Intermediate Physics for Medicine and Biology, 31–47 (2007)

Design and Implementation of a Machine Learning Model for Soccer Match Prediction Based on Player Statistics

Antonio Muñoz Barrientos, Humberto Muñoz Bautista(✉) 📷, Alejandro Padilla Díaz, and Francisco Javier Álvarez Rodríguez

Universidad Autónoma de Aguascalientes, Aguascalientes AGS 20100, México
`al229279@edu.uaa.mx, hmuntista@gmail.com`

Abstract. Football generates vast amounts of data daily on players, teams, and leagues, used for performance analysis. However, advanced analysis technologies are often inaccessible to lower-budget teams. This study developed a predictive tool to estimate the number of goals a team might score in a match, focusing on the UEFA Champions League with data from 2005 to 2022, totaling 4237 records. The model development included three stages: database creation, data preprocessing and analysis, and model construction. Data from the Fbref and FIFA Index, including player data from the EA FC video game, were collected using web scraping to build an expandable database. Additional variables, such as team origin league and momentum (average goals in the last five matches), were added to improve model accuracy. The model, a deep fully connected neural network evolved from a multi-layer perceptron, achieved 57% accuracy. The findings suggest that effective analysis tools can be created using data mining and machine learning, supporting lower-budget teams by providing more information and reducing competitive disparities.

Keywords: performance analysis · web scrapping · football predictive tool · neuronal network

1 Introduction

Sport has been part of society for hundreds of years and represents the competitive side of human beings. Throughout history, different tribes and societies have organized competitions to obtain a certain prize, title, or trophy. Currently, soccer is one of the most practiced sports in the world, with two competitions of great global impact (UEFA Champions League and the World Cup). The great amount of competition that exists makes it necessary to use tools that allow teams to prepare for the matches. The previous analysis provides enriching information to know the possibilities against a rival. Currently, technology provides us with tools that give us a great capacity for analysis, from visualized graphics to machine learning models that allow us to predict events. Taking advantage of these tools is key for all teams looking to have the most complete analysis and gain that advantage in favor of their planning.

However, access to analysis tools is not available to all institutions, and with data mining and machine learning we seek to build effective models that can predict the results of a soccer match to provide better and equal conditions for all teams.

2 Related Work

2.1 Model for Predicting Results in Colombian Professional Soccer

It is a model that suggests analyzing the number of goals scored by a team throughout the season. The number of goals scored by a team is taken as observations and divided into 9 different variables [3]. All variables depend on the goals scored and give some weight to the forecast, for example, the sum of the goals of the previous 5 games.

2.2 Predicting Outcome of Soccer Matches Using Machine Learning

This project makes use of machine learning techniques for outcome prediction, but the study seeks to emphasize more discrete variables. They study concepts such as refereeing injustices, fixed matches, poor match data, among others [4]. In the end they focused on a single league with predetermined variables and obtained a classification accuracy of 60%.

2.3 Explaining Soccer Match Outcomes with Goal Scoring Opportunities Predictive Analytics

The project consists of creating a model that can predict the possibility of a goal based on the circumstances of a play. It takes data on the origin of the play, distance from the goal, shooting angle, player quality, goalkeeper quality, among other variables. It uses 4 different methods of classification algorithms such as Random Forest, Logistic regression, decision trees and Ada-boost) [5].

The clear difference is that this project is more focused on match situations and not on predicting the outcome, but they can complement each other very well.

2.4 Algorithms

Prediction is a machine learning task that can be covered in different ways, using different algorithms. Random forest is an ensemble learning algorithm that can be used for classification, that is predicting a categorical response variable, and they can also be used for regression which involves predicting a continuous response variable [11]. Considered that the domain of our target data can be set in values from 0–10.

An approach for attacking this task is using regression. There are a lot of types of regression. Linear regression, that is used to find a linear relationship between one or more predictors [13]. Logistic regression is another option, permits the use of continuous or categorical predictors and provides the ability to adjust for multiple predictors [14].

Another type of model is the Artificial Neuronal Network (ANN). An Artificial Neural Network (ANN) is a mathematical model that tries to simulate the structure and functionalities of biological neural networks. Basic building block of every artificial neural network is artificial neuron, that is, a simple mathematical model (function) [12].

3 Methodology

At present, there are several ways to obtain information on soccer events, player statistics, match analysis, score predictions, etc. Unfortunately, these tools can only be acquired by a specific group of users who have the acquisitive power (professional clubs, bookmakers, etc.), having a great advantage over those who do not have these tools. The objective is to provide a viable solution for all those who cannot have access to expensive tools, to keep competitiveness and sportsmanship as even as possible.

The objective is to design a prediction model that can obtain the outcome of a soccer match involving teams from the 5 major European leagues.

The project was focused on the CRISP-DM methodology [1]. This methodology has 6 different phases:

Understanding of the Business, In this case the problem covered, which is the construction of a database with player's characteristics using techniques of data mining to recover data from web.

Understanding of the Data, Which helps to describe more precisely the characteristics of the players. In this part the selection of variables was made.

Preparation of the Data, Using techniques of data engineering, applying normalization and creating synthetic data that can enrich the information and add more elements that help to distinguish the data records.

Model, Create a model with the data that can receive input information and producing an output. We use assess different algorithms to study the data and analyze the behaviour of it. Neuronal networks are a common model of machine learning for train data that have complex patterns.

Evaluation, Determine the accuracy of the model on new data from which it was trained. We can use the accuracy and precision as metrics of a good model.

Deployment, Using the model in real environments. We can use connection points to a model in the cloud, using API's or putting data manually.

3.1 Database Building

Information Sources Selection. For the selection of the sources of information we decided to choose pages that had information that was easy to process through web scrapping. To obtain all the soccer matches we consulted the Fbref [9] website which contains a list of all the historical Champions League matches. This page contains statistics of all teams and players and has a large database. We use web scrapping to obtain all the matches with their respective results.

For more detailed information on each team, we consult FIFA Index [10] which is a site that provides statistical information on each team. The information is collected by expert scouts who describe the players, so that later mathematicians can translate those statistics into data. This platform provides us with different information on each team depending on the year that is consulted, which allows us to have a more accurate reference on how a team was in a certain year.

Database Building. We performed web scraping to the Fbref website to obtain all the matches of the Champions League seasons from 2005 to 2022. We extracted the name of the team, the score and whether they played away or home.

Each team was then searched on the FIFA Index page. The page provides statistics such as attack, midfield, and defense scores, as well as the overall team score. The idea is to turn all the team names into different columns of scoring data for their attack, midfield, and defense. Averaging these gives us the overall team score and we can rank teams based on their score. Teams with a score of 80+ are considered 5-star teams, while teams with a score of 65− are considered 2-star teams, and within this range the rest of the team values are defined. Thanks to this we can get a better idea of what kind of teams are facing each other.

Data Cleaning. In some situations, teams found in the FBREF data were not found in the FIFA Index data. In the database, the teams in these situations have their statistics set to 0, which causes a lot of inconsistency when applying machine learning algorithms. This happens for two main reasons:

-The team has a very different name between the two pages. This happened mainly with Greek, Russian, Turkish, etc. teams. Where their alphabet includes letters that are not used in the common alphabet, examples: Qarabağ FK, Oțelul Galați, Viktoria Plzeň. To correct this the team's names had to be changed to UTF-8 encoding to get rid of the special characters. The search was performed again with the corrected names to obtain the respective data.
-The team is not in EA FC videogame. EA Sports are in charge of the FIFA game, and they are the ones who choose which leagues and which teams are included in the game. There are situations where teams from an unrecognized league make it to the top European championship, this by a single season and even with a lot of luck. Since the teams are not recognized, EA Sports decided not to include them in the game, and therefore they are not in the FIFA Index. These teams being few, it was decided to give them random values to their statistics, but with the condition that their team quality must be low. Examples of teams: Sheriff Tiraspol, Maccabi Haifa, Trabzonspor.

3.2 Data Analysis and First Prediction Models

First Model Approach, Random Forest. We wanted to make an initial test with the Random Forest algorithm to try to classify the number of goals to be scored by a certain team with certain characteristics. Using the constructed dataset, we separated the dataset into 75% training and 25% testing. The algorithm was given freedom to expand to any number of nodes but with a maximum of 100 final leaves. In the end the algorithm had an accuracy of 28%.

Second Model Approach, Multi-layer Perceptron. By exhaustively analyzing the behavior of the Random Forest algorithm, it was concluded that its effectiveness as a classification algorithm could be compromised in certain cases, especially when faced with data sets with a large number of variations and similarities. This limitation becomes evident in situations where there are multiple instances of input data generating different outputs, which may complicate the algorithm's ability to generalize patterns and make accurate decisions.

The proposed neural network architecture is a multilayer perceptron [2]. "The multilayer perceptron is a unidirectional neural network consisting of three or more layers: an input layer, another output layer and the rest of the intermediate layers called hidden layers".

We chose to use more computational power to analyze as much data as possible. This approach allowed multiple tests in which the hyperparameters were varied, as well as the sizes of the training and test sets, to optimize the model's performance and ensure its ability to generalize effectively on unseen data. Results ranging from 28% to 32% were obtained with this approach.

Data Augmentation. After performing a set of tests with the multilayer perceptron model, it was concluded that the improvement in results is not necessarily linked to the specific architecture of the neural network, but rather to the quality of the input data. Consequently, it was decided to focus efforts on improving the quality of the data, as well as enriching the information provided to the neural network, without modifying the fundamental nature of the data itself.

The first step to improving accuracy and discrimination in classification was the inclusion of an additional variable designed to distinguish more clearly between the various cases present in the database. In the first iteration of the database, it was possible for discrepancies in the results to occur due to repetition of the same input data, generating different output values. This situation was due to the possibility of teams with similar characteristics from different leagues obtaining contrasting results. In order to mitigate these discrepancies and increase the specificity of the model, a strategic decision was made to introduce a new variable to identify which league each team belonged to. This additional variable would later be used to adjust the input data according to the corresponding league, thus allowing better differentiation between similar cases belonging to different contexts.

The major European leagues were listed to determine the league number. The order in which they are listed was done based on how competitive the teams from that league are in European competitions.

1. Spanish League
2. English League
3. Italian League
4. German League
5. French League
6. Dutch League
7. Rest of Europe

The idea is to make a proportion of the input data using the league to which they belong, following the formula.

$$X' = X * 1/league \qquad (1)$$

The higher the index a team has in its league, the lower its input data will be, and this will help us to better identify outliers. We performed tests again with the multilayer perceptron and obtained results of 43%.

Synthetic Data Generation. As a final strategic step, it was decided to incorporate an additional element aimed at providing a more accurate perspective on the state and dynamics of a team prior to a match. This concept was named "momentum" and is defined as the average number of goals scored by a team over the course of a season.

Momentum offers a standardized measure that reflects a team's scoring trend, regardless of whether it has played 30 games or just 10. In this way, momentum becomes a valuable tool for anticipating a team's offensive potential in an upcoming match, providing a solid basis for strategic decision-making for coaches and sports analysts alike.

The inherent variability of the data when considering only historical goals is notable, manifesting itself in the presence of significantly high peaks and notably low valleys in the graphical representation, see Fig. 1. However, in Fig. 2, a closer look at the momentum plot reveals a different pattern. Although fluctuations and variants are still present, their magnitude is considerably smaller compared to the raw goal data.

Although variations in a team's performance exist, momentum acts as a buffer, smoothing out the extremes and providing a more uniform representation of a team's performance over the course of the season. This distinctive feature of momentum not only makes it easier to interpret the data, but also provides a more accurate and reliable view of a team's overall status.

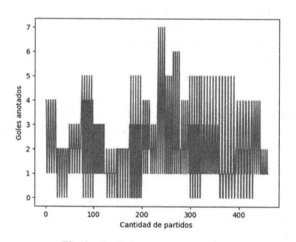

Fig. 1. Goals from a team over time

Thanks to this variable, it is much easier for the neural network to better identify the similarities between the number of goals scored by a team and its goal scoring potential. Testing with the perceptron obtained results of 50% of accuracy.

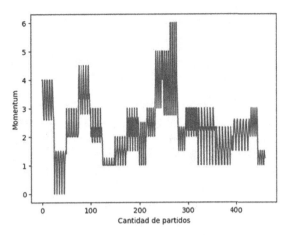

Fig. 2. Momentum from a team over time

3.3 Implementation of Deep learning

Deep learning[1] is a branch of machine learning that employs computational models composed of multiple layers of neural processing. These layers are designed to learn representations of data at successive layers of abstraction, enabling the identification of increasingly complex features and patterns in the input data [6]. This approach is noted for its ability to discover deep and meaningful structures in large-scale data.

The ability of the neural network to readjust its values once an output is obtained is very useful to adjust more accurately the detection of those different cases. With a good number of epochs[1], we can obtain a higher result in the accuracy of the algorithm because we would have a more detailed adjustment.

The proposed architecture is a balanced model where the hidden layers increase as the neural network advances and decrease as it approaches the output layer. The neural network values will have values between 10 and 50, this chosen arbitrarily to avoid overfitting.

3.4 Training of the Model

Configuring the model training with appropriate hyperparameters is crucial for achieving optimal performance. Factors such as the activation functions in neurons, training duration, loss function, and the number of epochs play vital roles in ensuring the model's effectiveness.

For the activation function of the neurons, we utilized ReLU (rectified linear unit), a widely adopted non-linear activation function in neural networks. One of the key advantages of the ReLU function is its ability to selectively activate neurons, thereby deactivating a neuron only when the output of the linear transformation is zero [17].

---

[1] Epoch: is an iteration from the neuronal network where the input data produces an output and then the weights and bias from the middle layers are update [16].

This is mathematically represented as:

$$f(x) = \max(0, x) \tag{2}$$

ReLU serves as an optimal activation function for preventing neurons from working with negative weights and values. Given our objective of predicting non-negative integers, it is imperative to avoid working with negative values.

The choice of loss function is critical for evaluating the model during training. In this case, we employed sparse categorical cross-entropy, a loss function specifically designed for multi-class classification tasks with integer labels. Sparse categorical cross-entropy measures the disparity between the true integer-coded labels and the predicted probability distributions across classes. Notably, it offers computational efficiency compared to one-hot encoding of the labels. For example, in intent classification scenarios where sentences are categorized into different intent groups, sparse categorical cross-entropy gauges the dissimilarity between predicted and true intent labels [18].

The training results indicate a progressive increase in accuracy and a simultaneous decrease in loss per epoch. This trend suggests that the model is not overfitting (Fig. 3).

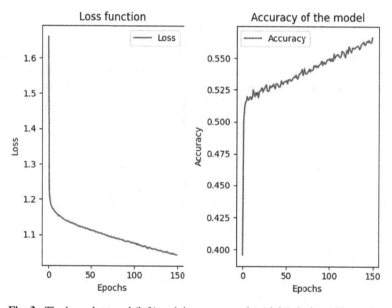

Fig. 3. The loss plot trend (left) and the accuracy plot (right) during 150 epochs.

3.5 Algorithms Resume

The reasons for choosing these algorithms were their ease of implementation and their ability to analyze the behavior of the data. The training of the networks is fast for the data available and requires little computational power. A summary of all the approaches mentioned above is now presented to provide more details (Table 1).

Table 1. A description of machine learning algorithms [15]

Algorithm	Learning type	Description	Remarks
Random Forest	Supervised	Group of trees with characteristics	The random forest model is a learning algorithm based on an ensemble of trees; that is, the algorithm averages the predictions of many individual trees. The individual trees are constructed from bootstrap samples rather than from the original sample [7]
Multilayer perceptron	Supervised	Process inputs through multiple connected layers	The multilayer perceptron is a universal function approximator. It can learn any type of continuous function between a set of input and output variables [8]
Deep neuronal network	Supervised	Multiple layers to extract features and learn patterns	Models that are composed of multiple layers of processing to learn data representations with multiple levels of abstraction. Discovers intricate structures in large data sets [6].

4 Results

We can observe that the algorithm had satisfactory results when employing a deep learning model and when incorporating additional variables. In this case, the prediction value reached 57%, which is useful to obtain a hypothetical result that is close to what can actually happen in practice (Fig. 4 and Table 2).

Although in this project we focused only on goals scored, we believe that there are other factors that could significantly improve the accuracy of the results obtained. These additional factors include goals against, which would provide a more complete picture of the team's defensive performance, and player availability, which can significantly influence the team's performance in each match.

Table 2. Accuracy performance of the different models and data

Model	Train/test data	Layers	Additional variables	Accuracy
Random forest	-	-	-	28%
Multi-layer perceptron	(80, 20)	150	-	30%
Multi-layer perceptron	(70, 30)	500	-	32%
Multi-layer perceptron	(70,30)	500	Leagues	43%
Multi-layer perceptron	(70,30)	500	Leagues and momentum	50%
Full connected neuronal network	(70,30)	(50,13,41,47,10)	Leagues and momentum	57%

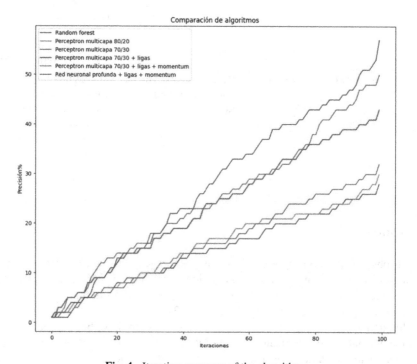

Fig. 4. Iteration accuracy of the algorithms

4.1 Results in a Real Context

The model was used to predict the results of the quarterfinals of the 23–24 season of the UEFA Champions League, giving the following results (Table 3):

Table 3. Comparative of the goals prediction of the model and the real outcome

Team	Model goals predicted	Goals scored
Arsenal	0	2
Bayern	0	2
Atlético de Madrid	1	2
Dortmund	**1**	**1**
Barcelona	0	3
PSG	0	2
Manchester City	**3**	**3**
Real Madrid	**3**	**3**

In the predictions of real matches, the model accurately predicted 3 of the 8 results. And of the 4 matches it was able to predict the outcome (win, lose or draw) of 2.

5 Conclusion

Predictions allow teams to gain a preliminary view of possible outcomes, which facilitates the development of more informed and effective strategies. In this context, the use of data mining techniques, such as web scraping, has been fundamental. These techniques have enabled the construction of a customized database that can be continuously updated, either manually, through APIs or using more web scraping. This database provides a solid foundation on which additional synthetic data can be generated, helping to more clearly visualize patterns in the collected data.

Neural networks have proven to be highly efficient structures for pattern recognition in complex data sets. Their analysis capacity varies according to their architecture, ranging from simple models to deep neural networks. In this study, the implementation of deep neural networks has successfully identified the patterns needed to predict the number of goals a team can score, based on the specific characteristics of its players and the team's conditions during the season.

It is important to keep in mind that soccer is a sport played by people, and human performance is inherently unpredictable and susceptible to a multitude of unexpected factors that can significantly affect its performance. Despite these uncertainties, the results obtained, with a prediction accuracy of 57%, are considered highly satisfactory within the context of the limitations and challenges inherent in predicting sporting events.

References

1. Gallardo, J.A.: Metodología para el desarrollo de proyectos en Minería de Datos CRIPS-DM (2018). Recuperado de http://www.oldemarrodriguez.com/yahoo_site_admin/assets/docs/Documento_CRISP-DM.2385037

2. Vivas, H.: Optimización en entrenamiento del perceptrón multicapa. Universidad del Cauca, Popayán (2014)
3. Gómez Bayona, C.A.: Modelo de predicción de resultados en el futbol profesional colombiano (2013). Retrieved from http://hdl.handle.net/10818/8996
4. Yezus, A.: Predicting outcome of soccer matches using machine learning. Saint-Petersburg University (2014)
5. Eggels, H., Van Elk, R., Pechenizkiy, M.: Explaining Soccer Match Outcomes with Goal Scoring Opportunities Predictive Analytics. In Mlsa@ pkdd/ecml (2016)
6. LeCun, Y., Bengio, Y., Hinton, G.: Deep learning. Nature **521**(7553), 436–444 (2015)
7. Schonlau, M., Zou, R.Y.: The random forest algorithm for statistical learning. Stand. Genomic Sci. **20**(1), 3–29 (2020)
8. Vidal González, M.: El uso del Perceptrón Multicapa para la clasificación de patrones en conductas adictivas (2014)
9. FBref. (s.f.). FBref.com. Retrieved de https://fbref.com/en/
10. FIFA Index. (s.f.). FIFAIndex.com. Retrieved de https://www.fifaindex.com/
11. Everingham, Y., Sexton, J., Skocaj, D., Inman-Bamber, G.: Accurate prediction of sugarcane yield using a random forest algorithm. Agron. Sustain. Dev. **36**, 1–9 (2016)
12. Krenker, A., Bešter, J., Kos, A.: Introduction to the artificial neural networks. Artificial Neural Networks: Methodological Advances and Biomedical Applications. In: Tech, 1–18 (2011)
13. Maulud, D., Abdulazeez, A.M.: A review on linear regression comprehensive in machine learning. J. Appl. Sci. Technol. Trends **1**(2), 140–147 (2020)
14. Weiß, C.: Notfall + Rettungsmedizin **21**(6), 516–518 (2018). https://doi.org/10.1007/s10049-018-0507-7
15. Bautista, H.M., Esparza, M.A.O., Arteaga, J.M., Villalba-Condori, K.: Use of Data Mining to Identify Preferences for Humanistic Courses and Support the Management of University Extension of the Autonomous University of Aguascalientes. In: CISETC, pp. 490–497 (2023)
16. Afaq S., Rao, S.: Significance of epochs on training a neural network. Int. J. Sci. Technol. Res. **09**(06), 1–4 (2020). https://www.ijstr.org/finalprint/jun2020/Significance-Of-Epochs-On-Training-A-Neural-Network.pdf
17. Sharma, S., Sharma, S., Athaiya, A.: Activation functions in neural networks. Towards Data Sci. **6**(12), 310–316 (2017)
18. Steele, B.: Feed Forward Neural Network for Intent Classification: A Procedural Analysis

Electric Scooters and Renewable Energy Integration Associated with Tourist Parks: A Dijkstra-Based Model for Smart Mobility Optimization

Gilberto Espadas-Baños[1], César Quej-Solís[1], Manuel Flota-Bañuelos[1], and Alberto Ochoa-Zezzatti[2(✉)]

[1] Facultad de Ingeniería, Universidad Autónoma de Yucatán, Av. Industrias No Contaminantes Por Periférico Norte, Apdo. Postal 150 Cordemex, C.P. 97310 Mérida, Yucatán, Mexico
[2] Doctorado en Tecnología, Universidad Autónoma de Ciudad Juárez, Av. Plutarco Elías Calles #1210 Fovissste Chamizal, C.P. 32310 Ciudad Juárez, Chihuahua, Mexico
alberto.ochoa@uacj.mx

Abstract. This research presents a novel strategy to enhance ecotourism in Yucatán's national parks by integrating electric scooters with solar energy systems, focusing on Los Petenes National Park, noted for its biodiversity and ecological significance. The study identifies 27 recreational activities within the park and proposes an optimization model aimed at improving tourist mobility, particularly for visitors unaccustomed to the region's high temperatures. This system not only reduces the risk of heat-related illnesses but also enhances the overall visitor experience through a convenient and sustainable transportation option. Utilizing Dijkstra's algorithm, a graph of the park's key recreational points was constructed to optimize tourist routes, minimizing waiting times and unnecessary travel. To select the most suitable electric scooter for these optimized routes, the Technique for Order Preference by Similarity to Ideal Solution (TOPSIS) was applied, evaluating multiple criteria such as performance, user experience, and sustainability. The GoTrax GXL V2 scooter was identified as the optimal choice, aligning with both operational requirements and visitor preferences. The integration of a photovoltaic system ensures that the energy demands of the scooter fleet are met through renewable sources, significantly reducing carbon emissions and operational costs. This renewable energy component directly supports the optimized transportation system, reinforcing the park's commitment to environmental conservation. The proposed model not only enhances the overall visitor experience by providing a convenient and sustainable transportation option but also contributes to the local economy by fostering economic opportunities in the region. The study offers a scalable and adaptable approach for other parks and tourist destinations seeking to implement eco-friendly technologies and sustainable practices in their operations.

Keywords: Ecotourism · Electric Scooters · Renewable Energy · Sustainable Mobility · Dijkstra's Algorithm · Hybrid Energy Systems · Tourism Activities Optimization · TOPSIS

1 Introduction

In his 1978 paper, Cohen examines the environmental impact of tourism, highlighting that while moderate tourism can support the preservation of attractions, mass tourism introduces significant environmental risks [1].

Ecotourism represents an alternative model of tourism that prioritizes the preservation of natural and cultural environments while promoting sustainable socio-economic development within host communities [2]. This approach highlights the active participation of local populations as essential social actors, fostering adaptable and varied ecotourism experiences that range from protected natural areas to urban and rural settings [3, 4]. The key principles of ecotourism include nature-based solutions, a sustainability dimension focused on conservation, the active involvement of local communities, educational initiatives and awareness-building programs aimed at environmental preservation, and the ethical promotion of indigenous products [5]. Ecotourism not only enhances the traveler's experience by offering authentic encounters with wildlife and local cultures but also serves as a protective measure against ecologically detrimental activities such as unsustainable mining, deforestation, and illegal wildlife poaching [5, 6].

Over the past decade, ecotourism in Mexico has generated approximately $740 MXN million in revenue, with the country's protected natural areas being the primary attractions. However, despite the growth in this sector and its increasing societal impact, there is still a lack of studies that thoroughly document the comprehensive environmental benefits and potential ecological constraints associated with this activity [7].

While ecotourism offers significant benefits, it is crucial to approach such projects with caution, as poorly planned initiatives have often led to protests from residents. In many cases, these ventures have resulted in environmental degradation and disruption of local livelihoods, undermining the very goals of sustainability and conservation [8]. Careful planning and active local participation are crucial to avoid these negative outcomes and ensure that ecotourism remains a positive force for both people and nature through sustainable development practices.

The Yucatan Peninsula comprises the Mexican states of Campeche, Quintana Roo, and Yucatan, as well as regions of northern Guatemala and Belize [9, 10]. The Mexican portion alone covers nearly 145,000 km^2 [9]. The region hosts 39 protected natural areas covering 2,977,752 hectares of land-based ecosystems. These areas account for 21.6% of the peninsula's total landmass and are vital for preserving extensive forests, tropical ecosystems, and their rich biodiversity [10]. Historically, local communities in this region have relied on sustainable agriculture, fishing, forestry, beekeeping, horticulture, and livestock farming [11]. Since the 1970s, the growth of tourism along Mexico's Caribbean coast has provided an additional source of income. By the 1990s, ecotourism emerged as a supplementary economic activity, supported by international, federal, and non-governmental organizations through funding, training, and resources [12].

A significant challenge for ecotourism in the Yucatan Peninsula is the region's intense climate. The area features a tropical climate with variations that range from warm and humid to sub-humid and semi-arid, with rainfall concentrated in the summer months, which can influence biodiversity and visitor experiences, necessitating adaptive management strategies to mitigate potential impacts [12]. The hottest temperatures typically occur between May and August, often exceeding 35 °C, which can impact both local ecosystems and the overall visitor experience [13]. High temperatures and humidity create significant difficulties for visitors, especially those unaccustomed to such conditions. Tourists from cooler regions often struggle with the intense heat, which can lead to exhaustion or heatstroke, necessitating adequate preparation and hydration strategies [14]. This limits their ability to fully enjoy the region's natural attractions. Transportation within national parks plays a vital role in shaping the cultural landscape, enhancing the visitor experience by improving travel efficiency and influencing how visitors interact with the park's natural and cultural features, while promoting sustainable practices [15].

Technological advancements have revolutionized various industrial sectors, including transportation, with innovations such as electric vehicles, autonomous systems, and improved logistics [16]. Generation Z (Gen Z), raised in a highly connected technological environment, exhibits distinct attitudes and lifestyles compared to previous generations, often emphasizing sustainability and a strong presence on digital platforms [17]. Findings suggest that while Gen Z are not universally more environmentally conscious than their counterparts in previous decades, they are more responsive to changes in the built environment, showing a greater propensity to adopt sustainable travel modes, such as public transportation, cycling, and eco-friendly accommodations [18]. This highlights Gen Z's potential to drive sustainable transportation initiatives, influenced by technological advancements and social media, such as promoting electric vehicles or advocating for public transit improvements through online platforms [19].

The need for an efficient and comfortable transportation system is crucial, enabling visitors to explore while avoiding harsh environmental conditions, thereby enhancing the ecotourism experience and minimizing ecological impacts. However, the long-term sustainability of ecotourism depends on understanding its effects on local biodiversity [20]. Poorly managed projects can disrupt sensitive ecosystems, leading to habitat loss, species displacement, and a decline in biodiversity [21]. To mitigate these risks, adaptive management strategies must be implemented, prioritizing conservation and closely monitoring the cumulative effects of tourism on natural habitats [22].

In this study, an innovative model is presented to improve mobility in Yucatán's national parks by integrating electric scooters powered by solar energy systems. Utilizing Dijkstra's algorithm, the model optimizes tourist routes and scooter distribution at key locations, enhancing both transportation efficiency and sustainability. Additionally, the Technique for Order Preference by Similarity to Ideal Solution (TOPSIS) method was applied to select the most suitable electric scooter based on multiple criteria, ensuring alignment with operational requirements and visitor preferences. The results demonstrate that the model significantly improves transportation systems in tourist parks, highlighting the potential of combining advanced optimization algorithms, multi-criteria decision-making methods, and renewable energy solutions. This approach illustrates how ecotourism can contribute to conservation and sustainable development through intelligent mobility optimization.

2 Literature Review

Dijkstra's algorithm, developed by computer scientist Edsger Dijkstra in 1959, is designed to solve the shortest path problem in graphs with positive edge weights. It computes the shortest path from a specified starting point to all other vertices in the graph, continuing until the shortest path to the target vertex is determined [23].

The application of Dijkstra's algorithm to transportation systems has become a key factor in optimizing urban mobility, enhancing route efficiency and resource allocation [24]. The use of Dijkstra's algorithm in cross-city transit systems has proven effective for optimizing route selection and travel efficiency, considering both time and cost parameters [25]. Previous studies have explored this optimization model aimed at improving the efficiency of transportation networks in densely populated metropolitan areas, leading to enhanced user satisfaction [26].

The integration of Dijkstra's algorithm in tourism optimization has gained attention in cities with numerous tourist attractions, facilitating improved route planning and visitor experiences [27]. More recent studies have applied the algorithm to recommend the shortest paths for tourists visiting scattered sites, improving both the experience and travel time [28, 29]. The combination of ecotourism with sustainable transportation has emerged as a central theme in urban development, promoting environmentally responsible practices and enhancing community well-being [30].

Sustainable urban transport can enhance tourist areas by preserving natural ecosystems and creating more attractive, environmentally friendly tourist destinations [31]. Studies have demonstrated that electric scooters are commonly used for recreational activities and short-distance travel, especially in regions that emphasize environmental sustainability initiatives [32]. Statistical evidence supports the suitability of electric scooters for ecotourism. Zero-emission scooters have been recognized as key contributors to sustainable urban mobility due to their potential to lower greenhouse gas emissions and enhance energy efficiency [33]. Research from China highlights how electric scooters can alleviate congestion and parking issues while offering flexible and reliable transportation options in both urban and natural conditions [34]. Through a generalized ordered logit model based on an online survey from various Spanish cities, a study identifies key factors driving the adoption and frequency of scooter-sharing with potential benefits like reducing noise, air pollution, and road congestion, thus enhancing urban livability [35]. Using data from a 2022 survey conducted in Malmö, Manchester, and Utrecht, a study explores the impact of shared micro-mobility on the perception of transport poverty across different income groups. The results indicate that e-scooters hold potential in improving mobility for low-income users, potentially fostering greater transport equity [36]. A study conducted in Austin and Minneapolis used spatial GIS hotspot analysis and negative binomial regression models to investigate e-scooter usage. The analysis revealed a positive correlation between a greater diversity of land uses and increased e-scooter activity in both cities [37]. Additionally, a study identifying spatial factors influencing e-scooter use suggests that strategic electric scooter distribution, land use, and connectivity can enhance micromobility services and contribute to sustainable urban development [38].

In general, electric scooters are considered an effective alternative for reducing carbon emissions compared to traditional vehicles, particularly in ecotourism contexts where sustainable practices are prioritized. Their low operational costs and compatibility with renewable energy sources, such as solar-powered charging stations, reinforce their importance in these environments [39]. These factors are essential in remote or protected ecotourism areas, where minimizing environmental impact and maximizing transportation efficiency and convenience are top priorities.

Site Description
The biosphere reserve known as "Los Petenes" is located within the facilities of the Center for the Conservation and Research of Wildlife at the Autonomous University of Campeche [Fig. 1]. It was declared a biosphere reserve on May 24, 1999, and registered as a Ramsar site in February 2004 due to its wetland value [40].

This site features a large, narrow coastline in northern Campeche state, located at 20°51′30″ N and 90°45′15″ W. It encompasses a land area of 100,866.52 ha and a marine area (considering coastal lagoons, estuaries, freshwater lakes, rivers and other bodies of water) of 181,991.10 ha [41].

Los Petenes is characterized by complex habitats, including islands with perennial freshwater availability. These unique Petenes ecosystems are found only in Mexico, Cuba, and Florida [42]. The dominant climate is warm sub-humid, characterized by summer rains and a mid-summer drought. Annual precipitation ranges from 729 to 1,049 mm, and the average temperature is 27.5 °C [43].

This protected area encompasses unique tropical wetlands and mangrove forests that are crucial for maintaining regional biodiversity and supporting various endemic species. The park's complex hydrological systems play a key role in preserving its rich fauna, including 45 species of fish [44]. Its diverse habitats, including lagoons and estuaries, provide essential resources for migratory birds and marine life [45]. Moreover, the park's landscapes and ecological significance attract researchers and tourists, promoting environmental awareness and ecotourism practices [46]. This dual role as both a conservation area and a tourist attraction underscore its significance for ecological preservation and economic development.

3 Methodology and Operation

This study employs a comprehensive optimization model to enhance mobility and energy efficiency within Los Petenes National Park. The methodology is structured into three main components: (1) optimization of tourist routes using Dijkstra's algorithm, (2) selection of the most suitable electric scooter using TOPSIS, and (3) integration of a solar renewable energy system for sustainable power generation.

Application of Dijkstra's Algorithm
The first step in this study was to develop an optimization model based on Dijkstra's algorithm. The aim was to enhance tourist mobility by minimizing travel time and improving electric scooter distribution throughout the park.

The park's map, containing 27 distinct recreational activities was modeled as a weighted, undirected graph

$$G = (V, E) \tag{1}$$

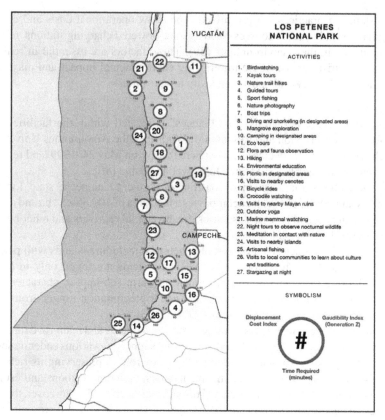

Fig. 1. Los Petenes Biosphere Reserve, Yucatan Peninsula, Mexico [41].

where nodes (V) represent key locations in the park where tourists engage in recreational activities, and edges (E) represent the paths between these locations, with weights corresponding to the travel distance and time between two nodes as shown in Fig. 1.

In addition to physical distances, a recreational value R was assigned to each activity, where the value was inversely proportional to the activity's popularity. This allowed the model to prioritize high-interest activities when calculating the most efficient tourist routes. The weights W_{ij} on the edges between nodes i and j were calculated as:

$$W_{ij} = D_{ij} + \left(\frac{1}{R_i} + \frac{1}{R_j}\right) \qquad (2)$$

where D_{ij} represents the distance between nodes, and R_i, R_j represent the recreational values of nodes i and j.

Dijkstra's algorithm was employed to determine the shortest paths between nodes, focusing on optimizing both travel efficiency and tourist satisfaction. The process began by selecting a source node s (e.g., the park's entrance), where all other nodes were initially assigned a tentative distance of infinity, except for s, which was set to zero. In the next step, for each unvisited neighboring node, the tentative distance was updated by adding the weight W_{ij} of the edge connecting the current node to its neighbor. The algorithm then selected the node with the smallest tentative distance, marking it as visited. This process was repeated iteratively until the shortest paths from the source node to all other nodes were calculated.

Selection of the most suitable electric scooter using TOPSIS

In this study, seven electric scooter models were evaluated based on a comprehensive set of criteria relevant to performance, user experience, and sustainability. The criteria were categorized into four main groups, as shown in Table 1.

Table 1. Criteria for evaluating electric scooter models

Performance	Physical attributes	User experience	Economic factors
Maximum speed (Km/hr)	Weight (Kg)	Handling	Price (USD)
Battery autonomy (Km)	Portability	Comfort	Maintenance costs
Charging time (Min)	Durability	Safety features	Availability of parts
Motor Power (W)	Quality construcition	-	-

Each criterion was assigned a weight reflecting its importance to the overall goal of enhancing tourist mobility while ensuring sustainability and user satisfaction. The weights were determined based on expert opinions and user preferences, particularly considering the expectations of Gen Z visitors.

Following the optimization of tourist routes with Dijkstra's algorithm, the next critical step was to select the electric scooter that best meets the requirements of these routes and the preferences of the visitors. To achieve this, we applied TOPSIS, a multi-criteria decision-making method that ranks alternatives based on their distance from an ideal solution.

To ensure that visitors can complete their selected routes without issues, the TOPSIS method was applied through several key steps.

Initially, a decision matrix was constructed, with each row representing an electric scooter model and each column representing a criterion. The performance ratings of each scooter with respect to each criterion were compiled into this matrix. To eliminate the effects of differing units of measurement and to facilitate comparability across criteria, the decision matrix was normalized using vector normalization. This process transformed the various criteria scales into a dimensionless scale, ensuring that no single criterion would disproportionately influence the outcome due to its scale of measurement.

After normalization, the matrix was weighted by multiplying each element by its corresponding criterion weight. This resulted in a weighted normalized decision matrix that accurately reflected the relative importance of each criterion in the decision-making process.

The next step involved determining the ideal best and ideal worst solutions. The ideal best solution comprises the highest performance values for beneficial criteria (such as maximum autonomy and minimum weight) and the lowest values for non-beneficial criteria (such as price and charging time). Conversely, the ideal worst solution consists of the lowest performance values for beneficial criteria and the highest values for non-beneficial ones. These ideal solutions serve as reference points against which all alternatives are compared.

The separation measures were then calculated by assessing the distance of each alternative from both the ideal best and ideal worst solutions. This was achieved by evaluating the differences between each weighted normalized value and the corresponding ideal values across all criteria. The overall distance reflects how close each scooter is to the ideal solution and how far it is from the worst-case scenario.

Finally, a relative closeness coefficient was computed for each scooter, representing its proximity to the ideal solution. This coefficient is calculated by comparing the distance of each scooter from the ideal best solution to its total distance from both the ideal best and ideal worst solutions. Scooters were then ranked based on their relative closeness coefficients, with higher values indicating a closer proximity to the ideal solution and, therefore, a more favorable option.

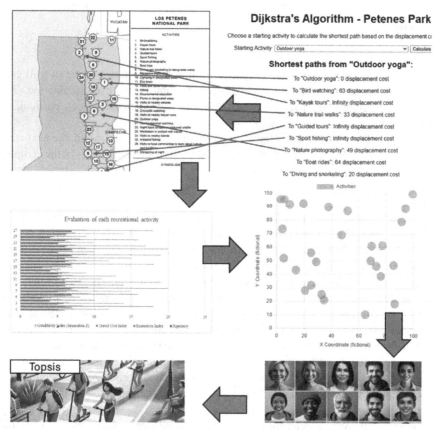

Fig. 2. Graphical representation of the optimization process using Dijkstra's algorithm and TOPSIS for recreational activity selection.

Figure 2 illustrates the integrated methodology for optimizing routes and selecting the most suitable electric scooter for visitors in Los Petenes National Park. This approach employs Dijkstra's algorithm, multicriteria analysis, and decision-making methods. Dijkstra's algorithm calculates the optimal routes between various activities in the park by minimizing displacement, resulting in a graph that maps the coordinates of these activities. Within the multicriteria analysis, the TOPSIS method is applied to evaluate and rank the electric scooters based on multiple performance criteria, ensuring that the selected model aligns optimally with visitor needs and the park's operational requirements.

Integration of Solar Energy into the Mobility System

To ensure the sustainability of the electric scooter system within Los Petenes National Park, a detailed analysis was conducted to integrate solar energy into the park's infrastructure. The objective was to design a renewable energy system capable of meeting the energy demands of the electric scooters, thereby reducing reliance on non-renewable energy sources and minimizing the carbon footprint associated with tourist mobility.

Solar panels were installed at key locations within the park to supply power for the scooters and other facilities. The design of the energy system included an analysis of the total daily energy consumption required for the scooters, calculated using the following equation:

$$E = N_s \times P_s \times T \tag{3}$$

where N_s represents the number of scooters, P_s is the power consumption per scooter, and T is the daily operating time.

Each scooter is equipped with a 1.5 kWh battery, providing an average range of 50 km per charge. With 20 scooters operating daily, the total daily energy demand amounts to 30 kWh. To meet this demand, a photovoltaic (PV) system was designed. The system's capacity was determined based on the average solar irradiance in Yucatán, which is approximately 5.5 kWh/m^2/day [47]. Monocrystalline PV panels with a rated power of 600 W and an efficiency of 21% were selected due to their high performance and suitability for the climatic conditions of the region.

The expected daily energy generation per panel (E_{panel}) is calculated by:

$$E_{panel} = P_{Panel} \times H_{Sun} \times \eta_{System} \tag{4}$$

where P_{Panel} is the panel's rated power (0.6 kW), H_{Sun} is the peak sun hours per day (5.5 h), and η_{System} is the system efficiency after accounting for losses (estimated at 87% due to a 13% loss factor).

By substituting the given values, the energy generated per panel is calculated to be 2.871 kWh per day. To meet the daily energy demand of 30 kWh, the required number of panels is determined by dividing the total energy demand by the energy generated per panel, yielding approximately 10.45 panels. Rounding up to the nearest whole number, a total of 11 panels are necessary to ensure sufficient energy generation.

The PV system includes a direct charging and battery swapping mechanism to maintain continuous scooter operation without the need for energy storage systems.

Solar charging stations are strategically located at key nodes identified by the optimized routes from Dijkstra's algorithm, facilitating efficient charging operations and minimizing additional travel distances.

4 Results

Optimization of tourist routes using Djikstra's Algorithms

Dijkstra's algorithm was employed to optimize tourist mobility within Los Petenes National Park by calculating the travel cost between locations, considering distance, travel time, and associated expenses. These factors are quantified into a Displacement Cost Index (DCI), as detailed in Table 2. This index categorizes routes from "low effort" to "highly challenging" based on the energy expenditure required to navigate between activities. By analyzing these costs, electric scooters can be strategically allocated and their real-time distribution optimized, minimizing both waiting times and redundant travel.

To further enhance the visitor experience, the Gaudability Index (GI) was introduced. This index quantitatively evaluates the enjoyment level of each activity, specifically tailored to reflect the preferences of Gen Z. By aligning activities with factors deemed engaging by this demographic, the index aims to sustain visitor interest over time. While the total number of activities tourists can engage in per day is contingent upon their budget, it is estimated that a maximum of seven activities can be feasibly completed within a single day.

Table 2. Displacement cost index (DCI) and classification in the Dijkstra's algorithm.

DCI	Classification	Description
0–5	Low effort	It is extremely simple and quick to reach other activities
6–10	Moderate effort	Reaching other activities requires little effort or time
11–15	Considerable effort	A moderate effort is required to travel to other activities
16–20	Challenging	Traveling between activities requires considerable time and effort
21–25	Very challenging	It is complicated to reach other activities, requiring significant effort
26 and above	Highly challenging	Reaching other activities is extremely complex and takes a lot of time and effort

For the analysis of results, the three activities exhibiting the highest GI were selected to evaluate the routes proposed by Dijkstra's algorithm. These activities include nature photography, diving and snorkeling, and nighttime star observation. By concentrating on these highly engaging experiences, the optimal pathways were identified to enhance overall visitor satisfaction.

In Fig. 3, various scatter plots are presented that illustrate the patterns of relative accessibility among different activities. The distribution of points reflects areas with higher levels of connectivity, indicated by clusters of points, and more remote or less accessible activities, represented by isolated points. Using Dijkstra's algorithm, each point represents the result of the shortest path calculation, depicting the Displacement Cost from an initial activity to others. Figure 3c illustrates optimal accessibility, showing a high concentration of points along low-cost travel routes, which indicates enhanced connectivity with other activities. In contrast, Fig. 3b presents moderate dispersion, indicating intermediate accessibility along routes with varying travel costs. Conversely, Fig. 3a depicts the least accessible activity, characterized by a broader distribution of points that reflects elevated travel costs and diminished connectivity.

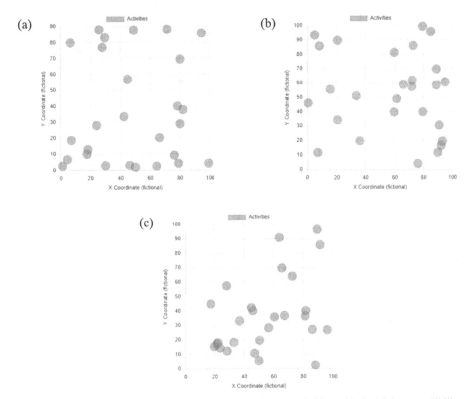

Fig. 3. Scatter plots of relative accessibility among the top activities with the highest gaudibility index, (a) nature photography, (b) diving and snorkeling, and (c) night stargazing.

Selection of the electric scooter using TOPSIS

Once the graphs generated by Dijkstra's algorithm were obtained, an analysis was performed to determine which scooters represent the most efficient option for tourist mobility along the routes.

To further refine the mobility solution, the TOPSIS multi-criteria decision-making method, as outlined in the Methodology, was employed to determine the most suitable electric scooter for visitors. Utilizing the previously established criteria and their assigned weights, the seven electric scooters were evaluated and ranked accordingly [Table 3].

The evaluation utilized a 7-point Likert scale to normalize the criteria and facilitate comparability across different units of measurement. The TOPSIS method systematically assessed each scooter against the criteria, calculating a relative closeness coefficient that represents its proximity to the ideal solution.

Based on these factors, the GoTrax GXL V2 was identified as the most suitable scooter, achieving the highest relative closeness to the ideal solution. This model offers an optimal balance across all evaluated criteria, demonstrating excellence in performance characteristics such as battery autonomy and charging time, as well as in user experience factors like handling and comfort. The Xiaomi Mi Electric followed in second place, with the Razor EcoSmart ranking third. Conversely, the Dualtron Thunder was considered the least suitable, occupying the seventh position in the evaluation.

The integration of TOPSIS into the optimization framework ensured that the selected electric scooter not only meets the technical demands of the optimized routes but also enhances the overall visitor experience and aligns with the park's sustainability goals.

By systematically considering multiple criteria and weighting them according to their importance, a comprehensive evaluation was achieved that supports informed decision-making in the selection of mobility solutions.

Table 3. TOPSIS ranking results of the top 7 electric scooter models based on a weighted sum of performance factors.

Brand	Weighted sum	Results (Top 7)
Xiaomi Mi Electric	0.324	2
Segway Ninebot	0.290	5
Razor EcoSmart	0.316	3
Apollo Explore	0.211	6
GoTrax GXL V2	0.350	1
Dualtron Thunder	−0.054	7
Unagi Model One	0.316	4

Performance of the solar energy system

The implemented photovoltaic system demonstrated effective support for the electric scooter operations within Los Petenes National Park. The average daily energy generation was measured at 31.58 kWh, slightly exceeding the required 30 kWh, which provided a buffer to accommodate variations in solar irradiance due to weather conditions.

The reliance on solar energy resulted in a significant reduction of carbon emissions. Assuming that grid electricity has an emission factor of 0.5 kg CO_2 per kWh, the solar energy system prevented the emission of approximately 15 kg of CO_2 per day, amounting to an estimated 5,475 kg of CO_2 annually.

Additionally, analyzing hybrid energy systems would be beneficial, as wind energy could effectively contribute to energy generation in this geographic area, enhancing the overall reliability and sustainability of the energy supply. Future work will examine the feasibility and logistics of this hybrid system, including its economic and environmental impacts.

Integration with mobility optimization

The solar energy system is directly linked to the optimized mobility framework established by Dijkstra's algorithm and the selection of the most suitable electric scooter through TOPSIS. The strategic placement of solar charging stations corresponds with the high-traffic nodes identified in the route optimization process. This alignment ensures that scooters can be recharged efficiently without deviating from the optimized paths, thereby maintaining the reduced travel times and increased visitor satisfaction achieved through the route optimization.

The use of solar energy directly supports the operational requirements of the selected electric scooter model, the GoTrax GXL V2, which was chosen based on criteria that include compatibility with renewable energy usage. The synergy between the energy supply and the mobility demand enhances the overall efficiency and sustainability of the transportation service within the park.

Integrating the photovoltaic system into the electric scooter mobility model not only meets energy needs but also reinforces the sustainability objectives of Los Petenes National Park. The direct relationship between the solar energy component and the mobility system is evident in operational efficiency, reduced environmental impact, and economic viability.

The photovoltaic system ensures operational efficiency by providing a reliable and continuous energy supply for the electric scooters, which is crucial for maintaining the optimized routes and schedules determined by Dijkstra's algorithm. Moreover, the significant reduction in carbon emissions aligns with the park's commitment to environmental conservation and offers a tangible benefit that can be communicated to visitors, thereby enhancing the ecotourism appeal. Additionally, the use of solar energy reduces operational costs associated with electricity consumption from the grid, contributing to the economic viability and financial sustainability of the mobility system.

By integrating the renewable energy system with the optimized mobility framework, the study demonstrates a holistic approach to sustainable transportation in ecotourism settings. The solar energy component is not an isolated addition but a fundamental element that directly influences the effectiveness and sustainability of the proposed mobility solution.

5 Conclusions

This research presents a sustainable smart mobility model for Los Petenes National Park, combining electric scooters with solar renewable energy systems to enhance eco-tourism and improve visitor mobility under challenging climatic conditions. By applying Dijkstra's algorithm, the study establishes an efficient optimization framework that reduces waiting times and unnecessary travel, while the Displacement Cost Index and Gaudability Index provide essential metrics for evaluating accessibility and visitor engagement, specifically tailored to Gen Z preferences.

The application of the TOPSIS method allowed for an objective and quantitative assessment of multiple electric scooter options, ensuring that the selected model aligns with both operational requirements and visitor preferences. The GoTrax GXL V2 was identified as the optimal choice, highlighting the importance of comprehensive criteria evaluation in the successful implementation of sustainable mobility initiatives in ecotourism settings.

Crucially, the integration of solar energy into the mobility system not only meets the energy demands of the optimized transportation network but also significantly reduces carbon emissions and operational costs, reinforcing the park's commitment to environmental conservation. This renewable energy component demonstrates the practical viability of combining clean energy solutions with advanced transportation models.

By integrating advanced optimization algorithms with multicriteria decision-making methods and renewable energy solutions, this study offers a scalable and adaptable model for other parks and tourist destinations seeking to implement eco-friendly technologies in their operations. The proposed approach contributes to sustainable development goals by enhancing visitor experiences, promoting environmental stewardship, and providing economic benefits to local communities.

References

1. Cohen, E.: The impact of tourism on the physical environment. Ann. Tourism Res, **5**(2), 215–237 (1978). https://doi.org/10.1016/0160-7383(78)90221-9
2. Martínez, Y.C., Baños, M.R., Monroy, H.C.: Ecotourism as a path to sustainable development in an isolated Magic Town. J. Tourism Anal.: Revista de Análisis Turístico (JTA) (2021). https://doi.org/10.53596/jta.v25i1.335
3. García, M.O.: Turismo masivo y alternativo. Distinciones de la sociedad moderna/posmoderna. Directory Open Access J. (2010). https://doaj.org/article/02235212e862484c8b3656592714c19d
4. Deery, M., Jago, L., Fredline, L.: Rethinking social impacts of tourism research: a new research agenda. Tour. Manage. **33**(1), 64–73 (2012). https://doi.org/10.1016/j.tourman.2011.01.026
5. Ouboter, D.A., Kadosoe, V.S., Ouboter, P.E.: Impact of ecotourism on abundance, diversity and activity patterns of medium-large terrestrial mammals at Brownsberg Nature Park, Suriname. PLoS ONE **16**(6), e0250390 (2021). https://doi.org/10.1371/journal.pone.0250390
6. Lee, T.H., Jan, F.-H.: Can community-based tourism contribute to sustainable development? Evidence from residents' perceptions of the sustainability. Tourism Manag. **70**, 368–380 (2019). https://doi.org/10.1016/j.tourman.2018.09.003
7. Donohoe, H.M., Needham, R.D.: Ecotourism: the evolving contemporary definition. J. Ecotour. **5**(3), 192–210 (2006). https://doi.org/10.2167/joe152.0
8. Kousis, M.: Tourism and the environment. Ann. Tour. Res. **27**(2), 468–489 (2000). https://doi.org/10.1016/s0160-7383(99)00083-3
9. Islebe, G.A., Calmé, S., León-Cortés, J.L., Schmook, B.: Biodiversity and Conservation of the Yucatán Peninsula (2015). https://doi.org/10.1007/978-3-319-06529-8
10. Durán-García, R., Méndez-González, M., Larqué-Saavedra, A.: The biodiversity of the yucatan peninsula: a natural laboratory. In: Cánovas, F.M., Lüttge, U., Matyssek, R. (eds.) Progress in Botany Vol. 78, pp. 237–258. Springer International Publishing, Cham (2017). https://doi.org/10.1007/124_2016_8
11. Doyon, S., Sabinot, C.: A new 'conservation space'? Protected areas, environmental economic activities and discourses in two yucatán biosphere reserves in Mexico. Conservat. Soc. **12**(2), 133 (2014). https://doi.org/10.4103/0972-4923.138409

12. Sonaglio, K.E.: Transdisciplinar o turismo: um ensaio sobre a base paradigmática making. PASOS Revista de turismo y patrimonio cultural **11**(1), 205–216 (2013). https://doi.org/10.25145/j.pasos.2013.11.013
13. INEGI: Anuario estadístico y geográfico de los Estados Unidos Mexicanos (2023)
14. Matsee, W., Charoensakulchai, S., Khatib, A.N.: Heat-related illnesses are an increasing threat for travellers to hot climate destinations. J. Travel Med. (2023). https://doi.org/10.1093/jtm/taad072
15. Youngs, Y.L., White, D.D., Wodrich, J.A.: Transportation systems as cultural landscapes in national parks: the case of Yosemite. Soc. Nat. Resour. **21**(9), 797–811 (2008). https://doi.org/10.1080/08941920801942065
16. Christian, M., Yulita, H., Girsang, L.R., Wibowo, S., Indriyarti, E.R., Sunarno, S.: The impact of cashless payment in application-based transportation on Gen Z user behavior in Jakarta. In: International Conference on IT Innovation and Knowledge Discovery (2023). https://doi.org/10.1109/itikd56332.2023.10100198
17. Contreras-Masse, R., Ochoa, A., Hernandez-Baez, I., Ronquillo, C., Garcia, H., Torres-Escobar, R.: The sustainable fashion revolution considering circular economy and targeting generation Z by reusing garments with Acrylan and Terlenka. Int. J. Combinatorial Optimization Probl. Infor. **15**(2), 72–84 (2024). https://doi.org/10.61467/2007.1558.2024.v15i2.472
18. Chen, X., Li, T., Yuan, Q.: Impacts of built environment on travel behaviors of Generation Z: a longitudinal perspective. Transportation **50**(2), 407–436 (2023). https://doi.org/10.1007/s11116-021-10249-6
19. Ma, Y., Li, Y., Han, F.: Interconnected eco-consciousness: Gen Z travelers' intentions toward low-carbon transportation and hotels. Sustainability **16**(15), 6559 (2024). https://doi.org/10.3390/su16156559
20. Samal, R., Dash, M.: Ecotourism, biodiversity conservation and livelihoods: Understanding the convergence and divergence. Int. J. Geoheritage Parks **11**(1), 1–20 (2023). https://doi.org/10.1016/j.ijgeop.2022.11.001
21. Hald-Mortensen, C.: The main drivers of biodiversity loss: a brief overview. J. Ecol. Nat. Resour. (2023). https://doi.org/10.23880/jenr-16000346
22. Pilliod, D., et al.: Chapter S. Adaptive management and monitoring, pp. 223–327 (2021)
23. Dijkstra, E.W.: A note on two problems in connexion with graphs. Numer. Math. **1**(1), 269–271 (1959). https://doi.org/10.1007/BF01386390
24. Bozyigit, A., Alankus, G., Nasiboglu, E.: Public transport route planning: Modified dijkstra's algorithm. In: 2017 International Conference on Computer Science and Engineering (UBMK) (2017). https://doi.org/10.1109/ubmk.2017.8093444
25. Sangaiah, A.K., Han, M., Zhang, S.: An investigation of Dijkstra and Floyd algorithms in national city traffic advisory procedures. Int. J. Comput. Sci. Mob. Comput. **3**(2), 124–138 (2014)
26. Tirastittam, P., Waiyawuththanapoom, P.: Public transport planning system by dijkstra algorithm: case study Bangkok metropolitan area. Zenodo (CERN European Organization for Nuclear Research) (2014). https://doi.org/10.5281/zenodo.1336370
27. Gunawan, E.P., Tho, C.: Development of an application for tourism route recommendations with the dijkstra algorithm. In: International Conference on Information Management and Technology (2021). https://doi.org/10.1109/icimtech53080.2021.9534998
28. Sipayung, L.Y., Sinaga, C.R., Sagala, A.C.: Application of dijkstra's algorithm to determine the shortest route from city center to medan city tourist attractions. J. Comput. Netw. Architect. High Perform. Comput. **5**(2), 648–655 (2023). https://doi.org/10.47709/cnahpc.v5i2.2699
29. Seran, K.J.T., et al.: Dijkstra algorithm to find shortest path of tourist destination in the border area of Indonesia – Timor Leste. In: AIP Conference Proceedings (2023). https://doi.org/10.1063/5.0154461

30. Dawood, S.A., Al-Hinkawi, W.S.: Sustainable transportation for healthy tourist environment: Erbil city-Iraq a case study. In: Calabrò, F., Spina, L.D., Mantiñán, M.J.P. (eds.) New Metropolitan Perspectives: Post COVID Dynamics: Green and Digital Transition, between Metropolitan and Return to Villages Perspectives, pp. 2496–2504. Springer International Publishing, Cham (2022). https://doi.org/10.1007/978-3-031-06825-6_238
31. Ioncică, D., Ioncică, M., Petrescu, E.-C.: The environment, tourist transport and the sustainable development of tourism. Directory of Open Access J. (2016). https://doaj.org/article/1797d8ee019844d0a9d962f45c3db7b0
32. Gössling, S.: Integrating e-scooters in urban transportation: problems, policies, and the prospect of system change. Transp. Res. Part D: Transp. Environ. **79**, 102230 (2020). https://doi.org/10.1016/j.trd.2020.102230
33. Hwang, J.J.: Sustainable transport strategy for promoting zero-emission electric scooters in Taiwan. Renew. Sustain. Energ. Rev. (2010). https://doi.org/10.1016/j.rser.2010.01.014
34. Liu, J., Guo, J., Li, R., Xu, J.: Perception, travel satisfaction associated with the use of electric moped scooters: evidence from Suzhou, China. Res. Square (2024). https://doi.org/10.21203/rs.3.rs-5016508/v1
35. Aguilera-García, Á., Gomez, J., Sobrino, N.: Exploring the adoption of moped scooter-sharing systems in Spanish urban areas. Cities **96**, 102424 (2020). https://doi.org/10.1016/j.cities.2019.102424
36. Guan, X., van Lierop, D., An, Z., Heinen, E., Ettema, D.: Shared micro-mobility and transport equity: a case study of three European countries. Cities **153**, 105298 (2024). https://doi.org/10.1016/j.cities.2024.105298
37. Bai, S., Jiao, J.: Dockless e-scooter usage patterns and urban built environments: a comparison study of Austin, TX, and Minneapolis, MN. Travel Behav. Soc. **20**, 264–272 (2020). https://doi.org/10.1016/j.tbs.2020.04.005
38. Hosseinzadeh, A., Algomaiah, M., Kluger, R., Li, Z.: E-scooters and sustainability: investigating the relationship between the density of E-scooter trips and characteristics of sustainable urban development. Sustain. Cities Soc. **66**, 102624 (2021). https://doi.org/10.1016/j.scs.2020.102624
39. Xu, L., Ao, C., Liu, B., Cai, Z.: Ecotourism and sustainable development: a scientometric review of global research trends. Environ. Dev. Sustain. **25**(4), 2977–3003 (2023). https://doi.org/10.1007/s10668-022-02190-0
40. Secretaría de medio ambiente y recursos naturales: Los Petenes, Reserva de la Biosfera (2024). https://www.gob.mx/semarnat/articulos/los-petenes-reserva-de-la-biosfera
41. Comisión Nacional de Áreas Naturales Protegidas. Los Petenes. Decretos, Programas de Manejo CONANP (2022)
42. Mas, J.-F., Sandoval, J.C.: Analysis of landscape fragmentation in the "Los Petenes" protected area, Campeche, Mexico. Investigaciones Geográficas (43), 42–59 (2000). http://www.scielo.org.mx/pdf/igeo/n43/n43a4.pdf
43. Roy, P.D., Torrescano-Valle, N., Islebe, G.A., Gutiérrez-Ayala, L.V.: Late Holocene hydroclimate of the western Yucatan Peninsula (Mexico). J. Quat. Sci. **32**(8), 1112–1120 (2017). https://doi.org/10.1002/jqs.2988
44. Can-González, M.J., Ramos-Miranda, J., Ayala-Pérez, L.A., Flores-Hernández, D., Gómez-Criollo, F.J.: Evaluating the fish community of "Los Petenes" biosphere reserve, Campeche, Mexico, through the characteristics of the environment and indicators of taxonomic diversity. Thalassas: An Int. J. Marine Sci. **37**(1), 331–346 (2021). https://doi.org/10.1007/s41208-020-00250-8
45. Oliva, M., Montiel, S., García, A., Vidal, L.: Local perceptions of wildlife use in Los Petenes biosphere reserve, Mexico: Maya subsistence hunting in a conservation conflict context. Trop. Conserv. Sci. **7**(4), 781–795 (2014). https://doi.org/10.1177/194008291400700414

46. Cinco-Castro, S., Herrera-Silveira, J.: Vulnerability of mangrove ecosystems to climate change effects: the case of the Yucatan Peninsula. Ocean Coast. Manag. **192**, 105196 (2020). https://doi.org/10.1016/j.ocecoaman.2020.105196
47. Buenfil Román, V., Espadas Baños, G.A., Quej Solís, C.A., Flota-Bañuelos, M.I., Rivero, M., Escalante Soberanis, M.A.: Comparative study on the cost of hybrid energy and energy storage systems in remote rural communities near Yucatan, Mexico. Appl. Energy (2022). https://doi.org/10.1016/j.apenergy.2021.118334

Use of Convolutional Neural Networks for the Recognition of Bird Species in Risk Categories in the State of Chihuahua

Jose Luis Acosta-Roman[1]([✉]), Alberto Ochoa-Zezzatti[1], and Martin Montes-Rivera[2]

[1] Universidad Autónoma de Ciudad Juárez, Ciudad Juárez, Chihuahua, Mexico
al237866@alumnos.uacj.mx
[2] Universidad Politécnica de Aguascalientes, Aguascalientes, Mexico

Abstract. In recent years, the state of Chihuahua has experienced overexploitation of natural resources, habitat fragmentation, contamination of soils and bodies of water, and desertification in several areas, in addition to the modification of the environment for urban expansion, which has led to the loss of biodiversity in the state. In view of this situation, there is a need for technological tools to evaluate and monitor the presence and absence of species in specific regions. Some of the most relevant organisms for this analysis are the birds of the state of Chihuahua, since they are excellent bioindicators of the state of the ecosystems and their presence is related to a high level of biodiversity. To address these situations of biodiversity loss, convolutional neural networks (CNN) are very useful, since these methodologies have demonstrated great effectiveness in different areas involving object recognition, due to their ability to learn automatically from the patterns, shapes and colors present in the images. In the context of population ecology, the application of CNNs allows the analysis of many images at the same time, in a fast way, which can contribute and facilitate the identification and monitoring of species, as well as the changes that occur in their populations, facilitating decision-making for conservation and sustainable development.

Keywords: Convolutional neural networks · Conservation · Chihuahua · Birds

1 Introduction

Biodiversity has always been important for human development, providing the materials or tools necessary for survival and later the development of technologies; however, in recent years the entire planet has experienced losses of natural resources, leading to the degradation of ecosystems, loss of ecosystems and changes in the normal distributions of species [1]. Some of the factors that have a major influence on biodiversity loss worldwide are agriculture, deforestation, and the expansion of urban areas for the creation of human settlements, which has led to the creation of eco-sustainable practices and models, which seek to reduce the impact of humans on the environment to try to mitigate and remedy the current situation of biodiversity loss [2, 3].

Thus, one of the most serious consequences of biodiversity loss is the loss of species, of course, and one of the most influential factors in this is population growth, which is due to the exploitation of the resources necessary to cover the basic needs of the people of a town or city, the latter together with poor planning of resource use or the lack of a plan aimed at sustainable development, can increase the rate at which biodiversity is lost in certain areas [4].

An example of all the above mentioned is the state of Chihuahua, this is the largest state in Mexico and therefore it is complicated to protect all the biodiversity that can be found in it, this is partly due to the orography and geographical position of the state, as these make it difficult to monitor the status of biodiversity and the presence or absence of species in the area, because if there is any disturbance in the environment organisms tend to seek a new area for their distribution, To this must be added that the state of Chihuahua serves as a transit site for migratory species, the most representative of which are birds, so that many species are affected over the years by the modification of their arrival environment during the migratory seasons or cause native species to look for a new area to inhabit [5, 6].

However, some different laws and strategies have emerged to try to mitigate the negative effects on ecosystems through techniques for evaluating the richness of ecosystems, such as sampling organisms, taking photographs to identify organisms and estimate population sizes, creating land use maps, and using satellite images to observe changes in the ecosystem over the years [7–9]. Naturally, with the development and advancement of technology, all these methods and techniques have had several improvements, allowing better monitoring of the state of ecosystems and better knowledge for developing new laws or modifications that will enable urban development without compromising the total resources of the future.

Therefore, the interest in biodiversity protection and the rapid advances in technology make it possible to implement more innovative tools such as monitoring technologies [10], object detection [11], drones [12], telemetry tools combined with tools such as deep learning and convolutional neural networks [13], all of which make it possible to recognize more quickly the causes of biodiversity loss and make decisions for the protection of ecosystems.

It is necessary to emphasize that, of the technologies mentioned above, those that involve deep learning and convolutional neural networks have shown great results in recent years when used for object detection, such is the case of their application in areas such as medicine, highlighting their potential for the recognition of tumors or any other medical condition [14], as well as in the automotive industry and other similar areas for the recognition of errors in parts during their production [15].

For that reason, this work has the objective of using convolutional neural networks for the recognition of species that are in some category of risk to contribute to the use of new methods in the recognition of these organisms in the state of Chihuahua and to contribute to decision-making for their protection, as well as to establish a relationship between the current status of the species and their environment.

2 Related Works

2.1 CNN in Species Conservation

Convolutional neural networks have proven to be very efficient for object detection in various areas and their application in species identification is no exception, but the efficiency of the model depends directly on the architecture being used, these architectures are chosen depending on the resources, time, number of images available and the purpose of the project [13, 16, 17].

Some architectures have shown great utility for the detection and identification of birds, for example, the EfficientNet-B0 architecture, which focuses on improving the efficiency of data processing, resulting in the use of a large number of images for model training, which improves learning and achieves better accuracy and efficiency for the recognition and classification of different species of birds through the use of a large number of images [18].

Another architecture is Mobilenet, which has an approach related to the lack of more limited computational resources, this is useful when seeking to reduce the number of parameters that are considered to perform an investigation, allowing the use of a large set of images separated into several classes and the models obtained are relatively lighter compared to other architectures, which even allows creating applications for mobile devices due to the high efficiency they show in the recognition of species [19].

VGG architectures are characterized by having several convolution layers, between 16 and 19, the function of these layers is to function as a filter that maintains the information that is presented in the atypical geometric shapes, and as the iteration of the model progresses it reduces the size of the filters during the interleaving [20].

3 Methodology

3.1 Datasets

In this work, a dataset of ten bird species in some category of risk, threatened or higher category was created. The selected species were searched in the iNaturalist database, taking as a basis the presence of the species in the state regardless of whether they are migratory or endemic species, which resulted in the species shown in Table 1.

Being species in some risk category, the number of images of these is lower than the amount that could be found of species that are not within these conditions, so the few-shot learning was added to the work, this technique allows the models to learn efficiently when there is a small number of samples for the study [21]. That said, the search for images resulted in obtaining 250 images belonging to the 10 different species of birds, the filter that was applied for their collection was that they present different positions in which it is possible to find the birds in their natural habitats such as perched on an object, several birds of the same species in an image, birds in flight, at a distance and proximity, as shown in Fig. 1. Of the 250 images, 200 were used for training the model, representing 80%, and 50 were used for testing, representing the remaining 20%, this allows the generated model to have enough information to learn from the images and avoid overfitting [22].

Table 1. Species of birds selected for this work and risk category in which they are found

Species No	Common name	Scientific name	Risk category
1	Golden eagle	*Aquila chrysaetos*	Threatened
2	White-rumped Hawk	*Buteo albonotatus*	Subject to special protection
3	Mottled owl	*Strix virgata*	Threatened
4	Painted bunting	*Passerina ciris*	Subject to special protection
5	Thick billed parrot	*Rhynchopsitta terrisi*	Endangered
6	Peregrine falcon	*Falco peregrinus*	Subject to special protection
7	Finsch's Amazon	*Amazona finschi*	Endangered
8	Mexican duck	*Anas diazi*	Threatened
9	Eared quetzal	*Euptilotis neoxenus*	Threatened
10	Rufous Hummingbird	*Selasphorus rufus*	Near threatened

Fig. 1. Examples of the images that were used for the study, taken as a reference the most common positions in which birds can be found in the wild, A. on some objects B. several birds C. birds in flight D. birds in the distance E. birds in proximity. F. behind objects.

3.2 Image Processing

This work was done in python programming language version 3.12.3 and a customized environment with CUDA 11.8, using a laptop with an intel core i7–6700 HQ 2.60 GHz processor, 16 GB of ram and a NVIDIA GeForce GTX 970M GPU.

Once the images of interest were collected, a code was developed in which the images of the birds were processed by applying magnifications, random rotations, and changes of scale so that the models would present greater robustness through the interpretation of variations and using 3 color channels (RGB), The images were resized to a size of 64x64 pixels to ensure the uniformity of all the images used, it is worth mentioning that reducing the size and quality of the images does not affect their processing and allows the analysis of the images to be faster [23] (Fig. 2).

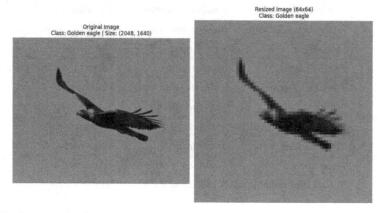

Fig. 2. Example of the resizing that was applied to all the images used in this work.

Subsequently, the images were resized again to 128 × 128 resolutions, allowing the model training to be applied employing multiple scales and data augmentation, which allows the model to be more complete and learn from features present in each image regardless of whether they are low or high resolution [24] (Fig. 3).

Fig. 3. Images representing the rotations, magnifications, and scale changes to which the images were exposed.

3.3 CNN Architecture

The CNN architecture used was tiny VGG, this architecture consists of the application of convolutional blocks combined with clustering layers followed by the use of ReLU, which are activation functions, and then a max-pooling layer is applied to reduce the features identified in the images. This convolutional neural network is a compressed version of the Visual Geometry Group model, which is used when limited computational resources are available. VGG models are characterized by using several 3 × 3 convolution layers, these layers have the function of filters that are applied to the images while the iteration is completed, allowing the extraction of as much information present for pattern recognition and by applying ReLU the model can learn from these patterns, these ReLU are also important because they allow the CNN can learn from complex images and recognize nonlinear patterns that are present [25, 26] (Fig. 4).

```
TinyVGG
─Sequential: 1-1
    └─Conv2d: 2-1
    └─ReLU: 2-2
    └─Conv2d: 2-3
    └─ReLU: 2-4
    └─MaxPool2d: 2-5
─Sequential: 1-2
    └─Conv2d: 2-6
    └─ReLU: 2-7
    └─Conv2d: 2-8
    └─ReLU: 2-9
    └─MaxPool2d: 2-10
─Sequential: 1-3
    └─Flatten: 2-11
    └─Linear: 2-12
```

Fig. 4. Image showing the functioning of the architecture Tiny VGG

3.4 Model Training and Evaluation

The model was trained by using cross entropy, this function allows the evaluation of the probability predicted by the model and the real probability represented by the images [27], additionally, 500 epochs were used in the images, indicating that the model made the path of that number of epochs to all the images of this work, in addition, the optimizer function was added to update the model parameters. Finally, batches of 32 images were used, to make the process more efficient and to take maximum advantage of the computational resources [28].

Once the model was trained, print functions were used within the code to obtain the confusion between the study species and the graph showing the maximum values of the test loss, to show how well the model performed in making predictions, and test accuracy, to show the total correct value of the predictions made using the study images with the trained model.

3.5 Image Based Prediction

Once the model was trained, we proceeded to use several images of the species used in this study to test the efficiency and success of the model for the correct identification of the species, this is done by loading an image not used for training the model, these images represent birds in flight, perched on objects, among branches or several birds in an image, because these are normally the most common situations in which a photograph can be taken of these organisms. Finally, each of these images is displayed with the name of the predicted species to which it belongs and the probability of that prediction.

4 Results

4.1 Graphics

Figure 5 shows the performance that the model had with the 250 images of the study applying the tiny VGG architecture, the graph on the left shows the train loss and test loss, these are used to show how good or bad is the model to make predictions by comparing them with the real labels that have the images used in the study, the maximum values obtained for the training loss were 2. 9347, while for the test loss were 2.1170, both curves of the graph show a downward trend, this indicates that the model minimizes the error with each passing epoch, however, there are many variations in the last part of the test loss graph, which may indicate overfitting of the data.

The graph on the right shows the learning performance of the model to be able to make correct predictions with the advance of the epochs, the maximum values obtained for the training accuracy were 0.2070 and test accuracy 0.3658, the curves show an upward trend, this means that the model is improving in its ability to predict and correctly classify the species of the study.

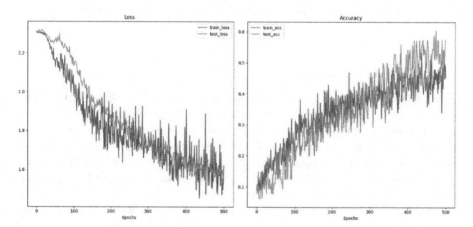

Fig. 5. Plots showing the train loss, test loss, train accuracy, and test accuracy of the model during the 500 epochs for model training.

4.2 Confusion Matrix

Below is the confusion matrix, Fig. 6, which evaluated the performance of the model obtained for the classification of the birds in the study, this allows visualizing how good the model is for classifying the different species of birds used in the study, on the Y-axis you can find the correct names of the species used in the study, while on the X-axis the model uses the same names, but for its predictions.

The matrix shows boxes with different colors, each of which contains the number of predictions that the model made for a given class; the darker colors represent a greater number of correct predictions while the lighter colors indicate that the species have been predicted very few times or that the model has made some errors during the classification of the organisms. The boxes on the diagonal, top left to bottom right, represent those cases in which the model correctly classified the species in the study, so it can be seen that the model is good at making predictions for species such as the eared quetzal, thick-billed parrot, and white-rumped hawk, which has a medium efficiency for the finsch's amazon and Golden Eagle, and for the Mexican duck, mottled owl, painted bunting, peregrine falcon, and rufous hummingbird it has more problems for its classification, this may be because some of the species have very similar characteristics. This is an indicator that the model shows a reasonable amount of accuracy in the predictions for some of the species, but adjustments are needed.

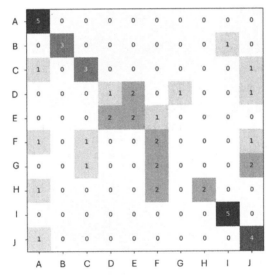

Fig. 6. Confusion matrix showing the classification performance of the model for species classification. A. Eared quetzal B. Finsch's amazon C. Golden eagle D. Mexican duck E. Mottled owl F. Painted bunting G. Peregrin falcon H. Rufous gumming bird I. Thick billed parrot J. White rumped hawk

4.3 Image Prediction

Table 2 shows the predictions made by the model loading new images of the species used in this work. As can be seen, this model presented higher prediction values for the species of thick-billed parrot, finsch's Amazon, and the peregrine falcon and, in the case of the first two, this may be because, of all the species used in the study, they are those in which the green color predominates while the rest of the species present darker colors, such as variations of brown or have small features of bluish colors, In the case of the peregrine falcon something similar happens, since it has morphological similarities with the golden eagle and white-rumped hawk, however, the colors and patterns in its plumage are very different.

As for the rest of the species, the predictions were correct but the values are low, this may be due to the similarity in the shades of the birds' plumage since most of them are variations of grayish or brownish colors, which makes it difficult for the model to distinguish the patterns present in the different species.

Table 2. Model prediction results using new images

Species No	Common name	Maximum prediction value	Predicted values in percent
1	Golden eagle	0.2801	28.01%
2	White-rumped hawk	0.3540	35.40%
3	Mottled owl	0.2619	26.19%
4	Painted bunting	0.3294	32.94%
5	Thick billed parrot	0.7455	74.55%
6	Peregrine falcon	0.7555	75.55%
7	Finsch's amazon	0.7187	71.87%
8	Mexican duck	0.2430	24.30%
9	Eared quetzal	0.4010	40.10%
10	Rufous hummingbird	0.1790	17.90%

5 Discussion

The results obtained in this work reflect the efficiency of the use of convolutional neural networks (CNN) for the recognition of endangered bird species in the state of Chihuahua. During the training, validation, and data acquisition process, a good capacity to recognize the complex patterns present in the images of the different species was observed, which allowed a correct classification of the species by entering new images; however, relatively low values can be observed in the recognition of the species, being the highest value 75.55%, while the studies carried out by Neeli, et al. in 2023 using Efficient net obtained an efficiency of 95.57% [18] and Islam, et al. [19] in 2019 in their study using mobilenet found an efficiency above 90%, this difference is because in those studies the number of images used was above 3000 photographs of various birds and the CNN architectures used are more complex and have object recognition features that are not suitable for low computational power, on the other hand, this study has only 250 images, this is due to the rarity of the species and the low availability of royalty-free images. One of the main contributions of this study was the use of data augmentation and few-shot learning techniques to mitigate the challenge associated with data sparsity; the application of rotation, magnification, and scaling of the images helped to increase the diversity of the input data, which significantly improved the model's ability to make identifications of birds under different visual conditions. This strategy is very useful when a limited dataset is available, as is the case for species that are in some risk category, so the use of the Tiny VGG architecture proved to be effective for this study. Although it is a compressed version of the VGG model, its performance was adequate for the proposed task, showing good efficiency between prediction accuracy and computational resources. e species and the low availability of royalty-free images.

However, although the model is robust, the graphs and the confusion matrix show that there is still room for improvement since the model shows difficulties in the recognition of certain species such as the mottled owl, painted bunting, and Mexican duck, this could be achieved by adding more images and increasing the number of epochs for the study.

6 Conclusion

The study showed that the use of convolutional neural networks (CNN) for bird recognition in the state of Chihuahua is beneficial and has great potential for its application in species conservation, however, modifications should be made, such as the addition of a greater number of images, increasing the number of epochs for the model or adding a segment to the code in which the modeling stops when it is detected that there is no significant improvement, all this to make better predictions. The Tiny VGG architecture was used in this work due to its balance between efficiency and computational resource consumption, allowing species identification even with hardware limitations.

All the species analyzed are in a risk category, which is why they were chosen for the study and were correctly classified using CNN, which denotes a broad potential for monitoring and protecting biodiversity in the state. One of the main achievements of this research was the incorporation of few-shot learning, which allowed for overcoming the limitation of having a limited number of images for model training. This turned out to be a key point to improve the performance of the model by having a smaller number of photos of species in some risk category since sightings of these species are less frequent.

In addition, image processing using data augmentation, rotations, and scale changes allowed the model to identify more complex patterns in diverse scenarios. This not only improved the accuracy of the classification but also served to add robustness to the model by adding and emulating the countries from which a photograph can be taken. Finally, this work shows that CNNs have great potential in conservation, serving as methods for monitoring and evaluating the status of species, which can contribute to decision-making for proper management of resources and animals in various areas, as well as the possible establishment of biodiversity protection areas.

7 Future Work

The future considerations of this work are the expansion of the dataset used, including a greater number of species, whether or not they are in any risk category, with the purpose of more accurately emulating the biological interactions of the species, this can serve to increase the accuracy of the model and the identification of common species of the areas or species that are in a different category. In addition, the integration of satellite images is planned to observe the changes in the regions over time and to establish whether or not urban developments or some other factors have a relationship with the establishment or movement of species in different zones.

Another aspect to consider would be the use of more complex architectures such as EfficientNet, ResNet, or Modelnet, which could improve the performance of the model without hardware limitations, in addition to the application of transfer learning to use other pre-trained models with other larger datasets, which will allow the efficiency of the model to be tested and compared.

Finally, we propose the development of an application based on CNN, which will allow citizens, volunteers, and people working directly on conservation projects to contribute to the collection of data on bird species in the area, which will allow having a constant source of information for monitoring the species, which in turn will facilitate the implementation of public policies for the conservation of endangered species in the state of Chihuahua.

Disclosure of Interests. The authors have no competing interests to declare that are relevant to the content of this article.

References

1. Geeta, R., Lohmann, L.G., Magallón, S., et al.: Biodiversity only makes sense in the light of evolution. J. Biosci. **39**, 333–337 (2014)
2. Caiado, R.G.G., de Freitas Dias, R., Mattos, L.V., Quelhas, O.L.G., Leal Filho, W.: Towards sustainable development through the perspective of eco-efficiency-A systematic literature review. J. Clean. Prod. **165**, 890–904 (2017)
3. Prakash, S., Verma, A.K.: Anthropogenic activities and biodiversity threats. Int. J. Bio. Innovations **4**(1), 94–103 (2022)
4. Zipperer, W.C., Northrop, R., Andreu, M.: Urban development and environmental degradation. In: Zipperer, W.C., Northrop, R., Andreu, M. (eds.) Oxford Research Encyclopedia of Environmental Science. Oxford University Press (2020). https://doi.org/10.1093/acrefore/9780199389414.013.97
5. De La Cerda Camargo, F.: Influence of orography on the weather patterns and water availability of a topographically complex Chihuahuan Desert region (2011)
6. Pool, D.B., Panjabi, A.O., Macias-Duarte, A., Solhjem, D.M.: Rapid expansion of croplands in Chihuahua, Mexico threatens declining North American grassland bird species. Biol. Cons. **170**, 274–281 (2014)
7. Hanisch, E., Johnston, R., Longnecker, N.: Cameras for conservation: wildlife photography and emotional engagement with biodiversity and nature. Hum. Dimens. Wildl. **24**(3), 267–284 (2019)
8. Turner, W., et al.: Free and open-access satellite data are key to biodiversity conservation. Biol. Cons. **182**, 173–176 (2015)
9. Pistocchi, A., Cassani, G., Zani, O.: Use of the USPED model for mapping soil erosion and managing best land conservation practices (2002)
10. Lahoz-Monfort, J.J., Magrath, M.J.: A comprehensive overview of technologies for species and habitat monitoring and conservation. Bioscience **71**(10), 1038–1062 (2021)

11. Van Gemert, J.C., Verschoor, C.R., Mettes, P., Epema, K., Koh, L.P., Wich, S.: Nature conservation drones for automatic localization and counting of animals. In: Computer Vision-ECCV 2014 Workshops: Zurich, Switzerland, 6–7 and 12 Sep 2014, Proceedings, Part I 13, pp. 255–270. Springer International Publishing (2015)
12. Hebblewhite, M., Haydon, D.T.: Distinguishing technology from biology: a critical review of the use of GPS telemetry data in ecology. Phil. Trans. R. Soc. B **365**(1550), 2303–2312 (2010)
13. Kaur, A., Kukreja, V., Chattopadhyay, S., Verma, A., Sharma, R.: Empowering Wildlife Conservation with a Fused CNN-SVM Deep Learning Model for Multi-Classification Using Drone-Based Imagery. In: 2024 IEEE International Conference on Interdisciplinary Approaches in Technology and Management for Social Innovation (IATMSI), vol. 2, pp. 1–4. IEEE (2024)
14. Yang, R., Yu, Y.: Artificial convolutional neural network in object detection and semantic segmentation for medical imaging analysis. Front. Oncol. **11**, 638182 (2021)
15. Carvajal Soto, J.A., Tavakolizadeh, F., Gyulai, D.: An online machine learning framework for early detection of product failures in an Industry 4.0 context. Int. J. Comput. Integrated Manuf. **32**(4–5), 452–465 (2019)
16. Mahmud, I., Kabir, M.M., Shin, J., Mistry, C., Tomioka, Y., Mridha, M.F.: Advancing wildlife protection: mask R-CNN for rail track identification and unwanted object detection. IEEE Access **11**, 99519–99534 (2023)
17. Kukreja, V., Kumar, M., Manwal, M., Jain, V.: Safeguarding biodiversity: remote forest poaching camp detection with CNN and spatial transformer networks. In: 2024 International Conference on Electronics, Computing, Communication and Control Technology (ICECCC), pp. 1–6. IEEE (2024)
18. Neeli, S., Guruguri, C.S.R., Kammara, A.R.A., Annepu, V., Bagadi, K., Chirra, V.R.R.: Bird Species Detection Using CNN and EfficientNet-B0. In: 2023 International Conference on Next Generation Electronics (NEleX), pp. 1–6. IEEE (2023)
19. Islam, M.R., Tasnim, N., Shuvo, S.B.: Mobilenet model for classifying local birds of bangladesh from image content using convolutional neural network. In: 2019 10th International Conference on Computing, Communication and Networking Technologies (ICCCNT), pp. 1–4. IEEE (2019)
20. Mascarenhas, S., Agarwal, M.: A comparison between VGG16, VGG19 and ResNet50 architecture frameworks for Image Classification. In: 2021 International Conference on Disruptive Technologies for Multi-Disciplinary Research and Applications (CENTCON), pp. 96–99. Bengaluru, India (2021), https://doi.org/10.1109/CENTCON52345.2021.9687944
21. Antonelli, S., et al.: Few-shot object detection: a survey. ACM Comput. Surv. **54**(11s), 1–37 (2022)
22. Muraina, I.: Ideal dataset splitting ratios in machine learning algorithms: general concerns for data scientists and data analysts. In: 7th International Mardin Artuklu Scientific Research Conference, pp. 496–504 (2022)
23. Semma, A., Lazrak, S., Hannad, Y., Boukhani, M., El Kettani, Y.: Writer identification: the effect of image resizing on CNN performance. ISPRS Int. Arch. Photogramm. Remote. Sens. Spat. Inf. Sci. **46w5**, 501–507 (2021)
24. Moreno, R.J., Avilés, O., Ovalle, D.M.: Evaluation of hyperparameters in CNN for detecting patterns in images. Visión electrónica **11**(2), 140–145 (2017)
25. Song, H., Lu, H.: Research on image recognition based on different depths of VGGNet. J. Image Process. Theory Appl. **7**, 84–90 (2024)
26. Ide, H., Kurita, T.: Improvement of learning for CNN with ReLU activation by sparse regularization. In: 2017 International Joint Conference on Neural Networks (IJCNN), pp. 2684–2691. IEEE (2017)

27. Ruby, U., Yendapalli, V.: Binary cross entropy with deep learning technique for image classification. Int. J. Adv. Trends Comput. Sci. Eng, **9**(10) (2020)
28. Ogundokun, R.O., Maskeliunas, R., Misra, S., Damaševičius, R.: Improved CNN based on batch normalization and Adam optimizer. In: International Conference on Computational Science and its Applications, pp. 593–604. Springer International Publishing, Cham (2022) https://doi.org/10.1007/978-3-031-10548-7_43

Exploring Deep Learning Applications in Neurodegenerative Diseases: A State-of-the-Art Review

Ayrton Santos, Claudia I. Gonzalez(✉), and Mario Garcia

Division of Graduate Studies and Research, TECNM/Tijuana Institute of Technology, 22414 Tijuana, Mexico
{ayrton.santos,mario}@tectijuana.edu.mx, cgonzalez@tectijuana.mx

Abstract. This paper presents a state-of-the-art review of the applications of deep learning in the study of principal neurodegenerative diseases, including Parkinson's disease, Alzheimer's disease, and Multiple Sclerosis. These diseases pose significant challenges in terms of early diagnosis and treatment due to their complex nature and progression. The aim of this review is to explore and highlight the main applications of deep learning techniques, including Convolutional Neural Networks (CNNs) in the classification and prediction of these neurodegenerative conditions. The review identifies and discusses the multimodal data employed in the experimentation, including medical imaging (MRI, PET), electrophysiological data (EEG), genetic data, and clinical assessments. We also address the challenges of integrating these diverse data modalities to enhance prediction accuracy and disease progression modeling. This review highlights the potential of deep learning models and their application in the area of neurodegenerative diseases, as well as identifying key areas for future research.

Keywords: Machine Learning · Artificial Intelligence · Neuronal Networks · Neurodegenerative Diseases · Alzheimer's disease · Parkinson's disease

1 Introduction

Neurodegenerative diseases are a group of age-related disorders that cause the death of specific types of neuronal cells. They are more common in the elderly population, making it a serious global health problem. Factors contributing to neurodegenerative diseases include genetic mutations, neuronal apoptosis, protein loss, and glial cell reactions [1]. There are different types of neurodegenerative diseases, which share symptoms. These are characterized by the progressive deterioration of physiological and cognitive functions. Diseases that impair the brain are varied, including Alzheimer's and Parkinson's disease [2]. Studying these neurodegenerative diseases is important because they pose a threat to health. The elderly population is the most affected, and it not only impacts them, but also generates high costs for health systems. In addition, the family and caregivers are also affected both emotionally and financially.

Artificial intelligence (AI), particularly deep learning, has emerged as a transformative technology in this domain. Deep learning algorithms, especially convolutional neural networks (CNNs) and recurrent neural networks (RNNs), have demonstrated remarkable success in tasks such as image recognition, natural language processing, and predictive modeling. These methods are now being repurposed to address the challenges of neurodegenerative diseases, offering novel solutions for diagnosis, progression monitoring, and therapeutic development. With the vast amounts of biomedical data generated from neuroimaging, genetics, and clinical assessments, deep learning has become indispensable for uncovering patterns and biomarkers that may remain hidden from traditional statistical methods.

This paper introduces the state-of-the-art in the implementation of deep learning applications and main datasets for the early detection of neurodegenerative diseases. Although a definitive cure has not yet been found, early diagnosis is critical, as it enables the timely administration of treatments that can help mitigate the onset of symptoms. The paper explores the use of advanced AI models in Neurodegenerative Diseases, including disease diagnosis, prognosis, and management. It also addresses critical challenges, such as the need for high-quality annotated datasets, model interpretability, and multimodal data integration. The search criteria for the paper review focus on filtering articles published between 2019 and 2024 in scientific journals. The review centers on neurodegenerative diseases, particularly Alzheimer's and Parkinson's, emphasizing the application of machine learning techniques for early diagnosis. Data sources such as OASIS and ADNI, which offer brain imaging and clinical information, are highlighted. Keywords used in the search include Machine Learning, Deep Learning, Artificial Intelligence, Convolutional Neural Networks, Neurodegenerative Diseases, Alzheimer's disease, and Parkinson's disease.

The work structure is as follows: Sect. 2 provides an overview of neurodegenerative diseases, detailing important statistics and the prevalence of Alzheimer's and Parkinson's disease worldwide. Section 3 presents a review of deep learning applications in diseases, focusing on specific approaches and models developed for the detection and diagnosis of both Alzheimer's and Parkinson's. Section 4 discusses the types of datasets used in deep learning for neurodegenerative diseases. Finally, the Conclusion summarizes the current state of deep learning applications in this field and outlines new directions for future research.

2 Overview of Neurodegenerative Diseases

Dementia is a decline in cognitive abilities that hinders a patient's ability to carry out daily life activities [2]. Here we present some of the main diseases, not a ranking, but the most prominent ones, which are detailed below.

Alzheimer's disease consists of a loss of cognitive ability, accompanied by memory failures, which progressively reduces memory retention effectiveness. One of the triggers of Alzheimer's is the abnormal production or accumulation of Aβ (beta-amyloid) in the brain [3]. The disease can last approximately 10 to 12 years, though this may vary depending on the patient. Currently, there is no single test to diagnose it; a series of medical evaluations is required to obtain confirmation [2].

The second most common after Alzheimer's, this neurodegenerative disease is caused by the accumulation and production of misfolded α-synuclein. Technically, Parkinson's disease presents with movement disorders, including resting tremors, rigidity, and instability, accompanied by the loss of brain cells known as neurons, which produce dopamine. By the time Parkinson's is evident, 70% of dopaminergic neurons are dead. In the terminal stages, more parts of the brain are involved [3].

Amyotrophic Lateral Sclerosis, this disease is fatal; its main symptoms include an irreversible loss of neurons responsible for movement, located in the spinal cord, brainstem, and cerebral cortex. The patient typically dies within 4 to 6 years. Currently, there is no medical treatment available. Associated genes have been found, but only in a minority of patients, who are classified as sporadic cases [3]. This means that it is not necessarily hereditary.

Neurodegenerative diseases are a broad range of progressive neurological problems that are age-dependent, affecting approximately 50 million people worldwide, especially older adults. In recent decades, the increase in the elderly population projects a mortality rate of 42%, representing 11.8% of all deaths. It is important to mention that some diseases are not fully understood, and for others, there is no cure or treatment, which ultimately leads to the patient's death [4].

In a study published in 2020, there were approximately 800,000 people affected by dementia. The current prevalence ranges between 7.9% and 9%, placing Mexico in 5th place with a high incidence. According to this study, it is estimated that by 2050, up to 3 million people will be affected by dementia [5]. As an interesting fact, the mortality risk ratio in urban areas is higher at 2.7. This is due to many factors, such as the quality of medical care, pollution, among other things, while in rural areas it is 1.6 [6].

3 Review of Deep Learning Applications by Disease

Currently, the development of technology, computer vision, neural networks, and pattern recognition has allowed us to expand these tools to other fields of research. Probably, artificial neural networks and artificial intelligence will be the future in addressing many health-related issues. AI, in this case, with its ability to recognize patterns, can help us predict which patients may be prone to developing some neurodegenerative disease. And although, as mentioned in previous sections, there is no definitive cure, treatment can be followed to prevent the loss of patients' quality of life from accelerating. Below, we will see a series of works focused on using machine learning techniques, a brief description of these are presented as follows and summarized in Table 1 for Alzheimer's Disease and in Table 2 for Parkinson.

3.1 Machine Learning Techniques in Alzheimer's Disease

Firstly, Dhakal et al. [7] proposed a dementia prediction system using machine learning through a project called OASIS (Open Access Series of Imaging Studies), provided by the Alzheimer's Research Center at Washington University. These data are preprocessed, and data transformation is performed to create a suitable set for training the model. Some of the main approaches used are AdaBoost, Decision Trees, Extra Tree, Gradient Boost,

K-Nearest Neighbor, Logistic Regression, Naïve Bayes (NB), Random Forests, and Support Vector Machines to combine features. Additionally, techniques such as Least Absolute Shrinkage and Selection Operator (LASSO) were applied. For this model, cross-sectional MRI images for young adults, middle-aged adults, and older adults, as well as longitudinal MRI images, were employed. The dataset consists of 150 patients aged between 60 and 96 years. The best result was achieved with the Support Vector Machine (SVM) algorithm, using all features reaching the highest accuracy, equivalent to 96.77%.

In the study by Chattopadhyay et al. [8], a 3D convolutional neural network was used to predict, based on MRIs, using biomarkers. The study consisted of 762 elderly patients: 459 healthy and 67 with mild cognitive impairment. The results showed that 236 patients had dementia, and the accuracy obtained was 76%.

In Li et al. [9] presented a multimodal feature selection of 3 types for Alzheimer's detection: magnetic resonance imaging, positron emission tomography, and cerebrospinal fluid analysis. Multimodal learning involves using different types of data and learning the best from that combined information, whether it is text, audio, or video. The issue is that common algorithms work with only one type of data at a time, which means that not all available information can always be utilized.

In Diogo et al. [10] the work is focused on Alzheimer's disease, early prediction of Alzheimer's and mild cognitive impairment is made using brain magnetic resonance imaging and machine learning. Two datasets were used: the Open Access Series of Imaging Studies (OASIS) and the Alzheimer's Disease Neuroimaging Initiative (ADNI). Some of the algorithms used included support vector machines, decision trees, random forests, extremely randomized trees, linear discriminant analysis, logistic regression, and logistic regression with stochastic gradient descent. The best results were obtained by the random forest and extremely randomized tree algorithms.

In Sarraf et al. [11], multiple statistical tools and machine learning algorithms were explored for diagnosing Alzheimer's disease in elderly individuals over 75 years old. As technology improved, it was applied to medical image classification. The samples were divided into 75% and 25%, used to train the algorithms, which were run on powerful GPUs. Using convolutional neural networks, the optimized vision transformer architecture called OViTAD was employed. These pipelines achieved average results of 94.32% and 97.88% for fMRI and MRI pipelines, respectively.

In Shamrat et al. [12] proposed AlzheimerNet framework, it is a classifier built on a convolutional neural network, designed to identify all stages of Alzheimer's, with one group of healthy individuals and another of affected individuals, using the ADNI MRI dataset. This was achieved by first applying preprocessing and data augmentation techniques. The models implemented were VGG16, MobileNetV2, AlexNet, ResNet50, and InceptionV3. Interestingly, InceptionV3 achieved high accuracy and was modified to create AlzheimerNet, with an RMSprop optimizer; the accuracy achieved was 98.67%.

In Jansi et al. [13] presented a model called Inception V3 which was enhanced with transfer learning, and it was employed to analyze MRI data and classify Alzheimer's patients and healthy individuals. Brightness adjustment was applied as part of the preprocessing to augment the dataset and improve accuracy. A technique called SMOTE was used to balance the classes. After preprocessing, the data was split into validation and training sets, achieving an accuracy of 87.69% with the OASIS dataset, which has been widely used in previous studies.

In Alamro et al. [14] research addresses feature selection, which produces better prediction performance than key gene sets. The five genes identified by both LASSO and Ridge methods achieved an area under the curve of 0.979. This work demonstrates how, with a reduced number of genes, Alzheimer's can be distinguished from healthy controls with high accuracy. The dataset used comes from Gene Expression Omnibus (GEO), filtered for Alzheimer's, and provides genetic data from brain tissue of healthy and affected patients within the same age range. Random Forest, Support Vector Machines, and convolutional neural networks were used.

In Thangavel et al. [15] an improved AdaBoost classifier technique is presented to diagnose Alzheimer's from MRI images, which reduces the construction time of a weak classifier using the PSO algorithm, replacing exhaustive search with an optimized PSO-based search in the weak classifier, which is a decision tree. The results show that the combination of PCA-PSO-AdaBoost is more accurate than other methods, outperforming PCA-AdaBoost and PSO-SVM-Cuckoo-AdaBoost by up to 16.47%.

In Franciotti et al. [16] presented a study that compares the performance of three machine learning algorithms: Random Forest, Gradient Boosting, and eXtreme Gradient Boosting, using multimodal biomarkers of subjects with mild cognitive impairment (MCI) obtained from the ADNI database. The prediction rate is related to the nature of the data, such as neuropsychological tests, Alzheimer's-related proteins, cerebrospinal fluid, MRI, among others. sMRI data alone had less accuracy (0.79), but when dealing with multimodal data, combining clinical and biological measures, higher accuracy was achieved (0.90).

3.2 Machine Learning Techniques in Parkinson's Disease

In McFall et al. [17] presented a study discussing how people with Parkinson's disease can develop dementia, but not all patients develop it gradually or at the same time. Data from 48 Parkinson's patients were collected, where 38 risk characteristics were evaluated, such as motor skills, cognitive abilities, blood molecules, among other factors. The Random Forest model, which is a machine learning algorithm, and a technique called Tree SHAP, which explains why certain factors are important for the model's predictions, were used. Random Forest was very accurate in classifying which patients developed dementia, with an area under the curve (AUC) of 0.84.

In Sotirakis et al. [18] six Inertial Measurement Unit (IMU) sensors were used to record the movements of patients with Parkinson's disease, applying linear regression algorithms and Random Forest, along with automated feature selection techniques. This resulted in seven models, where Random Forest was the most successful in predicting significant symptom progression over a 15-month period. These results indicate that the combination of sensors and machine learning models can provide a more accurate tracking of disease progression compared to traditional methods.

In a study presented in Hossain et al. [19], the voices of 525 people aged between 33 and 87 were recorded, and with these voice recordings, machine learning algorithms were able to distinguish between Parkinson's disease patients and healthy controls. A 10-fold cross-validation method was used across several models. During classification, the

Table 1. Summary of deep learning for Alzheimer's Disease.

Machine Learning techniques	Type of dataset	Authors	Year
AdaBoost, Decision Tree, Extra Tree, Gradient Boost, K-Nearest Neighbour, Logistic Regression, Naïve Bayes, Random Forest, Support Vector Machines	Mental test score, Degree of dementia, MRI image	Dhakal et al. [7]	2023
3D Convolutional Neural Networks	MRI images	Chattopadhyay et al. [8]	2023
Multi-modal neural networks, Anchor graph construction	MRI images, positron emission tomography, and cerebrospinal fluid analysis	Li et al. [9]	2022
Support Vector Machines, Decision Trees, Random Forest, Extremely Randomized Trees, Linear Discriminant Analysis, Logistic Regression	OASIS, Brain Imaging (MRI), Clinical Data, ADNI (Multimodal), Positron Emission Tomography (PET), CSF (Cerebrospinal Fluid)	Diogo et al. [10]	2022
Convolutional Neural Networks	fMRI, MRI	Sarraf et al. [11]	2021
Convolutional Neural Networks	ADNI (Multimodal), Positron Emission Tomography (PET), CSF (Cerebrospinal Fluid)	Shamrat et al. [12]	2023
Transfer Learning (Inception v3), SMOTE	OASIS, Brain Imaging (MRI), Clinical Data	Jansi et al. [13]	2023
Convolutional Neural Networks, Random Forest, SVM	Genetic Data of Healthy and Diseased Brains	Alamro et al. [14]	2023
AdaBoost, PSO (Particle Swarm Optimization), Convolutional Neural Networks	ADNI (Multimodal), Positron Emission Tomography (PET), CSF (Cerebrospinal Fluid)	Thangavel et al. [15]	2023
Random Forest, Gradient Boosting, Extreme Gradient Boosting	ADNI (Multimodal), Positron Emission Tomography (PET), CSF (Cerebrospinal Fluid)	Franciotti et al. [16]	2023

models with the implementation of a pipeline improved accuracy to 85.09%, with specific accuracy reaching 92%. The pipelines increased the accuracy of voice classification.

In Harvey et al. [20] a multivariate machine learning model was developed to predict cognitive outcomes in newly diagnosed Parkinson's disease cases using the PPMI dataset. Cognitive capacity assessments were conducted over an 8-year period to define two outcomes: cognitive impairment and conversion to dementia. The variables were organized, and machine learning algorithms were applied to them. The findings indicated that when machine learning algorithms were fed with multivariate information, they were more effective than their counterparts that used biofluid data and genetic variables.

On the other hand, the study presented by Hosseinzadeh et al. [21] aims to predict Parkinson's disease using hybrid machine learning systems such as multilayer perceptron, decision trees, and support vector machines, employing the PPMI dataset. By implementing cross-validation methods, the performance of the models was evaluated, highlighting the algorithms' ability to recognize patterns that identify Parkinson's disease.

In Islam et al. [22] 250 participants performed a standardized mobility task, specifically tapping a couple of times with their fingers in front of a webcam, to determine potential symptoms or the severity of Parkinson's disease symptoms.

In Govindu et al. [23] the voice measurement data from 31 people was collected, of which 23 are patients with Parkinson's disease, aged between 46 and 85 years, while normal readings come from people aged 23 years. An average of 195 phonations was recorded for each person, with a duration ranging from 1 to 36 s. Some of the tools used were logistic regression for classification, Random Forest classifier, and Support Vector Machine for the classification of Parkinson's disease, which offers a precision of 91.853% and a sensitivity of 0.95 for the Random Forest classifier. The results of the Random Forest model are good. The SVM and Random Forest models perform well for robust and outlier values.

In Alalayah et al. [24] a set of features from a person's voice is extracted and then early diagnostic methods for Parkinson's disease are employed. Subsequently, a series of machine learning algorithms are used to evaluate the selected features. This article applied two algorithms: t-distributed stochastic neighbor embedding (t-SNE) and principal component analysis (PCA). The experimental results demonstrated that the techniques had high success rates; the t-SNE algorithm achieved an accuracy of 97%, while its counterpart, PCA, reached 98%.

In the study by Abdullah et al. [25], a machine learning approach was presented for detecting Parkinson's disease through hand-drawn images. The proposed model achieved an accuracy of 95%.

4 Types of Datasets Used in Deep Learning for Neurodegenerative Diseases

The databases used in these studies consist of information collected worldwide by health organizations and hospitals. Although only some datasets are publicly available, others are gathered by scientists. These datasets are divided into three parts: those consisting of images, sound, and clinical variables of the patients. Table 3 presents a summary of these data set.

Table 2. Summary of deep learning for Parkinson's Disease.

Machine Learning techniques	Type of dataset	Authors	Year
Tree SHAP, Random Forest	Biomarcadores	McFall et al. [17]	2023
Automatic feature selection, Multivariate Linear Regression, Random Forest, Principal Component Analysis	Sensor Data	Sotirakis et al. [18]	2023
Support Vector Machine, Logistic Regression, K-Nearest Neighbors, Decision Tree, Random Forest, Multi-Layer Perceptron	Images from Audio spectrum	Hossain et al. [19]	2023
Random Forest, SVM, ElasticNet	PPMI	Harvey et al. [20]	2022
Extra Trees Classifier, XGBoost Classifier, K-Neighbors Classifier, Random Forest Classifier	PPMI	Hosseinzadeh et al. [21]	2023
Linear Regression, Decision Tree Regression, SVM	Domestic video Dataset	Islam et al. [22]	2023
Logistic Regression, Random Forest, SVM	Audio spectrum to image	Govindu et al. [23]	2023
Multilayer Perceptron, SVM, K-Nearest Neighbors, Decision Tree, Random Forest	Audio converted to image by Oxford University	Alalayah et al. [24]	2023
ResNet50, VGG19, Inception_v3	NewHandPD. Hand-drawn images by healthy patients and those with Parkinson's diseas	Abdullah et al. [25]	2023

4.1 Data Types

- **Imaging Datasets & Multimodal Datasets**

First, we have the OASIS [26] dataset (Open Access Series of Imaging Studies), which provides access to a magnetic resonance imaging (MRI) database focused on Alzheimer's disease.

Next, we have ADNI [27] (Alzheimer's Disease Neuroimaging Initiative), a collaborative effort that collects multimodal imaging data, such as MRI scans and positron emission tomography (PET) scans.
On the other hand, there is the UK Biobank [28], which includes computed tomography (CT) scans and MRI, and covers a wide range of risk factors, providing both imaging and genetic data.

- **Genomic and Biomarker Datasets**

Gene Expression Omnibus [29] is a public database that stores genetic data related to Alzheimer's and Parkinson's diseases. The available data allows for the analysis of how genes may vary between different groups of patients and their relationship to the progression of the disease.

- **Clinical and Behavioral Data**

MDS-UPDRS [30] is a clinical evaluation scale that measures motor and non-motor movements in patients with Parkinson's disease. This data helps assess the progression of the disease and the effectiveness of treatments, as it is one of the most deadly conditions.
PPMI [31] is a study that collects clinical and behavioral data from individuals with Parkinson's disease, focusing on identifying progression biomarkers. The collected data aids in research to combat the disease or develop therapies.

Table 3. Summary of Datasets for Alzheimer's and Parkinson's Disease.

Type	Name	Descriptions
Imaging Datasets	OASIS [26]	Size: 1000+ MRIs Image Type: MRI Resolution: 1 mm isotropic
	ADNI [27]	Size: 2,000+ subjects Image Type: MRI, PET Resolution: Varies by modality
	UK Biobank [28]	Size: 500,000 participants Image Type: MRI, CT Resolution: Varies
Genomic and Biomarker Datasets	ADNI Biomarkers [10, 11]	Processed genotype data can be in tabulated text formats such as CSV, TSV

(*continued*)

Table 3. (*continued*)

Type	Name	Descriptions
Clinical and Behavioral Data	GEO [29]	Description: A public database for gene expression data; contains various datasets related to Alzheimer's and Parkinson's
	MDS-UPDRS [30]	Size: Varied Description: Clinical assessment scale for Parkinson's disease
	PPMI [31]	Size: 4,000 + subjects Description: Clinical and behavioral data for Parkinson's disease
Multimodal Datasets	ADNI [27]	Description: Combines imaging, clinical, and biomarker data
	OASIS [26]	Size: 1000+ MRIs Image Type: MRI Resolution: 1 mm isotropic

5 Conclusions and Future Directions

Thanks to data recording and, in large part, digitalization, as well as increased awareness of neurodegenerative diseases, a number of innovations in libraries and tools have emerged. Supervised and unsupervised machine learning is achieving highly accurate results when analyzing MRI and CT scans. In the past, professionals had to perform these analyzes manually, which was prone to human error. Artificial intelligence should not be seen as a threat but rather as a supporting tool. AI models allow us to distinguish, in a matter of seconds, which patient might be at risk by making predictions from thousands of images. This is especially important as it is estimated that a large number of older people will be affected by neurodegenerative diseases, and it is crucial to raise awareness as this could have an impact on many areas of society. Future research should integrate various types of data, such as medical images, genetic data, and clinical records, to improve the prediction accuracy and diagnostic capabilities of neurodegenerative diseases. A major challenge in deep learning is the "black box" nature of the models. Future work should focus on creating more explainable models that allow medical professionals to better understand and trust the results. Additionally, fostering stronger collaborations between data scientists, neurologists, and medical professionals will lead to models that are better aligned with clinical needs and practices. Finally, as deep learning becomes more integral to healthcare, addressing ethical concerns and ensuring data privacy and security will be paramount, especially when handling sensitive medical data.

Acknowledgments. We thank TECNM/Tijuana Institute of Technology and CONAHCyT for support with the finances with the grant number CF-2023-I-555.

References

1. Candelise, N., Baiardi, S., Franceschini, A., Rossi, M., Parchi, P.: Towards an improved early diagnosis of neurodegenerative diseases: the emerging role of in vitro conversion assays for protein amyloids. Acta Neuropathol. Commun. **8**(1), 1–16 (2020)
2. Pao, P.C., et al.: HDAC1 modulates OGG1-initiated oxidative DNA damage repair in the aging brain and Alzheimer's disease. Nat. Commun. **11**(1), 2484 (2020)
3. Molteni, M., Rossetti, C.: Neurodegenerative diseases: the immunological perspective. J. Neuroimmunol. **313**, 109–115 (2017)
4. Qiao, J., et al.: Genetic correlation and gene-based pleiotropy analysis for four major neurodegenerative diseases with summary statistics. Neurobiol. Aging **124**, 117–128 (2023)
5. Pan American Health Organization: Dementia in Latin America and the Caribbean: prevalence, incidence, impact, and trends over time. PAHO, Washington, DC (2023). https://doi.org/10.37774/9789275326657
6. Romo-Galindo, D.A., Padilla-Moya, E.: Usefulness of brief cognitive tests for detecting dementia in the Mexican population. Archiv. Neurosci. **23**(4), 26–34 (2018)
7. Dhakal, S., Azam, S., Hasib, K.M., Karim, A., Jonkman, M., Al Haque, A.F.: Dementia prediction using machine learning. Procedia Comput. Sci. **219**, 1297–1308 (2023)
8. Chattopadhyay, T., Ozarkar, S.S., Buwa, K., Thomopoulos, S.I., Thompson, P.M.: Alzheimer's Disease Neuroimaging Initiative. Predicting Brain Amyloid Positivity from T1 weighted brain MRI and MRI-derived Gray Matter, White Matter and CSF maps using Transfer Learning on 3D CNNs. bioRxiv (2023). https://doi.org/10.1101/2023.02.15.528705. PMID: 36824826; PMCID: PMC9949045
9. Li, J., Xu, H., Yu, H., Jiang, Z., Zhu, L.: Multi-modal feature selection with anchor graph for Alzheimer's disease. Front. Neurosci. **16**, 1036244 (2022)
10. Diogo, V.S., Ferreira, H.A., Prata, D.: Early diagnosis of Alzheimer's disease using machine learning: a multi-diagnostic, generalizable approach. Alzheimer's Res. Ther. **14**(107), 1–15 (2022). https://doi.org/10.1186/s13195-022-01047-y
11. Sarraf, S., DeSouza, D. D., Anderson, J., Tofighi, G.: Alzheimer's Disease Neuroimaging Initiative. bioRxiv, 070441 (2017). https://doi.org/10.1101/070441
12. Shamrat, F.M.J.M., Alzahrani, F., Dajani, A., Alzahrani, A.: AlzheimerNet: an effective deep learning based proposition for Alzheimer's disease stages classification from functional brain changes in magnetic resonance images. IEEE Access **11**, 16376–16395 (2023). https://doi.org/10.1109/ACCESS.2023.3244952
13. Jansi, R., Gowtham, N., Ramachandran, S., Praneeth, S.: Revolutionizing Alzheimer's Disease Prediction using InceptionV3 in Deep Learning. In: 2023 7th International Conference on Electronics, Communication and Aerospace Technology (ICECA), pp. 1155–1160 (2023). https://doi.org/10.1109/ICECA59037.2023.10090239
14. Alamro, H., Thafar, M.A., Albaradei, S., Alzahrani, A.: Exploiting machine learning models to identify novel Alzheimer's disease biomarkers and potential targets. Sci. Rep. **13**(1), 4979 (2023). https://doi.org/10.1038/s41598-023-30904-5
15. Thangavel, S., Selvaraj, S.: Machine learning model and cuckoo search in a modular system to identify Alzheimer's disease from MRI scan images. Comput. Methods Biomech. Biomed. Eng.: Imag. Vis. **11**(5), 1753–1761 (2023). https://doi.org/10.1080/21681163.2023.2187239
16. Franciotti, R., Nardini, D., Russo, M., Onofrj, M., Sensi, S.L.: Comparison of machine learning-based approaches to predict the conversion to Alzheimer's disease from mild cognitive impairment. Neuroscience **514**, 143–152 (2023)
17. McFall, G.P., et al.: Identifying key multi-modal predictors of incipient dementia in Parkinson's disease: a machine learning analysis and Tree SHAP interpretation. Front. Aging Neurosci. **15**, 1124232 (2023). https://doi.org/10.3389/fnagi.2023.1124232

18. Sotirakis, C., Su, Z., Brzezicki, M.A., et al.: Identification of motor progression in Parkinson's disease using wearable sensors and machine learning. npj Parkinson's Disease (2023). https://doi.org/10.1038/s41531-023-00581-2
19. Hossain, M.A., Amenta, F.: Machine learning-based classification of Parkinson's disease patients using speech biomarkers. J. Parkinsons Dis. **14**, 95–109 (2023). https://doi.org/10.3233/JPD-223571
20. Harvey, J., et al.: Machine learning-based prediction of cognitive outcomes in de novo Parkinson's disease. NPJ Parkinson's Disease **8** (2022)
21. Hosseinzadeh, M., Gorji, A., Fathi Jouzdani, A., Rezaeijo, S.M., Rahmim, A., Salmanpour, M.R.: Prediction of cognitive decline in parkinson's disease using clinical and DAT SPECT imaging features, and hybrid machine learning systems. Diagnostics **13**(10), 1691 (2023)
22. Islam, M.S., et al.: Using AI to measure Parkinson's disease severity at home. NPJ Digital Medicine, 6.voting technique. Multim. Tools Appl. **83**, 33207–33234 (2023)
23. Govindu, A., Palwe, S.: Early detection of Parkinson's disease using machine learning. Procedia Comput. Sci. **218**, 249–261 (2023)
24. Alalayah, K.M., Senan, E.M., Atlam, H.F., Ahmed, I.A., Shatnawi, H.S.: Automatic and early detection of parkinson's disease by analyzing acoustic signals using classification algorithms based on recursive feature elimination method. Diagnostics **13**(11), 1924 (2023). https://doi.org/10.3390/diagnostics13111924
25. Abdullah, S.M., et al.: Deep transfer learning based parkinson's disease detection using optimized feature selection. IEEE Access **11**, 3511–3524 (2023)
26. [OASIS] Washington University in St. Louis. (n.d.). Open Access Series of Imaging Studies (OASIS). https://sites.wustl.edu/oasisbrains/
27. [ANRI] Alzheimer's Disease Neuroimaging Initiative. (n.d.). ADNI. https://adni.loni.usc.edu/
28. [ukbiobank] UK Biobank. (n.d.). UK Biobank: A major resource for health research. https://www.ukbiobank.ac.uk/
29. [GEO] NCBI. (n.d.). Gene expression omnibus (GEO). https://www.ncbi.nlm.nih.gov/geo/
30. [UPDR5] Movement Disorders Society. (n.d.). Movement disorders society. https://www.movementdisorders.org/
31. [PPMI] Parkinson's Progression Markers Initiative. (n.d.). Parkinson's progression markers initiative (PPMI). https://www.ppmi-info.org/

Automated Insights: LLMs in Neurodegenerative Disease Research and Comparison

Cesar Torres and Claudia I. Gonzalez(✉)

Division of Graduate Studies and Research, Tijuana Institute of Technology/TECNM, 22414 Tijuana, Mexico
`m21210008@tectijuana.edu.mx, cgonzalez@tectijuana.mx`

Abstract. The growing interest in large language models (LLMs) has led to exploring their potential in various domains, including medical research. This work focuses on using Llama-3 70B, an open-source LLM, to automate insights from neurodegenerative disease research papers. By applying zero-shot inference for hypothesis and result extraction, we aim to analyze multiple papers and find correlations in their methodologies. Our research seeks to enhance the understanding of complex findings, with an emphasis on accuracy and reliability. Currently, we are in the early stages, using a small open-source model to test a pipeline that shows promising results. We are evaluating a hand-picked set of highly cited papers from PubMed, published in the last four years, through a qualitative analysis comparing the model's output to our interpretations.

Keywords: Large Language Models · Neurodegenerative Disease · Inference for Hypothesis

1 Introduction

Large language models (LLMs) [1] are a class of deep learning models, based on Transformer architecture, a branch of Artificial Intelligence (AI) inspired by the understanding and generation of human-like text. They are trained on vast amounts of text in multiple categories to create models capable of performing feature extraction, understanding and generation of text in a reliable manner. Popular usages of LLMs include text generation, language services, summarization, and search and retrieval, such as conversational agents, translation, grammar correction, sentiment analysis, text condensation, question answering, and knowledge extraction.

The versatility of LLMs has led researchers to explore their applications in various domains, including healthcare. Neurodegenerative diseases, such as Alzheimer's disease, Parkinson's disease, amyotrophic lateral sclerosis, Huntington's disease, and multiple sclerosis, are a significant focus area. These diseases involve the gradual deterioration of the nervous system's structure and function, leading to the selective loss of structure and function of specific populations of neurons, ultimately resulting in neuronal death and/or significant impairment of motor and cognitive functions.

Recent studies have demonstrated the potential of LLMs in analyzing and extracting valuable information from medical research papers. For instance, Tang et al. [2] performed a systematic study on zero-shot medical evidence summarization in multiple domains, while Hey et al. [3] created a pipeline to identify neglected hypotheses in neurodegenerative disease research, highlighting the promise of automation and scalability in this field, L'ala et al. [4] designed PaperQA, a RAG based platform toperform question answering from research papers, exiding performance of simple LLM agents, another approach by Singhal et al. [5] developed MultiMedQA, a comprehensive benchmark for evaluating large language models (LLMs) in clinical applications, combining seven medical datasets. They evaluated the performance of PaLM and its variant Flan-PaLM on MultiMedQA, achieving state-of-the-art accuracy

Building on these findings, our research aims to investigate the viability of utilizing zero-shot prompting with state-of-the-art small LLMs to generate accurate hypothesis extraction from research papers focused on neurodegenerative diseases. This study seeks to contribute significantly to the growing body of research exploring the intersection of LLMs and healthcare, with far-reaching implications for medical research. By developing an efficient pipeline for hypothesis extraction from neurodegenerative disease reports, this research aims to empower medical researchers to swiftly and accurately extract insights from vast amounts of scientific literature. Ultimately, this will enhance the discovery of state-of-the-art knowledge, facilitate the identification of critical research gaps, and accelerate the translation of findings into clinical practice. By automating the time-consuming process of manual literature review, researchers will be able to focus on high-value tasks, such as hypothesis generation, experimental design, and therapeutic development.

2 Background

2.1 Natural Language Processing

The area of natural language processing (NLP) [6] derived from the necessity of allowing computers to convert text into a data structure. The area emerged from a multidisciplinary field that combines computer sciences, linguistics, and machine learning to enable computer to process, interpret, generate, and manipulate human language [7, 8]. Some popular NLP techniques to process text are the following.

Tokenization. Is the process of breaking down text into individual units (tokens). This could be reduced into words, sub words or characters, allowing to prepare the data for further analysis [9].

Lemmatization. Refers to the process of reducing words to their base form. Reducing dimensionality of the data and aggrupation of different words [10].

Word Embeddings. Creates a word representation that captures the semantic relationship between words in a high-dimensional vectorial space, aiming to map similar words close together in the vector space. Most of embedding techniques depend on the angle between pairs of word vectors as the primary method for determining the inherent quality of such a set of word representations. Some of the popular word embeddings are

Word2Vec from Mikolov et al. [1], GloVe by Pennignton et al. [11] and FastText from Bojanowski et al. [12].

Deep Learning Methodologies. Deep learning methodologies are utilized to process text given their abstraction capabilities, popular techniques include Recurrent Neural Networks (RNNs), Long Short-Term Memory (LSTM) and most recently Transformer architectures.

2.2 Transformer Architecture

The Transformer architecture from the work of Vaswani et al. [13] was a revolutionary idea that reformed the field of NLP [14]. The proposed attention mechanism allows the model to focus on specific parts of the token sequence.

The transformer was coined as a response to traditional convolutional neural networks with a self-attention mechanism. Provides an encoder-decoder with multiple layers of self-attention and feed-forward networks. Generating both a representation of the input and output sequence a token at the time. Allowing for parallelization, eliminating the need for recurrent processing. The basic structure and essential contents of Transformer architecture is illustrated in Fig. 1.

The key components of a transformer architecture are the following:

- **Positional Encoding**: Generates token embeddings by aggregating information on the position of each token in the sequence.
- **Token Embeddings**: Each token is embedded into a vector space, capturing its semantic meaning.
- **Attention mechanism**: Allows the model to concurrently focus on multiple segments of the input sequence assigning them weighs given their relative importance. The attention mechanism is denoted by Eq. 1.

$$Attention(Q, K, V) = softmax\left(Q * \frac{K^T}{\sqrt{d_k} * V}\right) \quad (1)$$

- Q (Query): represents the vector that is used to query the attention mechanism.
- K (Key): represents the vector that is used to compute the attention weights.
- V (Value): represents the vector that is used to compute the output of the attention mechanism.
- d_k (Dimensionality of the Key matrix): represents the number of dimensions in the key matrix.
- *Softmax*: function that normalizes the attention weights to ensure they sum up to 1.

- **Multi-Head Attention**: Attends to information from different subspaces, focusing each head on different parts of the sequence, defined by Eq. 2.

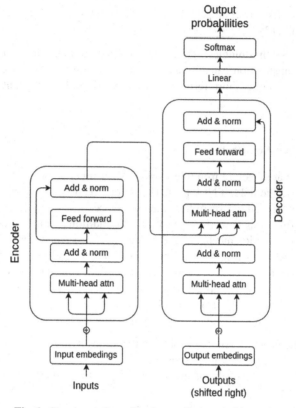

Fig. 1. Representation of basic transformer architecture.

$$MultiHead(Q, K, V) = Concat(head_1, head_2, \ldots, head_h) * W^0 \quad (2)$$

- $head_i$ (Attention Head): represents the output of each attention head.
- h (Number of Heads): represents the number of attention heads.
- *Concat*: a function that concatenates the output of each attention head.
- W^0 (Weight Matrix): represents a weight matrix that is used to linearly transform the concatenated output of the attention heads.

- **Residual Connection** (Add): Preserves original information by adding the output of the previous layer to the current layer.
- **Layer Normalization** (Norm): Normalizes to a mean of 0 and standard deviation of 1, stabilizing the training process.
- **Feed Forward Networks**: Traditional FFN allowing the model to learn complex contextual relationships.
- **SoftMax**: Generates a probability distribution over output tokens.

2.3 Large Language Models

Large Language Models (LLMs) are sophisticated NLP systems designed to comprehend, generate, and manipulate human language with extraodinary proficiency. These models are trained on extensive datasets, with billions or trillions of tokens/samples, which exposes them to a diverse range of linguistic data [15, 16]. LLMs are distinguished by their massive number of parameters, which enables them to capture intricate language relationships, nuances, and patterns. This capacity allows them to perform a wide array of tasks, including but not limited to:

- Text generation and completion.
- Language translation.
- Sentiment analysis.
- Question answering and information retrieval (RAG).
- Conversational dialogue and chatbots.

The primary objective of LLMs is to produce coherent and contextually relevant text, derived from the given prompt (input). Prompting refers to a string or sequence of tokens that instructs the model to generate a response. A prompt typically consists of three components. Given that P be a prompt represented as P = (C, T, I).

- Context: Background information for the task in question.
- Task specification: A concise description of the task to be addressed.
- Instruction: Guidance on format, tone, or style of the response.

Two prominent themes of prompting in LLMs are:

- Zero-Shot Prompting: Involves providing a query to the model without any prior training or fine-tuning on the given task [17]. The model relies on its knowledge base to perform an acurate respons.
- Few-Shot Prompting: Involves providing a prompt with a few examples of the task, typically 2-5, along with the task description (context). This allows the model to learn from the examples and adapt to the task with the given instruction.

2.4 Retrieval-Augmented Generation

Retrieval-Augmented Generation (RAG) is an NLP technique that combines retrieval-based methodologies with generative models to improve the relevance of generated text [18]. The RAG retrieves information from a large external corpus, which is utilized to augment the generation process. This approach enables LLMs to produce more informative and contextually relevant responses by leveraging external knowledge sources. Fig. 2 contains the basic structure of a RAG.

The architecture composes of the following components:

- **Retriever**: A retrieval system that searches a large external corpus or knowledge base to gather relevant information related to the input prompt.

The retreiver utilizes the prompt to query the system, producing a searches over the external corpus to gather relevant information related to the prompt.

The retrieved information is utilized to augment the input prompt, providing additional context afrom the corpus.

- **Generator**: A generative model, typically an LLM, that produces text based on the input prompt and the retrieved information. Produces text based augmentation with the given prompt, leveraging the exteran knowledge base to improve the quality of the response. The output may undergo to post-processing methods, such as flitering, refinement or ranking to ensure the accuracy and relevance.
- **External Corpus**: A vast repository of text data that serves as the knowledge source for the retriever (Fig. 2).

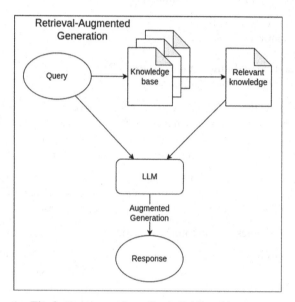

Fig. 2. Representation of basic RAG architecture.

3 Proposed Methodology

The current implementation is tailored to explore the capabilities of multiple backbones (LLMs) to perform key data extraction from multiple papers. In this case we focus our research in utilizing multiple open-source models with less than 14B parameters to perform the feature extraction. We implemented two approaches to perform this:

- The first implementation involved two prompts to streamline the hypothesis generation process. First, we create a short resume for each section of the paper, then we pass the hypothesis generation prompt with the resumes of the paper to the model to perform generation task.
- The second implementation was a simple but accurate NLP pipeline to perform extraction with a RAG interface utilizing the model.

3.1 Zero Shot Approach

In this approach we proposed the utilization of zero-shot prompting to obtain the hypotheses of the research papers. Due to the limitations presented by the model utilized Meta-Llama-3 70B (quantized 4 int) the model constraints us to a total of 8192 tokens combined for input and output. Due to this limitation we tackled to problem splitting the load into two separate tasks. Section summarization and hypothesis creation. The first step consisted in splitting the paper by its subsections and generating a resume of a maximum of 500 tokens per section. For this task we utilized the Prompt 1, we decided to pass the abstract as part of the prompt to contextualize the model and be able to link the generated resume to the overall topic of the paper.

```
Prompt 1 Summarization:
prompt = """
You are an AI expert at generating accurate and succinct summaries
for sections of scientific papers, with a focus on neurodegenera-
tive diseases. Your task is to summarize the provided section of the
paper with the utmost precision.
The summary should:
Be clear and concise, highlighting only the key points from the
section. Rely solely on the content of the section, title, key-
words, and abstract. Avoid any personal opinions, filler phrases,
or unnecessary information. Contribute to building a comprehensive
overview of the entire paper, precision is essential.
Start the summary with --START-- and end it with --END--.
Below is the relevant information for this section:
Paper Title: {title}
Abstract: {abstract}
Section Title: {section title}
Section Text: {section}
SUMMARY:
"""
```

After the summarization task we utilized the second prompt designed specifically to generate the hypothesis, based on the resumes. We also included the Title and full abstract and keywords with the resumes of the sections.

Prompt 2 Hypothesis generation:

```
Prompt = """
You are an AI expert at extracting accurate and succinct hypothe-
sis from scientific papers with a focus on neurodegenerative dis-
eases. You are given the title, full abstract, keywords and a resume
of each section of the scientific paper, it includes introduction,
background, results if available and conclusions. Your task is to
analyze the information and get the hypothesis from the paper based
on the abstract and the sections.
The hypothesis should:
Be clear and concise, highlighting the main idea of the paper.
Rely solely on the title, abstract, keywords, and sections
```

```
Avoid any personal opinions, filler phrases, or unnecessary infor-
mation.
Contribute to building a comprehensive overview of the entire
paper, making precision essential.
Start the hypothesis with --START-- and end it with --END--.
Below is the relevant information for this task:
Paper Title: {title}
Abstract: {abstract}
Keywords: {keywords}
Sections Text: {sections}
HYPOTHESIS:
"""
```

3.2 RAG Approach

For the second approach (Fig. 3) we implemented a single system prompt to pass in combination of the research paper separated by pages. Given the nature of RAG systems, we do not hit the limit of tokens for the model generation.

Prompt 3 RAG Prompt:

```
Prompt = """
System: Act as an expert academic researcher specialized in the
field of research paper summarization. I need your expertise to
provide an in-depth analysis of the Task provided by the user.
Please structure your response in a scholarly manner with cita-
tions, using a long-form format with bullets and examples.
Start the hypothesis with --START-- and end it with --END--.
Research paper: {research paper}.
Task: {task}.
"""
Tasks: {["What is the hypothesis of the paper?", "What are the
findings?"]}
```

4 Results

The current approach focused on the extraction and qualitative analysis of four research papers: Gene Therapy for Neurodegenerative Diseases [19], Ketogenic Diet: An Effective Treatment Approach for Neurodegenerative Diseases [20], NRF2 as a Therapeutic Target in Neurodegenerative Diseases [21], and the potential of hyperbaric oxygen as a therapy or neurodegenerative diseases [22]. In the subsequent sections we will display the outputs provided by the approaches previously described.

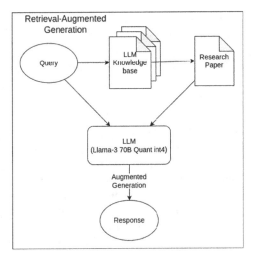

Fig. 3. Representation of basic RAG architecture.

4.1 Zero-Shot Approach

- **Gene Therapy for Neurodegenerative Diseases**: Gene therapy, utilizing advanced neurosurgical techniques such as interventional MRI-guided convection-enhanced delivery (iMRI-CED), has the potential to provide therapeutic benefit for neurodegenerative diseases, including Parkinson's, Huntington's, and Alzheimer's, by correcting pathogenic mechanisms, providing neuroprotection, and promoting neurorestoration.
- **Ketogenic Diet: An Effective Treatment Approach for Neurodegenerative Diseases**: A ketogenic diet may be an effective treatment approach for neurodegenerative diseases, including Alzheimer's disease, Parkinson's disease, and amyotrophic lateral sclerosis, by improving mitochondrial function, reducing oxidative stress and neuroinflammation, and regulating the intestinal microbiome.
- **NRF2 as a Therapeutic Target in Neurodegenerative Diseases**: NRF2 is a promising therapeutic target for neurodegenerative diseases, as its activation can mitigate mitochondrial dysfunction, oxidative stress, and neuroinflammation, and NRF2-activating compounds may be effective in treating neurodegenerative conditions such as Alzheimer's disease, Parkinson's disease, Huntington's disease, Friedrich's ataxia, multiple sclerosis, and stroke.
- **The potential of hyperbaric oxygen as a therapy or neurodegenerative diseases**: Hyperbaric oxygen therapy (HBOT) may be a viable treatment option for neurodegenerative diseases (NDDs) by reducing neuroinflammation, oxidative stress, and apoptosis, and improving microcirculation and neurogenesis, with potential benefits in epigenetic mechanisms.

4.2 RAG approach

- **Gene Therapy for Neurodegenerative Diseases**: The primary hypothesis of this research paper appears to be that focal neurosurgical delivery of gene therapy vectors,

specifically using adeno-associated virus (AAV) serotype 2, can provide a predictable and reproducible approach for treating neurological disorders by targeting specific brain regions or cell types.
- **Ketogenic Diet: An Effective Treatment Approach for Neurodegenerative Diseases**: A research paper summary! \n\nThe paper by Nadjar et al. (2016) investigated the effects of resolvin D1 and E1 on the resolution of inflammation in microglial cells in vitro.\n\n**Hypothesis:** The authors hypothesized that resolvin D1 and E1, which are pro-resolving lipid mediators, would promote the resolution of inflammation in microglial cells by inhibiting inflammatory responses.\n\nIn other words, they expected these compounds to help \"turn off\" the inflammatory process in microglial cells, which are immune cells involved in neuroinflammation.
- **NRF2 as a Therapeutic Target in Neurodegenerative Diseases**: Based on the provided text, it appears that the research paper explores the therapeutic potential of activating the Nuclear Factor Erythroid 2-related factor 2 (NRF2) pathway in neurodegenerative diseases and stroke.\n\nThe main hypothesis of the paper is likely that NRF2 activation can be a therapeutic target for preventing or treating neurodegenerative diseases, including stroke. The authors suggest that various compounds, including botanically derived and synthetic ones, can activate NRF2 and lead to beneficial effects on brain health and function.\n\nMore specifically, the authors seem to be exploring the idea that activating NRF2 can:\n\n1. Reduce infarct volume and cognitive impairments following stroke.\n2. Increase neuronal cell viability and reduce oxidative stress in the brain.\n3. Preserve blood-brain barrier integrity and prevent cerebral ROS levels from increasing.\n4. Attenuate oxidative stress and neuronal loss following ischemia/reperfusion injury.\n\nBy activating NRF2, the authors aim to develop new therapeutic strategies for neurodegenerative diseases, including stroke, and to identify potential natural products or synthetic compounds that can be used as treatments.
- **The potential of hyperbaric oxygen as a therapy or neurodegenerative diseases**: The hypothesis of this paper appears to be that Hyperbaric Oxygen Therapy (HBOT) may have therapeutic potential for Neurodegenerative Diseases (NDDs), and that sex and possibly race may impact the effect of HBOT. Additionally, the authors suggest that epigenetics may play a major role in understanding the effects of HBOT on NDDs.\n\nThe paper does not explicitly state a single hypothesis, but rather presents a body of research highlighting the potential benefits of HBOT for NDDs and suggesting that further study is needed to fully understand its therapeutic effects. The authors' goal appears to be to stimulate further research into the use of HBOT as a treatment for NDDs, with a focus on identifying potential sex and race-based differences in response to therapy.

5 Analysis

We can observe that the zero-shot approach yields more accurate and reliable hypothesis extractions compared to the RAG methodology, at least from a qualitative standpoint. While this study represents a first approach to exploring the viability of zero-shot prompting for hypothesis extraction, our preliminary findings suggest promise. Notably, the zero-shot approach generates concise and relevant hypotheses that effectively capture the essence of the research article. In contrast, the RAG output suffers from verbosity and a tendency to focus primarily on the abstract, keywords, and introduction.

However, it is essential to acknowledge that our current evaluation is based on qualitative assessments, and a comprehensive quantitative analysis is necessary to rigorously validate these findings. Future research should prioritize developing robust evaluation metrics to quantify the accuracy and reliability of hypothesis extractions. Despite this limitation, our initial results provide an encouraging foundation for continued exploration and refinement of the zero-shot approach. By addressing these methodological gaps, we can further assess the potential of zero-shot prompting to enhance hypothesis extraction in medical research.

6 Conclusions

This research marks the initial steps toward experimenting with hypothesis extraction from research papers on neurodegenerative diseases, demonstrating promising potential for leveraging Large Language Models (LLMs) to streamline literature review. Our qualitative analysis reveals that the zero-shot approach yields more accurate and reliable hypothesis extractions compared to the RAG methodology, despite requiring further refinement. To fully harness the capabilities of LLMs, future research will focus on developing robust evaluation metrics to accurately measure performance through statistical analysis, leveraging collaboration with medical researchers to create a meaningful metric. By addressing this methodological gap, we can rigorously validate the effectiveness of LLM-based hypothesis extraction, ultimately transforming the efficiency and efficacy of medical research. Utilizing open general models, such as Llama 3, provides a solid foundation for fine-tuning on domain-specific literature, paving the way for enhanced result quality and improved research outcomes.

Acknowledgment. We thank Tijuana Institute of Technology/TECNM and CONAHCyT for support with the finances with the grant number CF-2023-I-555.

References

1. Vaswani, A., et al.: Attention is all you need. In: Proceedings of the Neural Information Processing Systems (NeurIPS) Conference (2017)
2. Tang, L., Sun, Z., Idnay, B., et al.: Evaluating large language models on medical evidence summarization. npj Digital Medicine (2023). https://doi.org/10.1038/s41746-023-00896-7
3. Hey, S., Angle, D., Chatham, C.: Identifying neglected hypotheses in neurodegenerative disease with large language models. In: NeurIPS 2023 Generative AI and Biology (GenBio) Workshop (2023). https://openreview.net/forum?id=0gl0SJtd2E.
4. L'ala, J., et al.: PaperQA: Retrieval-Augmented Generative Agent for Scientific Research. arXiv abs/2312.07559, 2023. https://arxiv.org/pdf/2312.07559.
5. Singhal, K., Azizi, S., Tu, T., et al.: Large language models encode clinical knowledge. Nature **620**, 172–180 (2023). https://doi.org/10.1038/s41586-023-06291-2
6. Collobert, R., et al.: Natural Language Processing (Almost) from Scratch. NEC Labs America, Princeton, NJ (2009)

7. Devlin, J., et al.: BERT: pre-training of deep bidirectional transformers for language understanding. In: Proceedings of the North American Chapter of the Association for Computational Linguistics (NAACL) (2019).
8. Cho, K., et al.: Learning phrase representations using RNN encoder-decoder for statistical machine translation. In: Proceedings of the Conference on Empirical Methods in Natural Language Processing (EMNLP) (2014).
9. Manning, C.D., Surdeanu, M., Bauer, J., Finkel, J., Bethard, S.J., McClosky, D.: The stanford CoreNLP natural language processing toolkit. In: Proceedings of the 52nd Annual Meeting of the Association for Computational Linguistics, pp. 55–60 (2014)
10. Jurafsky, D., Martin, J.H.: Speech and Language Processing. Draft of August 20, 2024 (2019)
11. Mikolov, T., Sutskever, I., Chen, K., Corrado, G.S., Dean, J.: Distributed representations of words and phrases and their compositionality. In: Advances in Neural Information Processing Systems, pp. 3111–3119 (2013)
12. Pennington, J., Socher, R., Manning, C.D.: GloVe: global vectors for word representation. In: Proceedings of the 2014 Conference on Empirical Methods in Natural Language Processing, pp. 1532–1543 (2014)
13. Bojanowski, P., Grave, E., Joulin, A., Mikolov, T.: Enriching word vectors with subword information. Trans. Assoc. Comput. Linguist. **5**, 135–146 (2017)
14. Jurafsky, D., Martin, J.H.: Speech and Language Processing: An Introduction to Natural Language Processing, Computational Linguistics, and Speech Recognition. India: Dorling Kindersley Pvt, Ltd. (2014). https://web.stanford.edu/~jurafsky/slp3/ed3book.pdf.
15. Radford, A., Wu, J., Child, R., Luan, D., Amodei, D., Sutskever, I.: Language models are unsupervised multitask learners. In: Proceedings of the Neural Information Processing Systems (NeurIPS) Conference (2019). https://api.semanticscholar.org/CorpusID:160025533.
16. Devlin, J., Chang, M.-W., Lee, K., Toutanova, K.: BERT: pre-training of deep bidirectional transformers for language understanding. In: Proceedings of the North American Chapter of the Association for Computational Linguistics (NAACL) (2019). https://api.semanticscholar.org/CorpusID:52967399.
17. Kojima, T., Gu, S.S., Reid, M., Matsuo, Y., Iwasawa, Y.: Large language models are zero-shot reasoners. In: Proceedings of the 36th International Conference on Neural Information Processing Systems (NIPS'22), pp. 1613–1627. New Orleans, LA, USA (2024). https://doi.org/10.5555/3600270.3601883
18. Lewis, P., et al.: Retrieval-augmented generation for knowledge-intensive NLP tasks. ArXiv, vol. abs/2005.11401 (2020). https://api.semanticscholar.org/CorpusID:218869575.
19. Sudhakar, V., Richardson, R.M.: Gene therapy for neurodegenerative diseases. Neurotherapeutics **16**(1), 166–175 (2019)
20. Ye, T., Sean, L.X., Haiyan, Z.: Ketogenic diet: an effective treatment approach for neurodegenerative diseases. Curr. Neuropharmacol. **20**(12), e300822208211 (2022). https://doi.org/10.2174/1570159X20666220830102628
21. Brandes, M.S., Gray, N.E.: NRF2 as a therapeutic target in neurodegenerative diseases. ASN Neuro. (2020). https://doi.org/10.1177/1759091419899782
22. Mensah-Kane, P., Sumien, N.: The potential of hyperbaric oxygen as a therapy for neurodegenerative diseases. GeroScience **45**, 747–756 (2023). https://doi.org/10.1007/s11357-022-00707-z

Convolutional Neural Network Models for Classifying of Peach (Prunus persica L)

Flossi Puma-Ttito[1], Carlos Guerrero-Mendez[1](✉), Daniela Lopez-Betancur[1](✉), Tonatiuh Saucedo-Anaya[1], Rafael Castaneda-Diaz[1,2], and Luis Martinez-Ytuza[1]

[1] Universidad Autónoma de Zacatecas, Unidad Académica de Ciencia y Tecnología de la Luz y la Materia (Lumat), Zacatecas, Mexico
{guerrero_mendez,danielalopez106}@uaz.edu.mx
[2] Instituto Politécnico Nacional, Unidad Profesional Interdisciplinaria de Ingeniería Campus Zacatecas (UPIIZ), Zacatecas, Mexico

Abstract. In this research, a group of convolutional neural network (CNN) architectures for peach image classification task were compared. The models were trained to identify the categories of a set of images that could be found in a sorting machine, such as "Healthy" or "Damaged", which correspond to the separation of peaches with or without damage. Specifically, five ResNet models (ResNet-18, ResNet-34, ResNet-50, ResNet-101, and ResNet-152) from the Torchvision library were evaluated for their ability to categorize peach images. The models were trained employing feature extraction in a transfer learning approach and evaluated using established metrics. The evaluation results indicate that ResNet-18 is the optimal CNN model for this task. The metrics for ResNet-18 are as follows: Accuracy – 95.68%, Precision – 95.74%, Recall – 95.68%, Specificity – 90.81%, F1-score – 95.61%, Geometric Mean – 93.07%, and Index of Balanced Accuracy – 87.04%. By using the AI-based grading system, peach producers could optimize their classification production and have each product evaluated according to international standards.

Keywords: Image classification · ResNet architecture · Deep Learning · Peach classification · Convolutional Neural Networks

1 Introduction

The peach (*Prunus persica L.*) belongs to the Rosaceae family, and is a fruit consumed worldwide. This fruit can be consumed fresh, sliced, preserved, in syrups and in desserts at any time of the day. The consumption of this fruit provides vitamin A, B1, B2, C, phosphorus, calcium, among other essential elements and vitamins [1]. Likewise, peaches are highly perishable fruits due to their short shelf life after harvest. This is primarily attributed to their high-water content. Transportation and post-harvest storage can lead to losses of 15 to 25% of the total peach production [2].

Peach sorting is a vital process in agriculture that directly influences the quality of the final product and consumer satisfaction. Traditionally, this process has been executed manually, which can be time-consuming, costly and susceptible to errors. The quality of peaches depends on several factors, such as size, color and flavor. A peach is considered damaged when it has physical damage, pests, diseases or any visible foreign material. The Mexican Official Norm NMX-FF-060-SCFI-2009 [3] indicates that peaches to be marketed must meet certain requirements such as: being whole, without critical defects, harvested manual and carefully. However, traditional marketing does not consider the quality of the peaches; in the market they are classified and sold according to the experience of the seller. The advancement of grading technologies may be a very attractive area of interest for many peach marketers.

In contrast, the human brain can process information, solve problems, make decisions and evaluate information from both its environment and its own body [4]. From these human capabilities, Artificial Intelligence (AI) has been developed, which is the ability of a computer system to simulate the behavior of the human brain to learn through training based on the data provided [5]. This ability is applied in various fields such as education, science and technology.

Currently, AI has practical applications to solve human problems more efficiently. Convolutional Neural Networks (CNN) are mathematical algorithms that replicate how humans learn through computational blocks and multiple layers of neurons that approximate any continuous function. These are mainly used for image analysis through classification and detection. In agriculture, CNNs have been used for classification and detection of fruits [6], diseases [7], sizes [8], colors [9], etc. The CNNs offer a promising solution due to their ability to automatically learn relevant features from image data.

A subfield of AI is Deep Learning (DL). It allows computational models, composed of multiple processing layers, to learn data representations with multiple levels of abstraction. Some examples of these representations are voice recognition, visual object recognition, object detection, among others. CNN have made significant advances in processing images, videos, voice, and audio for classification or identification [10]. Using AI and DL algorithms, state-of-the-art sorting machines have been developed that analyze images to detect abnormal or defective products with high precision, without the need for human intervention [11, 12]. These machines can distinguish between different samples of peaches based on characteristics such as size, color, weight and other product properties. The development of these technologies is of great interest to agricultural producers, as it allows them to classify their products more efficiently.

Due to the fundamental importance of peach classification, various studies have been carried out and innovative methods and techniques have been developed. In research carried out at the Universidad Autónoma del Estado de México, a study of peach classification was carried out using CNN. 960 images were recorded with a professional Nikon D3500 camera, using 360 images of ripe and unripe peaches and 600 images of healthy and damaged peaches, of which 80% were for training and 20% for validation. The authors designed a simple CNN model that achieved a classification accuracy of 95.31% for choosing ripe and unripe peaches and an accuracy of 92.18% for classifying healthy and damaged peaches [13].

Yao et al. [14] in their research applied deep network models (Mask R-CNN and Mask Scoring R-CNN) for segmentation and recognition of peach diseases. The number of acquired images for the diseases: brown rot, anthracnose, scab, bacterial shot hole, gummosis, powdery mildew, and leaf curl disease were 94, 157, 654, 427, 91, 50, and 87, respectively. When using ResNet50 as a backbone network based on Mask R-CNN, the segmentation accuracy of segm_mAP_50 increased from 0.236 to 0.254. When using ResNetx101 as the backbone, the segmentation accuracy of segm_mAP_50 increased from 0.452 to 0.463.

Akbar et al. [15] investigated bacteriosis, it is one of the most prevalent and deadly infections affecting peach crops globally. In this paper, a new lightweight CNN model (WLNet) based on Visual Geometry Group (VGG-19) was proposed to detect and classify images into bacteriosis and healthy images. The deep knowledge of the proposed model is used to detect bacteriosis in peach leaf images. They used a dataset consisting of 10000 images: 4500 are bacteriosis and 5500 are healthy images. The authors trained the WLNet model for leaf classification. The model was compared with four different CNN models: LeNet, Alexnet, VGG-16 and the simple VGG-19 model. The proposed model obtained an accuracy of 99%, which is higher than LeNet, Alexnet, VGG-16 and the simple VGG-19 model. The results obtained indicate that the proposed model is more effective for detecting bacteriosis in peach leaf images, compared to existing models.

Based on the previous research, the authors in this study are motivated to evaluate the performance of several advanced Torchvision architectures (an open-source computer vision package for the Torch machine learning library) in classifying peach quality. For this purpose, a comparison for five Torchvision architectures available in PyTorch was conducted to identify the most effective one for classifying peach quality and their potential for implementation in a peach sorting machine. The evaluation and comparison of these models are performed using a set of statistical metrics commonly applied in DL.

2 Methods, Techniques and Instruments

This section describes the use of five CNN models for peach image classification. The equipment, methods, and performance metrics used in this research are detailed.

2.1 CNN Models

CNNs are a type of artificial neural network architecture composed of blocks that work together to process images. The main function of their layers is to identify relevant features of an image. These architectures consist of stacked convolutional layers, and over time new ways of building these layers have been developed to improve learning efficiency.

In classification tasks, CNN-based models are widely used. Among the architecture families with high accuracy, ResNet stands out according to Torchvision reports. ResNet uses residual blocks to train deep networks with great efficiency in image classification. By comparing different models within the ResNet family, such as ResNet-18, ResNet-34, ResNet-50, ResNet-101, and ResNet-152, seeks to identify the optimal architecture that balances accuracy, complexity, and processing time in classifying peaches.

ResNets represent an innovative approach in the field of computer vision. Unlike conventional networks, where each layer learns to directly map an input to an output, ResNets focus on learning to map the input to the difference between the desired output and the original input. This difference, known as 'residual', allows ResNets to build deep architectures without suffering from performance degradation. Figure 1 shows the general architecture of a ResNet model. In a variety of tasks, such as image classification, object detection, and segmentation, ResNets have proven to be highly effective [16].

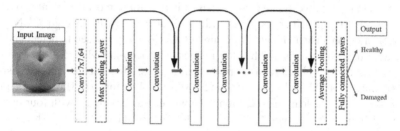

Fig. 1. A schematic view of ResNet architecture.

- **Resnet-18**

The ResNet-18 model architecture, with its 18 deep layers, despite its relatively small size, has proven to be highly effective in terms of training time and computational resource usage for image classification.

- **Resnet-34**

ResNet-34 is an extremely powerful and versatile CNN architecture with 34 deep layers, suitable for a wide variety of computer vision tasks. Its ability to capture complex features and its prowess in identifying objects in images, such as animals, vehicles, and food, are factors that position it above ResNet-18 in terms of training performance.

- **ResNet-50**

CNN architecture with 50 layers, it stands out for its depth and its ability to learn highly complex representations of images. It has proven to be exceptionally effective in tasks such as large-scale image classification, object detection, and image segmentation. In fact, ResNet-50 is one of the most widely used and best-supported architectures in deep learning libraries.

- **ResNet-101**

ResNet-101 is composed of 101 layers, which makes it a deeper network compared to its predecessors such as ResNet-18, ResNet-34, and ResNet-50. This 101-layer architecture can capture more complex patterns and features in images. Due to its depth, ResNet-101 typically performs better on challenging image classification tasks. However, it is important to note that its higher number of parameters also implies a higher requirement for computational resources for both training and running.

- **ResNet-152**

ResNet-152, with its 152 layers, stands out as one of the deepest neural networks of its time. This impressive depth allows it to learn extremely complex and abstract features present in images. Its architecture is especially suitable for tasks that require exceptional precision. Indeed, ResNet-152 has demonstrated outstanding performance in image classification, reaching very high levels of accuracy on widely known datasets such as ImageNet. However, it is important to keep in mind that this increased power comes with a significant computational cost.

2.2 Transfer Learning

Training a CNN model from scratch requires many labeled images. To reduce computational costs and training time, the best option is to use Transfer Learning (TL). This technique trains only a part of the pre-trained model to perform the task of classifying images.

In this research, ResNet architecture with pre-trained weights on ImageNet [15] will be used and retrained for peach classification. The network layers will be frozen, except for the final layer, which will be modified to have the two necessary outputs and trained using the peach database. Table 1 presents the total parameters of the CNN models and the parameters that will be trained after freezing the network. In addition, the disk space that each model occupies is detailed.

Table 1. Size and total of trainable parameters in ResNet models.

CNN model	CNN with TL (Trainable parameters)	CNN pre-trained with ImageNet (Total parameters)	Size on disk (MB)
ResNet-18	1026	11689512	44.7
ResNet-34	1026	21797672	83.3
ResNet-50	4098	25557032	97.8
ResNet-101	4098	44549160	170.5
ResNet-152	4098	60192808	230.4

2.3 Data Acquisition

The Mexican Official Standard NMX-FF-060-SCFI-2009 establishes the requirements that peaches (Prunus persica L.) in their different varieties must meet for their fresh marketing in Mexico. According to this standard, the fruits must be healthy and intact, free of pesticides and with the organoleptic characteristics of their variety.

To create an image database, 316 peaches were collected from a local grower in Zacatecas, México. Image capture was carried out in a controlled environment with white light. Each peach was placed at a fixed distance in front of an Atvio A489 camera, on a stationary white background. The resulting photographs had a resolution of 420 × 400 pixels.

The final dataset consisted of 3378 images, divided into two categories: healthy peaches (2476 images) and damaged peaches (902 images). 80% of the images were used to train a computer vision model and the remaining 20% to validate it. Figure 2 shows an example of each class. Additionally, data augmentation techniques were applied to mitigate overfitting and enrich the data set.

Fig. 2. Number of peach images of each class.

2.4 Data Augmentation

Data augmentation consists of generating new images from raw data by applying mathematical operations. These new images expand the overall dataset, which in turn improves the training of CNN models.

In this research, a numerical transformation was implemented to generate new images within the dataset. For this, the RandomHorizontalFlip module from the PyTorch library was used. This transformation allows to flip raw images horizontally in a random manner, which further diversifies the dataset used to train the model.

2.5 Training Parameters

Hyperparameters are key and adjustable components in CNN models, playing a crucial role in the training process by influencing their behavior and performance on specific tasks. Appropriate selection of these hyperparameters is essential to maximize the effectiveness of a CNN model. Among the relevant hyperparameters, optimization is critical. The optimization algorithm Adam (Adaptive Moment Estimation) is widely used in fruit classification [17–19] and will be employed in this research. Adam updates the model parameters at each iteration by estimating the moment and magnitude of gradients, and its ability to automatically adjust the learning rate contributes to its outstanding effectiveness in classification tasks.

One of these key parameters is the "learning rate", which adjusts the speed at which the model calculates the gradient of the loss function. In other words, it controls how much the architectures predictions change as it updates its results based on the model error. If we select a high learning rate, the model will make rapid changes to its parameters, while a low learning rate will adjust slow. To find the right balance, it is necessary to choose a value that allows the error to decrease effectively and for the model to converge towards the minimum error in the fewest number of epochs. The hyperparameter "epoch" refers to the number of times the entire training dataset is passed through the CNN model during the tuning process. Each epoch represents a complete training cycle. Additionally, "batch size" must be considered. Models are trained using batches of data. The batch size determines how much data is processed simultaneously before updating the model parameters. Finally, the "momentum" hyperparameter is used to speed up gradient descent. Basically, momentum considers a fraction of previous gradients when updating the current parameters. The hyperparameter selection for the research was carried out through iterative testing and adjustments, the values of which proved to provide good results in terms of accuracy and efficiency in the models used. It was also carried out impartially, without favoring any specific model. The hyperparameters used are lists in Table 2.

Table 2. Tuning hyperparameters in the training process

Hyperparameter	Value
Optimization algorithm	Adam
Learning rate	0.001
Bach size	16
Epochs	50
Seed	40
Run	1

The computer specifications for the training process were as follows: an 11th Gen Intel® Core™ i7-11700K processor, 32 GB of RAM, an NVIDIA RTX 3060 12 GB graphics card, and the Windows 11 Pro operating system. The implemented algorithms were executed using Python 3, an open-source programming language. Furthermore, the CNN models were trained using the PyTorch library, specifically version 1.9.1. In addition, to optimize the training and evaluation process, we leveraged the Torchvision package, which provides a vast array of pre-trained models and is indispensable for developing advanced computer vision applications.

2.6 Performance Evaluation

- **Confusion Matrix**

To summarize the results of the training process, it is essential to use a confusion matrix. This tool is a table that allows evaluating the performance of a model by providing a detailed view of its predictions. In the confusion matrix, columns represent the true classes, while rows reflect the classes predicted by the model. The elements on the main diagonal indicate the number of correctly classified cases, while the elements off the diagonal represent the misclassified cases. Analyzing the confusion matrix provides us with key information to adjust and select the optimal set of hyperparameters for the CNN model, thus facilitating the improvement of its performance on specific tasks.

In the confusion matrix, there are four relevant terms: True Positives (TP), which indicate the number of correctly classified cases; True Negatives (TN), which refer to the number of items correctly identified as not belonging to the class; False Positives (FP), which are the items incorrectly classified as belonging to the class when in fact they are not; and False Negatives (FN), which are the cases in which the model incorrectly predicts that an item does not belong to the class when it does. These metrics are essential to calculate the model performance and tune the hyperparameters effectively [20].

These terms are essential for evaluating the quality of model predictions and for calculating key performance metrics such as accuracy, precision, recall, specificity, F1-score, G-mean, and Index of Balanced Accuracy (IBA) [21].

The accuracy in a model measures the proportion of correctly classified samples compared to the total number of samples.

$$Accuracy = \frac{TP + TN}{(TP + TN + FP + FN)}. \tag{1}$$

Precision indicates the fraction of true positives among all positive predictions made by the model. It is represented by Eq. (2).

$$Precision = \frac{TP}{TP + FP}. \tag{2}$$

Recall or sensibility evaluates the ability of the model to detect all real positive cases. Using Eq. (3) it can be calculated.

$$Recall = \frac{TP}{TP + TN}. \tag{3}$$

Specificity measures the ability of the model to correctly identify negative cases. Eq. (4) is used to calculate specificity.

$$Specificity = \frac{TN}{TN + FP}. \tag{4}$$

The F1-score is a metric that combines accuracy and sensitivity into a single value, calculated as the harmonic mean of both, as shown in Eq. (5). This metric evaluates the balance between the model's ability to correctly identify positive cases and its ability to detect all actual positive cases.

$$F1 - score = \frac{2TP}{(2TP + FP + FN)}. \tag{5}$$

The geometric mean combines performance metrics such as accuracy and sensitivity into a single value. By calculating the square root of the product of these metrics it provides a balanced evaluation of the model. This is defined in Eq. (6).

$$G_{mean} = \sqrt{\frac{TP}{TP + FN} \times \frac{TN}{TN + FP}}. \tag{6}$$

The Index of Balanced Accuracy measures the ability of the model to correctly classify all classes, especially in unbalanced sets. A weighting factor of 0.1 was used [11, 22]. Eq. (7) can be used to calculate the IBA.

$$IBA = \left(1 + 0.1\left(\frac{TP}{TP + FN} - \frac{TN}{TN + FN}\right)\right) \times \left(\frac{TP}{(TP + FN)} \times \frac{TN}{(TN + FP)}\right). \tag{7}$$

The metrics previously described in this section were used to evaluate and compare the performance of the models in peach classification.

3 Results and Discussion

The validation run was used to calculate the validation for each model. Based on these results, the best model for classifying peaches was identified. In the run, 80% of the image set was assigned for training and the remaining 20% for validation.

The results obtained in this study demonstrate the effectiveness of ResNet models for the classification of healthy and damaged peaches. The average performance of the models exceeded 94%, evidencing their ability to discriminate between both classes. The analysis revealed that ResNet-18 is the most suitable CNN model for this task and could be implemented on a peach sorting machine, as it obtained an overall performance with a mean accuracy value of 95.68%. Furthermore, Table 3 lists the validation accuracy results for each model CNN. ResNet-18 obtained the highest individual value, indicating high generalization ability and accuracy in classifying peaches. Other models such as ResNet-34, ResNet-50, ResNet-101 and ResNet-152 showed inferior performance compared to ResNet-18. Although these models also reached high accuracy values, their mean value was lower, indicating a lower generalization capacity.

All these pre-trained models were evaluated using various metrics: accuracy, precision, sensitivity, specificity, F-Score, G_mean and IBA, using the same hyperparameters. According to Table 3, which lists average performance results for the metrics used, ResNet-18 achieved better results than the other architectures. Furthermore, ResNet-152, which is one of the deepest CNN models, shows poor performance. In the same way, the performance measures were calculated for each class, in order to know which class obtained the lowest error percentage, based on: sensitivity, specificity, F-Score and precision, which is because these measures represent model performance; It is possible

Table 3. Performance metrics

CNN model	Accuracy	Precision	Recall	Specif	F1-score	G_mean	IBA	Time (min)
ResNet-18	0.9568	0.9574	0.9568	0.9081	0.9561	0.9307	0.8704	10.58
ResNet-34	0.9424	0.9425	0.9424	0.8874	0.9415	0.9126	0.8374	14.23
ResNet-50	0.9424	0.9424	0.9424	0.9171	0.9424	0.9293	0.8658	18.30
ResNet-101	0.9496	0.9501	0.9496	0.935	0.9498	0.9421	0.8889	27.40
ResNet-152	0.9496	0.9494	0.9496	0.9201	0.9494	0.9342	0.8753	10.58

Where Specif.: Specificity.

to determine that the classes with the least error were: ResNet-18 and ResNet-101; The class with the largest error was ResNet-152. These were all based on the precision metric. Although all models showed good performance, slight differences were observed. For example, ResNet-50 obtained the best results in precision and F1-score, while ResNet-34 presented the lowest results.

From Table 3, it can be stated that as the depth of the network (number of layers) increases, a significant increase in training time (min) is observed.

Theoretically, it is suggested that there is a trade-off between accuracy and computational efficiency. Training time increased significantly with model complexity, with ResNet-152 being the slowest model and ResNet-18 being the fastest.

In this research, it is observed that the training time of ResNet models increases. This is because the number of parameters directly influences the time required to train a model. In other words, models with more parameters require greater computational resources to calculate the weights and biases of the neurons during the training process. Therefore, models with fewer parameters tend to train faster. If the ResNet-18 model is implemented on a classifier machine, performance like that lists in Table 3 is expected and if this task is inserted into a production process, the classification of peaches would be immediate.

4 Conclusions

This research demonstrates the exceptional capability of convolutional neural networks (CNNs), in particular ResNet-18, which is the best model for accurately and efficiently sorting peaches. The model achieved an impressive accuracy of 95.68%, outperforming traditional sorting methods. This breakthrough paves the way for the development of automated sorting systems that can significantly revolutionize the peach industry. By automating the sorting process, we can expect increased efficiency, lower post-harvest losses, and ultimately higher quality products reaching consumers. ResNet-18's computational efficiency makes it an ideal candidate for deployment on sorting machines, where speed and accuracy are paramount. The findings in this research agree with Yao et al. [14].

References

1. Africano, K.L., Almanza-Merchán, P.J., Balaguera-López, H.E.: Fisiología y bioquímica de la maduración del fruto de durazno [*Prunus persica* (L.) Batsch]. Una Revisión. Rev. Colomb. Cienc. Hortic. **9**(1), 161 (2015). https://doi.org/10.17584/rcch.2015v9i1.3754
2. Duraznos para consumo en fresco en el sur de Santa Fe : ¿cómo definir su momento óptimo de cosecha? Consultado: el 3 de junio de 2024. [En línea]. Disponible en: https://repositorios latinoamericanos.uchile.cl/handle/2250/2675748
3. DOF – Diario Oficial de la Federación: Consultado: el 17 de septiembre de 2024. [En línea]. Disponible en: https://www.dof.gob.mx/nota_detalle.php?codigo=5089982&fecha=12/05/2009#gsc.tab=0
4. Corvalán, J.G.: Inteligencia artificial: retos, desafíos y oportunidades – Prometea: la primera inteligencia artificial de Latinoamérica al servicio de la Justicia. Revista de Investigações Constitucionais **5**(1), 295 (2018). https://doi.org/10.5380/rinc.v5i1.55334
5. Flores, F.A.I., Sanchez, D.L.C., Urbina, R.O.E., Coral, M.Á.V., Medrano, S.E.V., Gonzales, D.G.E.: Inteligencia artificial en educación: una revisión de la literatura en revistas científicas internacionales. Apunt. Univ. (2021). https://doi.org/10.17162/au.v12i1.974
6. Naranjo-Torres, J., Mora, M., Hernández-García, R., Barrientos, R.J., Fredes, C., Valenzuela, A.: A review of convolutional neural network applied to fruit image processing. Appl. Sci. **10**(10), 3443 (2020). https://doi.org/10.3390/app10103443
7. Lopez-Betancur, D., et al.: Comparación de arquitecturas de redes neuronales convolucionales para el diagnóstico de COVID-19. Computación y Sistemas (2021). https://doi.org/10.13053/cys-25-3-3453
8. Medina, G., Chuk, O., Luna, A., Bertero, R.: Redes neuronales convolucionales para determinar redondez en)partículas de arena (2022)
9. Méndez Almansa, J.E., Silva Salamanca, J.S.: Desarrollo de una aplicación móvil para el reconocimiento de la madurez de un grupo de frutas a través del análisis de imágenes por medio de redes neuronales", abr. 2022, Consultado: el 13 de junio de 2024. [En línea]. http://repository.unipiloto.edu.co/handle/20.500.12277/11498
10. LeCun, Y., Bengio, Y., Hinton, G.: Deep learning. Nature **521**(7553), 436–444 (2015). https://doi.org/10.1038/nature14539
11. Navarro-Solís, D., et al.: Analysis of convolutional neural network models for classifying the quality of dried chili peppers (*Capsicum Annuum* L). In: Calvo, H., Martínez-Villaseñor, L., Ponce, H., Cabada, R.Z., Rivera, M.M., Mezura-Montes, E. (eds.) Advances in Computational Intelligence. MICAI 2023 International Workshops: WILE 2023, HIS 2023, and CIAPP 2023, Yucatán, Mexico, November 13–18, 2023, Proceedings, pp. 116–131. Springer Nature Switzerland, Cham (2024). https://doi.org/10.1007/978-3-031-51940-6_10
12. Lopez-Betancur, D., et al.: Evaluating CNN models and optimization techniques for quality classification of dried chili peppers (*Capsicum annuum* L.). Int. J. COP Infor. **15**(2), 13–25 (2024). https://doi.org/10.61467/2007.1558.2024.v15i2.462
13. Zenteno, M.D.A., Castilla, J.S.R., de la Vega, J.A.: Clasificación de frutos del durazno en maduros, no maduros y dañados hacia la cosecha automatizada. CIBA Revista Iberoamericana de las Ciencias Biológicas y Agropecuarias **10**(19), 39–53 (2021). https://doi.org/10.23913/ciba.v10i19.107
14. Yao, N., Ni, F., Wu, M., Wang, H., Li, G., Sung, W.-K.: Deep learning-based segmentation of peach diseases using convolutional neural network. Front. Plant Sci. (2022). https://doi.org/10.3389/fpls.2022.876357
15. Akbar, M., et al.: An effective deep learning approach for the classification of Bacteriosis in peach leave. Front. Plant Sci. (2022). https://doi.org/10.3389/fpls.2022.1064854

16. He, K., Zhang, X., Ren, S., Sun, J.: Deep Residual Learning for Image Recognition. Presentado en Proceedings of the IEEE Conference on Computer Vision and Pattern Recognition, pp. 770–778 (2016). Consultado: el 3 de junio de 2024. [En línea]. Disponible en: https://openaccess.thecvf.com/content_cvpr_2016/html/He_Deep_Residual_Learning_CVPR_2016_paper.html
17. Rawung, B.H., Djamal, E.C., Yuniarti, R.: Classification of lemon fruit ripe using convolutional network. AIP Conf. Proc. **2714**(1), 030029 (2023). https://doi.org/10.1063/5.0129395
18. Mathur, M.: A comparative analysis of deep learning algorithms for fruit disease classification. J. Electr. Syst. **20**(7s), 2621–2633 (2024). https://doi.org/10.52783/jes.4098
19. Saedi, S.I., Rezaei, M., Khosravi, H.: Dual-path lightweight convolutional neural network for automatic sorting of olive fruit based on cultivar and maturity. Postharvest Biol. Technol. **216**, 113054 (2024). https://doi.org/10.1016/j.postharvbio.2024.113054
20. Lopez-Betancur, D., et al.: Convolutional neural network for measurement of suspended solids and turbidity. Appl. Sci. **12**(12), 6079 (2022). https://doi.org/10.3390/app12126079
21. Guerrero-Mendez, C., Saucedo-Anaya, T., Moreno, I., Araiza-Esquivel, M., Olvera-Olvera, C., Lopez-Betancur, D.: Digital holographic interferometry without phase unwrapping by a convolutional neural network for concentration measurements in liquid samples. Appl. Sci. **10**(14), 4974 (2020). https://doi.org/10.3390/app10144974
22. Maeda-Gutiérrez, V., et al.: Comparison of convolutional neural network architectures for classification of tomato plant diseases. Appl. Sci. **10**(4), 1245 (2020). https://doi.org/10.3390/app10041245

Cleaning Binary Distortion on MNIST Dataset

Rafael Castaneda-Diaz[1,2], Daniela Lopez-Betancur[1(✉)],
Carlos Guerrero-Mendez[1(✉)], Efrén González Ramírez[1],
Salvador Gómez-Jiménez[1], and Flossi Puma-Ttito[1]

[1] Universidad Autónoma de Zacatecas, Unidad Académica de Ciencia y Tecnología de la Luz y la Materia (LUMAT), Zacatecas, Mexico
{danielalopez106,guerrero_mendez}@uaz.edu.mx
[2] Instituto Politécnico Nacional, Unidad Profesional Interdisciplinaria de Ingeniería Campus Zacatecas (UPIIZ), 98160 Zacatecas, Mexico

Abstract. Dealing with noise in digital images is an important task in various scientific research processes. Noise often has different characteristics, with one of the key factors being its randomness. Distorted images many times are treated with techniques to remove distortion and improve image quality. However, in some cases, simulated distortion itself could be useful for testing hypothesis about the performance of the cleaning process or specific patterns. This paper presents an analysis of the performance of a convolutional autoencoder (CAE), proposed by us. The CAE reconstructs images from the MNIST dataset that were distorted with three increasing levels of binary noise: 10% (79 pixels), 20% (157 pixels), and 30% (235 pixels), respectively. In the validation stage, SSIM metric was used to compare the reconstructed images with original ones, showing that SSIM's value decreased as distortion level increased: 0.918 (10%), 0.940 (20%), 0.900 (30%). Furthermore, the loss function throughout 150 epochs has shown: 0.008, 0.009, 0.113, respectively. In the experimental work with this CNN architecture images were reconstructed with very good visual quality, highlighting the potential for practical applications in imaging technology.

Keywords: Autoencoder · convolutional · neural network · noise · images · SSMI

1 Introduction

In a wide variety of scientific research, dealing with noise in digital images is one of the most important tasks for obtaining accurate results. In many cases noise is considered as corruption or distortion, rendering a low-quality image. Distortion could be an obstacle to a good interpretation of such information contained in images obtained from experimental scenarios. Alternatively, noise patterns could be just a source of criteria for making inferences about visual perception of images. Basically, noise is defined as a random variation of the values of some pixels, i.e. random variations in intensity or brightness [1].

From this point of view, variational patterns in intensity or brightness came from different real applications, e.g., environmental factors for cameras, functions of sensors, vibrations, dusty scenarios, humidity, chemical compositions, etc. [3–5]. Nevertheless, to test some hypothesizes about these types of variational patterns, there is the possibility of downloading available datasets and adding to them some noise under specific characteristics.

In Deep Learning theory some algorithms came up to clean noise from images. The aim of these algorithms is to reconstruct corrupted information. Some of algorithms are the autoencoders (AE). An AE is a neural network, with architecture and functionality, which has the capacity to reconstruct corrupted information. For this work, three levels of noise were induced to each image from MNIST dataset. The alteration process consisted of replacing the proportions of 10%, 20% and 30% of the total of pixels in each image with completely black pixels, i.e., set the value pixel to 1. In consequence, three new datasets were created for the experimental work developed during this study. Especially, these datasets were used to evaluate the capacity of a convolutional autoencoder (CAE), proposed by us, to clean images. The experiment implementation of this architecture reconstructed images with very good visual quality.

The outline of this paper is as follows. In Sect. 2, autoencoders are described in the mathematical sense and their functionality with images. Also, the state of the art about some techniques for denoising images are given. In Sect. 3, the ways datasets were generated and organized are described in detail. Also, the proposal of autoencoder and its implementation, resources and its performing evaluation are described. Section 4 contains the results of the model's application as well as the discussions about the SSIM index score generated after each running time. Sections 5 and 6, contain some conclusions and suitable work for the future.

2 State of the Art

Mathematical Foundation

The origin of autoencoders (AE) dates back to1980's. The foundation is the learning procedure called *back-propagation*. This procedure consists of repeated adjustments for weights of the connections in the neural network in such way that it minimizes a measure of the difference between the actual output vector of the network and the desired output vector. These adjustments in weights and their interactions, especially in hidden layers of the network, which are independent of input and the output, represent important meaningful features and regularities in the task domain of the network [6].

In mathematical words, an AE is a network that has the same number of input units as output units, and it is trained to generate an output x' that is close to the input x. The structure of this model may be viewed as the composition of two functions, f and g, that is $g \circ f$, where $f : \mathbb{R}^n \to \mathbb{R}^m$ is the encoder and where $g : \mathbb{R}^m \to \mathbb{R}^n$ is the decoder, respectively. Usually, $m \leq n$ because of the functionality of AE as a compressor of information or as a reductor of the high dimensionality of input data. Thus, the input $x \in \mathbb{R}^n$ is mapped into a code $h = f(x)$, where $h \in \mathbb{R}^m$ represents meaningful features of x. Furthermore, $\mathbb{R}^m$ is defined as the latent space. On the other hand, the decoder g maps x into $g(f(x)) = g(h) = x'$, where $x' \approx x$. Therefore, $x' \in \mathbb{R}^n$ is defined as the reconstruction of x from h [7].

To measure the degree of mismatch between input x and its reconstruction x', the weights w of the network are determined by minimizing an error function, e. g., the sum-of-squared-error with the form

$$E(w) = \frac{1}{2} \sum \|x'(x, w) - x\|^2. \tag{1}$$

Equation (1) is called the loss function. $E(w)$ is the difference between x and x'. It's more desirable to avoid having an AE as an identity mapping, which maps x' to x such that $x \approx x'$. Alternatively, one approach is to force the model to extract important and meaningful features from input data x. In this way, a useful model is a *denoising autoencoder (DAE)*. If so, before training a DAE, input data x is corrupted with some kind of noise to give a modified input data $\tilde{x}$. In consequence, the output $x'(\tilde{x}, w)$ depends on $\tilde{x}$ and w [8]. Similarly, the degree of mismatch is given by the error function:

$$E(w) = \frac{1}{2} \sum \|x'(\tilde{x}, w) - x\|^2. \tag{2}$$

Image denoising is considered an ill-posed problem. Owing to the difficulties in finding a unique solution, some techniques have been developed. A very simple and intuitive mathematical definition for noise on images is stated as follow:

$$\tilde{x} = x + n. \tag{3}$$

where $\tilde{x}$ is the distorted image, n is some additive noise added to x. Equation (3) represents the transformation of x into $\tilde{x}$, which are in the same Euclidean space. In a practical sense, denoising $\tilde{x}$ implies to process $\tilde{x}$ to obtain x', which is the best estimation of x. Such conceptualization is a starting point for understanding and dealing with noise as a relevant fact among different study fields. According to literature reviews, the most discussed types of noise are Gaussian noise, impulse noise, quantization noise, Poisson noise, speckle noise and salt-pepper noise. The presence of noise in images and its dealing with imply areas like medical imaging, remote sensing, military surveillance, biometrics and forensics, industrial, agricultural automation and many others. The solution to decreasing noise problem is treated with a wide variety of proposed methods and algorithms. Denoising methods are classified as spatial domain methods, transform domain methods and CNN-based denoising methods. Spatial domain methods are divided into spatial domain filtering and variational denoising methods, which aim is to remove noise by calculating the gray value of each pixel based on the correlation between pixels/image patches in the original image. In contrast, transform domain methods include Fourier transform, cosine transform, wavelet domain, block-matching and 3D filtering. In general, with these methods, the characteristics of image information and noise are different in the transform domain. Finally, CNN-based methods are classified in two categories: multilayer perceptron (MLP) and deep learning methods. Fundamentally, CNN-based methods attempt to learn a mapping function by optimizing a loss function on a training set that contains distorted images [5].

In a detailed way, in [9] CNN methods are described as more flexible and with more capacity respect to spatial and transform domain methods. CNN methods are divided into two approaches: the first one, for denoising general images; and the second one, for denoising specific images. In this case, authors refer to general images those which represent general purpose and for specific images that were created in specific fields like medicine, remote sensing, infrared sensing, etc. According to the authors, they concluded that different techniques related to CNN methods can remove all king of noise from images, and additionally, CNN architecture can be modified to remove bottleneck of vanish gradient.

In fact, according to consulted literature reviews on methods for image denoising using CNN architectures, it's possible to set up experimental work with a simple CNN model, or even with implementing more advanced CNN models reported in the literature. Furthermore, there are different CNN architectures for different applications, e. g., classification, denoising information, generative modeling, anomaly detection in medicine, missing value imputation, image compression, dimensionality reduction, and so forth. In [1, 2], two different models were used for denoising and compressing images. In both publications, a simple fully connected autoencoder (SAE) and a convolutional autoencoder (CAE) were proposed as DAE's. Both concluded that CAE had better performance than the SAE. Inspired by these efforts, we have proposed a similar CAE architecture with the aim of training it to reconstruct contaminated images. Also, we focused on the performance of the CAE in the light of induced noise patterns.

3 Materials and Methods

For this study, MNIST dataset of handwritten digits was downloaded from keras.io; MNIST for Modified National Institute of Standards and Technology. This dataset contains 70,000 grayscale images with values from 0 (completely white) to 1 (completely black), each with a size of $28 \times 28 \times 1$ pixels, that is, 784 pixels.

3.1 Dataset Description and Preprocessing

Initially, for default the dataset is divided into two subsets: 60,000 images for training and 10,000 for testing or validating the model. Subsequently, both subsets were corrupted by altering a portion of the pixels—10% (79 pixels), 20% (157 pixels), and 30% (235 pixels), respectively. The alteration process involved replacing these proportions of pixels with completely black pixels, set to a value of 1. As a result, this noise addition process created three new distinct datasets, each containing 70,000 corrupted images, and divided as the MNIST dataset: 60,000 for training and 10,000 testing the CAE model, respectively. In Fig. 1, one element from validation subset of MNIST is shown with the corresponding one in each new subsets for validation, respectively.

3.2 Distortion Levels on Dataset

After the distortion process of MNIST dataset with three different levels, the resulting datasets were saved and labeled as follow: DL-1 for 10% (70 of 784 pixels), DL-2 for 20% (157 of 784 pixels) and DL-3 for 30% (235 of 784 pixels). Also, as it was with MNIST dataset, DL-1, DL-2 and DL-3 are available in the same directory on the PC.

Distortion levels in 784 pixels

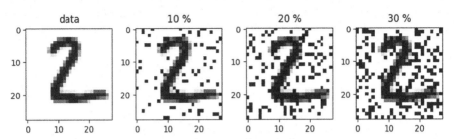

Fig. 1. Visual distortion levels: from 0% (data of MNIST) to 10%, 20% and 30%.

3.3 The Model Architectures

Mostly, convolutional autoencoders (CAE) are structured on two basic parts: the first one, an encoder, which extracts features from input data and generates a representation into lower dimensions space (*latent space*); the second one, a decoder, that will take the latent representation to reconstruct the input data. In Fig. 2 model is shown.

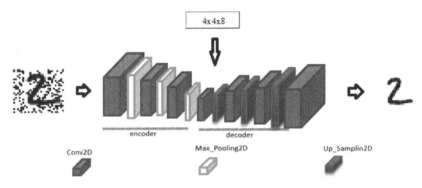

Fig. 2. Convolutional autoencoder architecture (CAE).

A detailed description is given below:

a) The encoder, which takes input images of size $28 \times 28 \times 1$, an element x of $\mathbb{R}^{784}$, processes x through 4 convolutional layers and 3 max pooling layers, mapping them into a latent space of dimension $4 \times 4 \times 8 = 126$, that is, it maps x to an element of $\mathbb{R}^{126}$. Notice that after each convolutional layer there is a rectified linear layer (ReLU).

b) The decoder takes the code (latent values) and process them through 3 convolutional layers and 3 transposed convolutional layers to reconstruct the original image. The resulting output images should have $28 \times 28 \times 1$ dimensions.

The model was designed to denoise images from three datasets with some degree of mismatch degree to MNIST dataset images. For this reason, the model was cloned and save separately three times with different names; it guarantees that weights of each version of the model are adjusted according with the distortion levels during training stage. Table 1 shows in more detail the type of layer and its output dimensions as the number of learnable parameters.

Table 1. Convolutional autoencoder layer details.

Layer	Output	Learnable Parameters
conv2d (Conv2D)	[(None, 28, 28, 32)]	320
max_pooling2d (MaxPooling2D)	[(None, 14, 14, 32)]	0
conv2d (Conv2D)	[(None, 14, 14, 8)]	2312
max_pooling2d (MaxPooling2D)	[(None, 7, 7, 8)]	0
conv2d (Conv2D)	[(None, 7, 7, 8)]	584
max_pooling2d (MaxPooling2D)	[(None, 4, 4, 8)]	0
conv2d (Conv2D)	[(None, 4, 4, 8)]	584
Up_Sampling2D	[(None, 8, 8, 8)]	0
conv2d (Conv2D)	[(None, 8, 8, 8)]	584
Up_Sampling2D	[(None, 16, 16, 8)]	0
conv2d (Conv2D)	[(None,14, 14, 32)]	2336
Up_Sampling2D	[(None,28, 28, 32)]	0
conv2d (Conv2D)	[(None,28, 28, 1)]	289
Total of learnable parameters:	7009	

3.4 Experiment Setup

The model was programmed using TensorFlow and Keras' libraries. Table 2 shows the characteristics of the computer resources implemented for this experiment.

Table 2. Specifications and setup

	Specifications
Memory	16 GB
Processor	13th Gen Intel(R) Core(TM) i7-13650HX 2.60 GHz
Graphics	NVIDIA GeForce RTX 4050 Laptop GPU
Operating Systems	Windows 11 Home Single Language

3.5 Training Details

For the created datasets DL-1, DL-2, DL-3, the same hyperparameters values for each cloned model were set up for training stages. Each training phase took 150 epochs with the following hyperparameters values: learning rate $= 0.001$, Adam's parameters $\beta_1 = 0.9$, $\beta_2 = 0.999$ and $\epsilon = 1 \times 10^{-8}$. These key Adam's parameters are set for default on Keras API. For the random initialization of weights was using a random seed equal to 42.

3.6 Performing Evaluation

Evaluating model performance was done in two ways: the first one, was through the estimation of loss function Mean Squared Error (MSE) which calculated the average of the squared difference errors between images from MNIST validation subset and the reconstructed images taken from DL-1, DL-2, DL-3 validation subsets, respectively. The second way was through the implementation of the metric Structural Similarity Index Metric (SSIM), which can assess the structural similarity between two images. The SSIM is an algorithm that captures from two signals x and y from images, three aspects: luminance $l(x, y)$, contrast $c(x, y)$ and structure $s(x, y)$. These three aspects are combined as follow:

$$SSIM(x, y) = l(x, y)c(x, y)s(x, y) = \frac{(2\mu_x\mu_y + C_1)(2\sigma_{xy} + C_2)}{(\mu_x^2 + \mu_y^2 + C_1)(\sigma_x^2 + \sigma_y^2 + C_2)}, \quad (4)$$

where x and y are discrete signals, μ_x and μ_y are the respective mean intensity. σ_x and σ_y are estimations of the signals contrast, respectively. σ_{xy} is the correlation coefficient between x and y. The constants C_1 and C_2 in Eq. (4) are included to avoid instability when $\mu_x^2 + \mu_y^2$ and $\sigma_x^2 + \sigma_y^2$ are close to cero. SSIM index score ranges from -1 to 1. If the score is close to 1, that indicates that similarly is almost total. If it is close to 0, then similarity is low. And for negative values there is not any structural similarity.

In this work reconstructed (denoised) images from validation subsets and the corresponding one from MNIST testing subset must be compared to have criteria about the model's performance. To see more about SSIM applications in detail, refer to [5, 9–11].

4 Results and Discussions

The implementation of this model for reconstructing images demonstrated good performance levels. In this work, we confirmed that CNN-based models have very good abilities to reach this outcome. In [1, 2], both papers, a very similar CAE and a full-connected autoencoder were compared on denoising images. CAE was better than fully connected. However, as it's shown in Fig. 3, the validation loss over the 150 epochs tends to increase as the noise level in the images rises. It indicates that higher distortion levels make it more difficult to the model adjust its weights. For example, in [1] for a training of 20 epochs, the CAE reached a validation loss of 0.0871 and for fully connected autoencoder a validation loss of 0.2305 and with a training time of 100 s.

The SSIM index score obtained in this experiment showed contradictory results, specifically with DL-1 and DL-2. According to the visual perception of images with

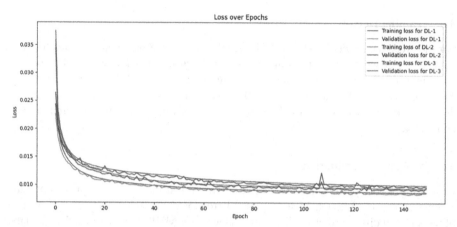

Fig. 3. Training loss vs. validation loss throughout 150 epochs. The training and validation loss for DL-1, DL-2 and DL-3, are shown respectively.

Table 3. Validation loss and SSIM index

Metrics	DL-1	DL-2	DL-3
Validation Loss	0.0082	0.0084	0.0093
SSIM index	0.918	0.940	0.900
Training time	821 s	820 s	825 s
Validation time	0.66 s	0.66 s	0.68 s

higher distortion level, SSIM for DL-2 should be lower than that for DL-1, but it was not. In contrast, the SSIM index for DL-3 is lower than the first two levels. In Fig. 4, in row b), one can see that reconstructed images are visually very similar respect to that one of MNIST validation subset, but with scrutiny one can notice many differences. Additionally, the results show that the computation time for training and validation model also increases with higher noise levels. Refer to Table 3 for details.

In [11], SSIM index is described as popular measure in many different scientific projects for almost two decades. However, in this paper it was demonstrated that this index or its linear transformation is not a metric in the mathematical sense. Also, is discussed the fact that SSIM cannot correctly determine the similarity between two images, just only similarity of visually close images of the same scene. Authors in [12] compared SSIM index with the Pierson correlation coefficient (PCC) and concluded that PCC is a more accurate measure of similarity of the compared images than the SSIM index. As a factual element for discussion about using SSIM, for example, in [5] SSIM and Peak Signal-to-Noise Ratio (PSNR) metrics were implemented to evaluate the denoising results on Lena image that was corrupted with Gaussian noise with standard deviation 30 and cleaned with some filtering methods: Weiner filtering (PSNR = 27.81 dB; SSIM = 0.707); Bilateral filtering (PSNR = 27.88 dB; SSIM = 0.712); PCA method (PSNR = 26.68 dB; SSIM = 0.596); Wavelet transform domain method (PSNR = 21.74 dB; SSIM

= 0.316); Collaborative filtering: BM3D (PSNR = 31.26 dB; SSIM = 0.845). Hint: dB for decibels. Authors concluded that Block-Matching and 3D Filtering (BM3D) method is better than the other ones and considered it with a big potential for noise reduction and edge protection. Evidently, authors based their argument on PSNR and SSIM values.

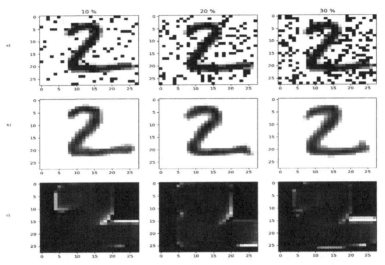

Fig. 4. a) images from validation dataset in DL-1, DL-2 and DL-3; b) reconstructed images with CAE; c) images resulting from resting reconstructed images and that one from MNIST validation dataset. Note that images in row c), visually are coincident in some regions. However, is not possible to give concluding arguments about that.

5 Conclusions

According with the results obtained in this experiment, we have concluded that:

- Despite the simplicity of CAE architecture, it is capable for extracting fundamental features from corrupted images. Even so, denoising (reconstructing) corrupted images, CAE achieves as results images that are visually very similar to the original MNIST images.
- On the other hand, SSIM showed structural similarity after reconstruction. Considering that SSIM scores 0.918 (10% of pixels), 0.940 (20% of pixels) and 0.900 (30% of pixels) are proximate to 1, so it's possible to conclude that CAE has a big potential for cleaning contaminated grayscale images similar to MNIST images.

6 Future Work

Visually, the random induced noise in MNIST images has some regularities. In this sense, implementing those types of noise mentioned in the state-of-the-art section would be interesting to explore in future work with the proposed CNN model. On the other hand, color images datasets could be suitable for experimenting under similar conditions and whit the same CAE or with a more refined one. Considering the arguments given in the consulted literature reviews, for future experiments, is suitable for implementing the PCC and PSNR in order to compare with SSIM index measure after to a denoising process.

Acknowledgements. This project is supported by CONAHCYT and El Instituto Politécnico Nacional de México.

Disclosure of Interests. It is now necessary to declare any competing interests or explicitly state that the authors have no competing interests.

References

1. Venkataraman, P.: Image Denoising Using Convolutional Autoencoder. el 24 de julio de 2022. arXiv: arXiv:2207.11771. https://doi.org/10.48550/arXiv.2207.11771
2. Zhang, Y.: A better autoencoder for image: convolutional autoencoder. In: ICONIP17-DCEC (2018). https://users.cecs.anu.edu.au/~Tom.Gedeon/conf/ABCs2018/paper/ABCs2018_paper_58.pdf. Accessed 23 Mar 2017
3. Lopez-Betancur, D., Moreno, I., Guerrero-Mendez, C., Gomez-Melendez, D., de J. Macias P., M., Olvera-Olvera, C.: Effects of colored light on growth and nutritional composition of tilapia, and biofloc as a food source. Appl. Sci. **10**(1), 362 (2020). Consultado: el 22 de septiembre de 2024. https://www.mdpi.com/2076-3417/10/1/362
4. Guerrero Méndez, C., Saucedo Anaya, T., Araiza Esquivel, M.A., Balderas Navarro, R.E., López Martínez, A., Olvera Olvera, C.A.: Measurements of concentration differences between liquid mixtures using digital holographic interferometry (2017). Consultado: el 22 de septiembre de 2024. http://148.217.50.3/jspui/handle/20.500.11845/1447
5. Fan, L., Zhang, F., Fan, H., Zhang, C.: Brief review of image denoising techniques. Vis. Comput. Ind. Biomed. Art **2**(1), 7 (2019). https://doi.org/10.1186/s42492-019-0016-7
6. Learning representations by back-propagating errors | Semantic Scholar. Consultado: el 12 de septiembre de 2024. https://www.semanticscholar.org/paper/Learning-representations-by-back-propagating-errors-Rumelhart-Hinton/052b1d8ce63b07fec3de9dbb583772d860b7c769
7. Goodfellow, I., Bengio, Y., Courville, A.: Deep Learning. MIT Press, Cambridge (2016)
8. Vincent, P., Larochelle, H., Bengio, Y., Manzagol, P.-A.: Extracting and composing robust features with denoising autoencoders. In: Proceedings of the 25th International Conference on Machine Learning - ICML 2008, pp. 1096–1103. ACM Press, Helsinki (2008). https://doi.org/10.1145/1390156.1390294
9. Methods for image denoising using convolutional neural network: a review—Complex & Intelligent Systems. Consultado: el 10 de octubre de 2024. https://doi.org/10.1007/S40747-021-00428-4
10. Wang, Z., Bovik, A., Sheikh, H., Simoncelli, E.: Image quality assessment: from error visibility to structural similarity. IEEE Trans. Image Process. **13**, 600–612 (2004). https://doi.org/10.1109/TIP.2003.819861
11. Nilsson, J., Akenine-Möller, T.: Understanding SSIM. el 29 de junio de 2020 arXiv: arXiv: 2006.13846. Consultado: el 18 de septiembre de 2024. http://arxiv.org/abs/2006.13846
12. Старовойтов, В.В.: Индекс SSIM не является метрикой и плохо оценивает сходство изображений. *Системный анализ и прикладная информатика*, vol. 0, núm. 2, Art. núm. 2, ago. 2019. https://doi.org/10.21122/2309-4923-2019-2-12-17

Rule-Based Expert System with Bayesian Theory and Fuzzy Inference for Vocational Guidance: A Tool to Prevent School Dropouts

Cynthia Cristina Martinez Padilla[✉]

Master of Science in Engineering, Postgraduate and Research Department, Universidad Politécnica de Aguascalientes, Aguascalientes, Mexico
mc230019@alumnos.upa.edu.mx

Abstract. Vocational guidance has a high impact on the lives of students. Its application within high school programs should be considered almost mandatory for young people to successfully complete a university degree and embark on their professional life with pleasure, integrating into the labor area in something that truly satisfies them and encourages them to always be innovating and seeking to exceed their own expectations. This work seeks to develop new technological tools and algorithms for career choice as the main axis of success in higher education. By applying techniques such as Bayesian theory and Fuzzy inference, we seek to create an expert system based on rules that guides young people in this important decision. Likewise, the probability of success or failure in the chosen career is highlighted based on issues such as the experience and aptitudes of the student, the type of educational demand that the career has and the Sufficient probability (LS) and Required probability (LN) using fuzzy sets to approximate the result. The accuracy obtained in most of the areas evaluated corresponds to more than 50% of the correct answers in which the expert system correctly classified the students if we consider those areas that obtained the highest number of students in their classification. The people who took the test, who were mostly men and women between 16 and 35 years old, said that they had not had any kind of guidance, and that they chose their university career based on their own tastes, leaving aside the skills in which they excel. 50% agreed with the result obtained in the test, they were correctly classified, 30% commented on other interests, since they were not classified correctly, and 20% showed a neutral position with the result. Of the total number of people surveyed, 70% of them said that it is valuable information for choosing their educational and professional path.

Keywords: Expert Systems · Inference engine · Fuzzy logic · Bayesian Theory · Forward chaining · Backward chaining · Dropout · Vocational Guidance · Bayesian probability

1 Introduction

Talking about Dropping out of school, it is a problem that affects not only students but also different institutions around the world, with consequences for society. It can be defined as a person who withdrew from an educational institution before receiving a degree or a person who enrolled in the previous year and decided not to transfer to another school and not continue in their classes [1]. In principle they see this as a learning program until the student leaves school. On the other hand, the spatial connection of concepts occurs when students change subjects in the same institution, when they change universities, even when they leave the educational system, but continue in the future. Whether it is the same house or different houses [2].

The causes of school dropouts can be categorized into three main factors: economic, political, and social. Economic reasons, such as a lack of material resources or the need to work full-time, along with personal factors like learning difficulties, disinterest, and lack of motivation, are among the primary reasons for students leaving school prematurely. In addition, family causes, including early pregnancy, unions, and psychological or emotional family problems, social factors, such as economic and social inequality, further contribute to this issue. All these factors result in a multitude of undesirable consequences, including decreased employment opportunities, inadequate financial resources, students may experience adverse consequences affecting their self-esteem and confidence, ultimately resulting in challenges related to emotional and mental well-being [3].

In addition, it can have an adverse impact on society at large, it can aggravate the shortage of skilled workers in their branches, affecting the workforce and contribute to crime, among other social problems.

There are four steps in this educational guidance process:

1. **General phase of professional training:** From childhood to primary, build character traits, tolerance, self-reflection, and use all the resources that are good for the development of thinking. (Enhanced reception).
2. **Preparatory phase for choosing a job:** In secondary school, intellectual interests, knowledge, and special skills related to a project or field of work appear the person presents special needs or real risks. (Communication and opening doors).
3. **The stage of creation and development of interests and professional skills:** In the first year of the degree, you determine the interests, knowledge and skills that will make the program suitable for your work. (Scientific Society)
4. **The phase of integration of professional needs:** In the last years of work and in the first two years of work for the process of professional adaptation [4].

In some countries, it is so difficult to guide the work. First, most teachers lack the psychology expertise needed to guide students through the career problems they feel they have. Secondly, the tools available to teachers, such as career guidance courses or practical materials, are insufficient or ineffective to support students in strengthening and developing decision-making abilities. This means that university jobs are often chosen quickly without the necessary education [5].

This indicates that one of the main causes of university dropout is the lack of career guidance, it also shows that many students enter college without really knowing what career they want to pursue and may end up making the wrong decision [5].

1.1 Contribution

The main problem or axis in this work is to avoid as far as possible the dropout of school among young university students due to the lack of information and guidance. We seek to implement the use of technologies at high school levels to carry out successful vocational guidance. The objective of the research is whether developing new technological tools and algorithms such as expert systems trained with inference rules will reduce this gap, correctly profiling and classifying young people in the areas in which they have better aptitudes and chances of success. The basis for this expert system is the already used CHASIDE test, which is detailed in Sects. 3 and 4 respectively.

1.2 Distribution

In Sect. 2, a brief exploration of works related to vocational guidance is made, as well as works in which expert systems focused on education were used. In Sect. 3, reference is made to the theoretical framework used in carrying out this work. In Sect. 4, the methodology that was developed, how it was applied and the data set with which it was worked are explained. In Sect. 5, the main results found are shown as well as a comparison of them. And finally, in Sect. 6, the conclusion reached with the use of the expert system.

2 State of the Art

A review of the literature, as presented by Gottfried Career [2], on career decision making and school dropout, shows that students drop out of school for many reasons, including personal factors, issues related to lack of motivation, interpersonal relationships, teenage pregnancy, etc., and financial issues related to family income level; know strategies, activities, resources, and educational evaluation. Educational institutions and parents must participate. This is a great support for children, because it has a positive effect on motivation, behavior and even learning.

Other studies related to dropping out of school as presented by Terrence [6] about the long-term impact of supporting students in elementary school, have found it efficient to take courses during their stay at the academic institution to improve their grades and reinforce what they have learned, thus students have greater chances of completing their studies.

Other studies such suggested by Guerra y Morales [7], that career guidance influences college dropout because the decision-making process for selecting a college career should be guided by professionals who encourage young people to continue their studies.

Discussion of leadership and professional practice and counseling became part of education when psychology was legalized in America in the late 19th century. First, in the revolution of the 20th century, there was a change, in particular the goal was to create and achieving human happiness, and the program focused on helping, counseling and caring for people to live peacefully, even contentedly, the best among the country's population. Yalandá and Trujillo in their article on the importance of vocational guidance in school education [8] mention at the educational level, institutional guidance in primary and secondary education plays an important role in ensuring smooth transitions between secondary and tertiary education. It is important to prepare instructions and techniques for self-awareness of mental and emotional capacities for the training of professional skills.

In the research conducted by Wulansari [9], an expert system software was developed to assist students in recognizing their potential. This technology serves as a tool for students to use in identifying and harnessing their capabilities. The development of this expert system utilized the Software Development Life Cycle (SDLC) methodology. The system design was based on Multiple Intelligence theory and implemented using the Unified Modeling Language (UML). The outcome is an expert system that is practical for utilization.

In the study conducted by Kabathova and Drlik [10] regarding the utilization of various Machine Learning methodologies, an analysis was performed on the performance metrics of multiple learning classifiers using academic data. The findings indicated that these results were pertinent in forecasting course finishers, rather than course completers. Some of the classifiers utilized in this study were: Logistic regression, Decision tree, and Naïve Bayes classifier.

As discussed by Oybek and Nasiba in their study on Technological Innovation in Education [11], the utilization of educational technology holds significant promise in enhancing the overall quality of education and facilitating personalized learning. Additionally, expert systems have the capability to offer students timely and individualized feedback, thereby encouraging a more profound level of learning. In their study, the researchers delve into the advantages of educational technology and expert systems in relation to augmenting student engagement, fostering critical thinking and problem-solving abilities, and enhancing learning outcomes.

By comprehending the potential advantages and obstacles, institutions can develop efficient strategies to utilize these tools for enhancing the quality of education and offering students a more personalized learning experience.

Studies like Makri's [12] refers to the ability of educational technology and expert systems to enhance student engagement as a potential benefit. Additionally, these technologies are seen as valuable tools for improving the overall learning experience. Digital tools, such as online discussion forums, virtual simulations, and interactive games, have the potential to enhance the interactive and engaging nature of learning for students. These tools provide opportunities for students to actively participate in their learning process by encouraging collaboration, problem-solving, and critical thinking skills. For instance, virtual simulations can offer students the opportunity to engage with real-world scenarios to apply and enhance their knowledge and skills.

As González Trejo mentions in his study on technological tools in virtual environments [13], digital tools could grant students access to a plethora of resources, such as online lectures, textbooks, and multimedia materials, that can significantly improve their comprehension of course content. Similarly, expert systems can assist students in recognizing their deficiencies in knowledge and skills, while offering tailored feedback to help remedy these shortcomings.

Also, Syifa, Barliana and Rahmanullah in their study [14] mention that the infrastructure is a crucial challenge, as institutions must ensure that they possess the necessary technology and resources to support these tools. Faculty readiness and training are also important. Instructors must be adequately prepared to integrate these tools into their teaching and learning practices. Additionally, student adoption and usage can present challenges, as certain students may lack familiarity with these tools or exhibit resistance towards utilizing them.

3 Theorical Framework

3.1 Expert Systems

Are powerful tools for automating cognitive tasks, modeling the behavior of human experts in a specific knowledge domain, and attempting to transfer the skills and experience of experts to systems [15]. Non-experts can improve their problem-solving skills or experts can help them in their work [16].

3.2 Knowledge Base

Is the key to solving problems. Identify all knowledge about the main parts of the problem. In other words, it is an indication of the expert's knowledge [17].

3.3 Data Base

The database is the part that contains all the implemented facts, the first facts when the system starts, and the facts obtained at the end [17].

3.4 Inference Engine

An inference engine is a component that contains the reasoning and techniques of a modeling system used by experts. This method analyzes a specific problem and finds the best answer or solution. The decision engine starts monitoring by matching the rules in the knowledge base with the facts in the database. There are two modern methods of modeling in which the inference starts from a hypothesis and works backwards to conclude the truth contained in this hypothesis. And forward chaining, as opposed to backward chaining, starts with a set of data and goes all the way to the end [16].

3.5 Rule-Based Systems

Are an effective tool for dealing with problems governed by decision rules, the simplest methods used in expert systems. The knowledge base contains a set of variables and rules that define the problem, and the inference engine applies classical logic to these rules to arrive at a decision. A method is a logical process that consists of two or more things and has two parts: premises and decision. The system receives input information and then uses rules to process and analyze the information and produce output [18] Most expert systems require rules because they are rule-based systems, that is, knowledge is stored in the form of rules [16].

3.6 Forward Chaining

Involves teaching the first step in a chain of behavior and adding steps one at a time. For example, you must first complete step 1 individually before moving on to step 2. If step 1 is performed all the time, step 2 will be added after the previous job is completed and continue until all parts of the chained job are completed. Are completed [19].

3.7 Backward Chaining

Involves teaching a step and then adding steps to the sequence. But training begins with the last step in the behavior chain. For example, you need to complete the last step first to access the other steps. After the last step is completed each time and individually, the second step in the sequence is added. This sequence continues until all stages are individually completed from start to finish [19].

3.8 The CHASIDE Test

Was developed by Dr. John L. Holland. This test is a tool that allows you to identify the needs and desires of the individual as well as the capacity and ability to work. Or work. The test consists of 98 items developed with YES-NO options, allowing the user to decide which way to go [20] The field of education is divided into [20]:

- C (Management and Accounting)
- H (Humanities, Law, Social Sciences)
- A (Arts)
- S (Health Sciences)
- I (Engineering), Technical and IT Careers)
- D (Defence and Security)
- E (Agriculture and Natural Sciences.

3.9 Bayesian Theory

Arose from using probability theory to solve problems presented by legal systems. However, high computing costs made progress impossible. To increase fit, dependent and independent correlations were constructed between the variables. This type of system is called a stochastic network. Every action and fact in the environment have a probability associated with it, and these probabilities lead people to make decisions or actions, which are modified by observations and connections with other facts that are true [21]. Bayesian analysis has been used in a variety of areas, including ecology, the social sciences, medicine and and artificial intelligence and genetics. This includes using data to evaluate the evidence for contradictory claims and hypotheses. This can be done using the Bayes factor (Eq. 1), which is the ratio of the posterior probabilities to the prior probabilities of the independent hypotheses [22].

$$p(A|B) = \frac{p(B|A) \times p(A)}{p(B)} \qquad (1)$$

3.10 Sufficient Probability (LS)

Represents a measure of the expert's belief in hypothesis H in the presence of evidence E [23]. The appropriate probability (Eq. 2) is defined as the ratio of p(E|H) to p(E|¬H):

$$LS = \frac{p(E|H)}{p(E|\neg H)} \qquad (2)$$

3.11 Required Probability (LN)

Is the measure of hypothesis H in the absence of evidence E [23] and is defined as (Eq. 3):

$$LN = \frac{p(\neg E|H)}{p(\neg E|\neg H)} \qquad (3)$$

3.12 The Prior Probability of an Outcome

In rule-based expert systems, becomes the prior probability of p(H), (Eq. 4) [24]:

$$OH = \frac{p(H)}{1 - p(H)} \qquad (4)$$

The prior probabilities are only used to initially adjust for the uncertainty of the results. So, to get the posterior probability, LS updates the prior probability if the prior of the rule is true (ie, evidence), (Eq. 5), and LN if the prior is false (Eq. 6).

$$O(H|E) = LS \times O(H) \qquad (5)$$

$$O(H|\neg E) = LN \times O(H) \qquad (6)$$

Then the posterior probability is obtained using the posterior probability. If the prior of the rule is true (ie, evidence), (Eq. 7), and LN if the prior is false (Eq. 8):

$$p(H|E) = \frac{O(H|E)}{1 + O(H|E)} \qquad (7)$$

$$p(H|\neg E) = \frac{O(H|\neg E)}{1 + O(H|\neg E)} \qquad (8)$$

3.13 Fuzzy Inference Systems (FIS)

Are flexible computational methods that allow the training of human expertise in IF-THEN rules. Fuzzy models represent the experience of an expert in controlling these systems. Optimization algorithms allow the adaptation and control of various fuzzy processes such as the black box problem. Python is the fastest programming language in the field in recent years. There are many benefits to using open-source libraries, including saving time and resources [25].

Inference systems include rules and consequences for those rules which are processed and evaluated within the following phases: preprocessing, fuzzification, rule base, inference engine, defuzzification and postprocessing [25]. (Fig. 1):

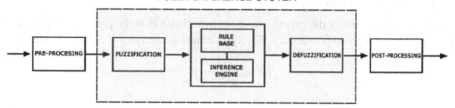

Fig. 1. Processing cycle of an FIS

3.14 Confusion Matrix for Classification with Accuracy

A common way to assess classification quality in data is through predictive decision classes and a confusion matrix, also known as an error matrix. The matrix is a tool that contains only the information contained in the data. The sum of the main diagonal elements of a confusion matrix is widely used to measure the success of classification based on algorithms or human observations [26]. Utilizing metrics customized from classification can aid in assessing feature attribution models in a standardized and interpretable manner. We acknowledge that certain metrics may hold increased significance in particular situations. For example, one may prioritize feature attribution methods that minimize the number of false positives (i.e., high attribute precision) or prioritize attribution that minimizes false negatives (i.e., high attribute recall) [27].

Let us first define True Positives as the set of images belonging to the target class within the mosaic, True Negatives as the set of images not belonging to the target class, and All Samples as the feature attributions. To begin, for each mosaic, we establish the following definitions:

- True Positive evidence (TP) = $\sum_{i \in T} |max(0, \alpha_i)|$
- False Positive evidence (FP) = $\sum_{i \in N} |max(0, \alpha_i)|$
- True Negative evidence (TN) = $\sum_{i \in N} |min(0, \alpha_i)|$
- False Negative evidence (FN) = $\sum_{i \in T} |min(0, \alpha_i)|$

The indicators are combined in the matrix, thus enabling the assessment of the feature attribution performance. Thanks to this link, we are able to borrow and extend the metrics that have been developed for the classification case in order to evaluate feature attribution. Therefore, as an illustration, we can establish the following definitions (Eq. 9) [27]. This functionality enables practitioners to determine the most appropriate option for their specific needs.

$$Accuracy = \frac{TruePositives + TrueNegatives}{All\ Samples} \quad (9)$$

4 Methodology

The application used to implement the test is developed using the Python language, and the expert library for building expert systems.

4.1 Data Set

The test was administered electronically to 103 high school students from the city of Aguascalientes, Mexico, who were in their fifth semester of their studies and were between the ages of 16 and 17; and to 7 students of the Master of Engineering Sciences who were in their third semester and between the ages of 27 and 35. A representative sample of approximately 47.3%, equivalent to 52 students, was taken from these students. Each student was asked to answer yes or no to each of the questions posed. At the end, they were shown the result of the area or areas with the highest score, some of the characteristics that people in that area should have, and recommendations of some related careers (Table 1).

Table 1. Distribution of students who supported the test

	HIGH SCHOOL	MASTER	
Women	28	1	
Men	75	6	
Total	103	7	**110**

4.2 Data Analysis

The test used consists of a questionnaire of 21 questions taken from the CHASIDE test that identify the person's aptitude and interests for selecting a university career. The 21 items are grouped into the 7 professional categories mentioned above [28] and have a value of 1 point each so that the maximum score is 3 points in each of the 7 areas; The area with the highest score is then considered to be the one in which the student has the greatest chance of achieving academic success [29]. The remaining 9 rules correspond

Table 2. Example of Facts created to support the rules

```
1   class AccountingArea(Fact):
2       """Is Accounting"""
3       pass
4
5   class AccountingArea(Fact):
6       """Is Accounting"""
7       pass
8
9   class AccountingArea(Fact):
10      """Is Accounting"""
11      pass
12
13  class AccountingArea(Fact):
14      """Is Accounting"""
15      pass
```

```
16  class Is5120(Fact):
17      """Coordinate work"""
18      pass
19
20  class Is5120(Fact):
21      """Coordinate work"""
22      pass
23
24  class Is46(Fact):
25      """Distribute schedules"""
26      pass
27
28  class IsTest(Fact):
29      """Coordinate work"""
30      pass
```

to the validations through which the result will be displayed on the screen. To create the Database, facts were declared for each of the rules that will be evaluated, 7 for the hypotheses, 14 for the questions that will use the forward method and 9 more for the printing of results (Table 2).

To create the Knowledge database, yes/no inference rules were used, the first 7 questions correspond to the Backward chaining method [19] in which, 7 hypotheses are consolidated and from which continuity will be given to the following questions (Table 3).

Table 3. Hypotheses formulated for the Backward chaining method

```
1    #ACCOUNTING
2    @Rule(NOT(Es15(es_15=W())))
3        def question_es15(self):
4            self.es_15 = int(input("Do you organize your money so that it
             lasts until your next payment? (yes=1/no=0)"))
5            increment_counter1(self.es_15)
6            self.declare(Es15(es_15=self.es_15))
7
8    #MEDICAL AND HEALTH
9    @Rule(NOT(Es69(es_69=W())))
10       def question_es69(self):
11           self.es_69 = int(input("Do you think that public health should
             be a priority, free and efficient for all? (yes=1/no=0)"))
12           increment_counter4(self. es_69)
13           self.declare(Es69(es_69=self.es_69))
```

If in any of them the person answers 'YES', then the system will continue asking the following questions related to the area in process. If the person answers 'NO', the system then discards that area and continues with the next one using the Forward chaining method (Table 4).

Table 4. Example of rules created for the Forward chaining method

```
1    self.today_weather = Es100(state=input("**100.** Are you currently
     working? Answer yes/no ***"))
2    self.declare(self.today_weather)
3
4    self.thats_another_thing7=Its200(it_is_average=float(input("**200.**
     What is your current fiancé? ***")))
5    self.declare(self.thats_another_thing7)
```

The process ends when the 7 areas are finished being evaluated. If during the process any of the areas reaches its maximum score, then the result is displayed on the screen (Table 5).

The flow chart shown below explains the behavior of the rules used (Fig. 2).

Once all the questions of the CHASIDE test have been applied, a validation is carried out which returns the area with the highest score. To obtain the probabilities of

Table 5. Rules to validate the printing of results

```
1   @Rule(EsContable(es_contable=W()) & EsContable(es_contable='si'))
2   def cuenta_pregunta(self):
3       print("Your area is: ACCOUNTING AND ADMINISTRATION.")
4       print("Characteristics of people in this area: Persuasive, Objective, Practical, Tolerant, Responsible, Ambitious.")
5       self.esotracosa7 = Es200(es_Promedio=float(input("**200.** What is your current fiancé? ***")))
6       self.declare(self.esotracosa7)
7
8   @Rule(IsHumanist(is_humanist='yes') & (Is100(state='yes') | (Is100(state='no'))) & Is200(is_Average=W()))
9   def question_probhumanistif(self):
10      LS = 15.52 #10.06
11      LN = 20.44 #5.03
12      PH = 0.72 #0.65 #p(h)
13      OS = PH/(1-PH)
14      if (self.today_weather['state']=='yes'):
15          OS_T = OS*LS
16          OS = OS_T/(1+OS_T)
17      if (self.today_weather['state']=='no'):
18          OS_N = OS*LN
19          OS = OS_N/(1+OS_N)
20      self.tomorrow_weather = EsHumanistas(OSr=OS)
21      self.declare(self.tomorrow_weather)
22      ComoEgreso = Egresados_Inference.fuzzy_system_sim([self.esotracosa8['es_Promedio'], 60])
23      print("With an Income Average %f" %self.esotracosa8['es_Promedio'], " you have a probability of success of: %f" %self.tomorrow_weather['OSr'])
24      print(f"of graduating from a LOW DEMAND degree with an approximate average of: {ComoEgreso[0][0]} %")
        print("Related courses: Psychopedagogical counseling, Law, Social Work, Communication.")
25
```

sufficiency and need, LS and LN respectively, a database from a University in Argentina was used, belonging to 2000 students, the items used in this set were degree received, student average and degree program. One more academic center field was added to this database, classifying the majors into 6 main areas according to classification courses of a University of Aguascalientes. 6 new inference rules were created to calculate for each area LS and LN respectively.

To calculate the student's success or failure in the university course given as optimal in the CHASIDE evaluation, the student's average at the time of admission and the current demand for the course, based on the number of students it offers, were taken as a reference. A scale of high, medium and low average was considered; likewise, a scale of high, low and medium demand was considered (Table 6).

With these characteristics, the Mamdani implication method will be used to obtain the max-min composition through the UPAFuzzySystem library [25] and thus estimate

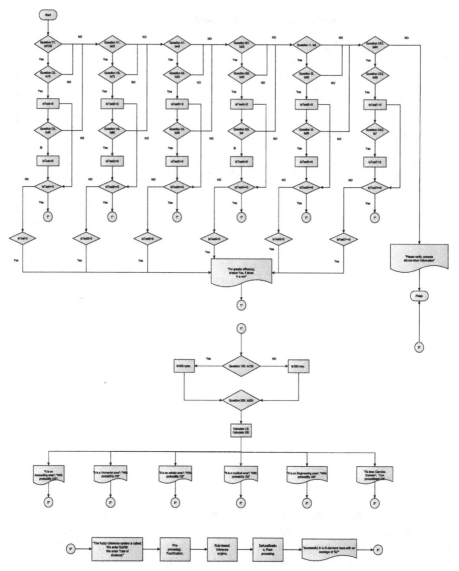

Fig. 2. Flow chart of the process

whether the student will graduate or not from the course and whether he will do so with honors or fail and is represented by the following inference rules (Table 7).

And the graphs of the universes of input and outputs of discrete and continuous type (Fig. 3). The fuzzy sets for fuzzy inference system are: a) Is the first input universe with average student income defined in terms of low, middle and high income; b) Is the second input universe with demand for the career, defined in terms of low, middle and high demand; c) Is the final input universe with the results of the system, in terms of failed (not graduated), graduated and honors (graduated with honors).

Table 6. Values used for average ranges and career demands

		Values
Average	High	8.0–10.0
	Middle	6.0–7.99
	Low	0–5.99
Demand of students	High	150–200
	Middle	90–149
	Low	50–89

Table 7. Inference Rules for probability of success

Average student income	Student demand for the career	Graduation result
High	High	Graduate
High	Middle	Graduated with Honors
High	Low	Graduated with Honors
Middle	High	Graduate
Middle	Middle	Graduate
Middle	Low	Graduated with Honors
Low	High	Reprobate
Low	Middle	Reprobate
Low	Low	Graduate

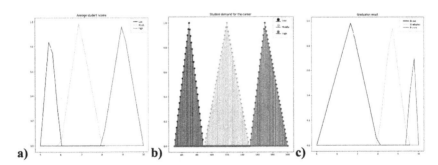

Fig. 3. Fuzzy sets for fuzzy inference system

To estimate the probability of success and the type of graduation that the student will have, the following rules were considered (Table 8) and were created under the scheme shown in the table (Table 9):

Table 8. Rules for determining the result

Average student income	High	8 to 10
	Middle	6 to 7.99
	Low	5 to 5.99
Student demand for the career	High	150 to 200
	Middle	90 to 149
	Low	50 to 89
Graduation result	Graduated with Honors	9.5 to 10
	Graduate	8 to 9.49
	Reprobate	5 to 7.99

Table 9. Example of rules for fuzzy inference engine

```
1  Graduates_Inference.add_rule([['Average Income', 'Low'], ['Demand
   for Degree', 'Low']], ['and'], [['Final Result', 'Graduate']])
2  Graduates_Inference.add_rule([['Average Income', 'Low'], ['Demand
   for Degree', 'Medium']], ['and'], [['Final Result', 'Failed']])
3  Graduates_Inference.add_rule([['Average Income', 'Low'], ['Demand
   for Degree', 'High']], ['and'], [['Final Result', 'Failed']])
4
5  Graduates_Inference.add_rule([['Average Income', 'Medium'], ['De-
   mand for Degree', 'Low']], ['and'], [['Final Result', 'Honors']])
6  Graduates_Inference.add_rule([['Average Income', 'Medium'], ['De-
   mand for Degree', 'Medium']], ['and'], [['Final Result', 'Gradu-
   ate']])
7  Graduates_Inference.add_rule([['Average Income', 'Medium'], ['De-
   mand for Degree', 'High']], ['and'], [['Final Result', 'Gradu-
   ate']])
8
```

5 Results

After applying the tests to people, the Fig. 4 shows examples of the questions that correspond to the Chaside test and the way in which they were answered. (Fig. 4).

The result that was shown on the screen to the students was like the following example (Fig. 5).

The results obtained for each category are shown (Table 10).

Of the total number of students evaluated, a sample of approximately 47.3% was taken, equivalent to 52 students. Using the Minimum Square Error technique and the accuracy for each area [27] was obtained as shown in the Table 11 using the Eq. 9.

It was found the students taking the sample commented that 50% said they agreed with the recommended area, since it was the one they identified with the most and even the career that interested them at university. 30% said that the area shown interested them but not enough to dedicate themselves professionally to it, they even studied another

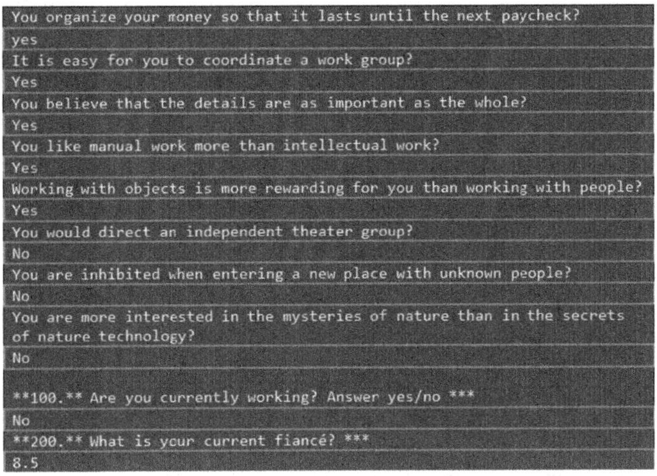
You organize your money so that it lasts until the next paycheck?
yes
It is easy for you to coordinate a work group?
Yes
You believe that the details are as important as the whole?
Yes
You like manual work more than intellectual work?
Yes
Working with objects is more rewarding for you than working with people?
Yes
You would direct an independent theater group?
No
You are inhibited when entering a new place with unknown people?
No
You are more interested in the mysteries of nature than in the secrets of nature technology?
No

100. Are you currently working? Answer yes/no ***
No
200. What is your current fiancé? ***
8.5

Fig. 4. Example of applied CHASIDE test questions

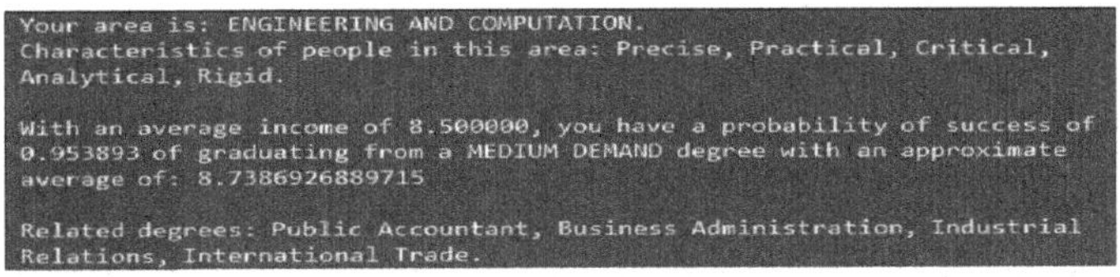
Your area is: ENGINEERING AND COMPUTATION.
Characteristics of people in this area: Precise, Practical, Critical, Analytical, Rigid.

With an average income of 8.500000, you have a probability of success of 0.953893 of graduating from a MEDIUM DEMAND degree with an approximate average of: 8.7386926889715

Related degrees: Public Accountant, Business Administration, Industrial Relations, International Trade.

Fig. 5. Example of results displayed

Table 10. Distribution of test results classified by men and women

AREA	MHS	WHS	MM	WM
C	20	9	0	1
H	0	4	1	0
A	0	1	0	0
S	20	7	0	0
I	25	7	5	0
E	10	0	0	0
Total	75	28	6	1

MHS = Men's high school, MM = Men's master's, WHS = Women's high school WM = Women's master's

career not in line with their interests; the remaining 20% showed apathy when doing the exercise, so they did not consider it relevant.

Table 11. CHASIDE Test applied using an expert system

AREA	Total	TP	TN	AC
C	30	22	8	0.5769
H	5	3	2	0.0961
A	1	0	1	0.0192
S	27	18	9	0.5192
I	37	23	14	0.7115
E	10	7	3	0.1923
Total	*110*			

TP = True Positives, AC = Accuracy for 52 students, TN = True Negatives

The accuracy obtained in most of the areas evaluated corresponds to more than 50% of the correct answers in which the expert system correctly classified the students if we consider those areas that obtained the highest number of students in their classification.

The final relationship of the possible results to be obtained, which is applied in the fuzzy inference system made up of two premises: Average school entry (AES), Demand for university career (DUC); and a conclusion: Final school average (FSA) is shown in the following Confusion Matrix [27] (Fig. 6). This graph can be interpreted as a bilateral relationship, where the higher the average and the lower the demand for the chosen career, then the higher the probability of success will be with better grades in it. If the relationship between the average income and the demand for the career is equitable, that is, with average values, there is also a high probability of success although it would not be considered with honors. And finally, if the average income is low and the demand for the career is in high demand, then the probability of success would be considered low or almost zero.

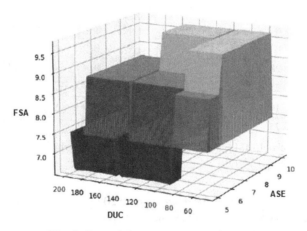

Fig. 6. Fuzzy inference system results graph

6 Conclusion

Vocational guidance has a high impact on the lives of students, its application within the educational programs of upper secondary education should be considered almost obligatory for young people to successfully complete a university degree. The people who were given the test commented that they do not have or had at one time any type of guidance, however, due to their interests they chose their university degree and master's degree that they are currently pursuing. Adding the probability of success to this study using Bayesian theory, fuzzy logic and inference rules also helped the examinees understand that having prior knowledge in an area of interest significantly improves their academic performance. Not having prior knowledge despite its low probability of success also motivates young people to venture into new areas of opportunity.

Comparing our work with some work related to expert systems in education we can mention the following characteristics regarding our research work. Our system is similar in accuracy to the reviewed systems, therefore, it is considered reliable for predicting success or failure based on the characteristics described such as the average income and the demand for the career as shown below (Table 12).

Table 12. Comparison of this work with the reviewed literature

Author	Year	Development	Technique	Objective	Result
Wulansari et al. [9]	2021	Expert System	Software Development Life Cycle (SDLC), Multiple Intelligence theory, Unified Modeling Language (UML)	Serves as a tool for students to use in identifying and harnessing their capabilities	Only assesses using the theory of multiple intelligences by Howard Gardner. So that it can be used as a guide for future researchers who will conduct similar research
Kabathova and Drlik [10]	2021	Analysis was performed on the performance metrics of multiple learning classifiers using academic data	Logistic regression, Decision tree, Naïve Bayes classifier	Utilization of various Machine Learning methodologies in forecasting course finishers, rather than course completers	the classification models could be considered trustworthy enough to make a proper prediction, The prediction accuracy varied between 77 and 93%

(*continued*)

Table 12. (*continued*)

Author	Year	Development	Technique	Objective	Result
Cynthia Cristina Martinez Padilla	2024	Expert System	Rule-based Systems, Forward and Backward chaining, Bayesian Theory, Fuzzy inference systems	Develop a system for career choice, profiling and classifying young people in the areas in which they have better aptitudes and chances of success	The accuracy obtained in most of the areas evaluated corresponds to more than 50% of the correct answers in which the expert system correctly classified the students

References

1. Moscoviz, L., Evans, D.K.: Learning Loss and Student Dropouts during the COVID-19 Pandemic: A Review of the Evidence Two Years after Schools Shut Down Abstract. Center for Global Development. 609 (2022)
2. Gottfried, M.A., Plasman, J.S.: Linking the timing of career and technical education course-taking with high school dropout and college-going behavior. Am. Educ. Res. J. **55**, 325–361 (2018). https://doi.org/10.3102/0002831217734805
3. Torres, J.A.O., Santiago, A.M., Izaguirre, J.M.V., Garduza, S.H., Garcia, M.A., Alejandro, G.F.: Multilayer fuzzy inference system for predicting the risk of dropping out of school at the high school level. IEEE Access (2024). https://doi.org/10.1109/ACCESS.2024.3425548
4. Barroso, M., et al.: Orientación Vocacional de los estudiantes del Técnico Superior de Ciclo Corto en Enfermería. JCM_UCM_SSp (2023)
5. Cisneros-Bravo, B.E., Rodríguez-Aguilar, R.M., Niño-Membrillo, Y.E., Cuevas-Rasgado, A.D.: Falta de orientación vocacional como factor en la deserción universitaria. Caso de estudio: zona Oriente del Estado de México. RIDE Revista Iberoamericana para la Investigación y el Desarrollo Educativo **14** (2023). https://doi.org/10.23913/ride.v14i27.1715
6. Terrence, T.J., et al.: The long-term impact of systemic student support in elementary school: reducing high school dropout. AERA Open **4** (2018). https://doi.org/10.1177/2332858418799085
7. Erazo Guerra, X.F., Rosero Morales, E. del Rocio Rosero Morales, E.: Orientación vocacional y su influencia en la deserción universitaria. Horizontes. Revista de Investigación en Ciencias de la Educación **5**, 591–606 (2022). https://doi.org/10.33996/revistahorizontes.v5i18.198
8. Felipe, L., Tombe, Y., Vianey, A., Rodríguez, T.: La importancia de la orientación vocacional en la formación escolar 1. Pedagogía y Educación **8** (2021). https://doi.org/10.31948/rev.fedumar
9. Wulansari, R.E., Sakti, R.H., Ambiyar, A., Giatman, M., Syah, N., Wakhinuddin, W.: Expert system for career early determination based on Howard Gardner's multiple intelligence. J. Appl. Eng. Technol. Sci. **3**, 67–76 (2022)

10. Kabathova, J., Drlik, M.: Towards predicting student's dropout in university courses using different machine learning techniques. Appl. Sci. (Switz.) **11** (2021). https://doi.org/10.3390/app11073130
11. Eshbayev, O., Nasiba, M.: Innovating higher education by using education technology and expert systems: pathways to educational innovation. Int. J. Soc. Sci. Res. Rev. **6**, 112–120 (2023). https://doi.org/10.47814/ijssrr.v6i6.1217
12. Makri, A., Vlachopoulos, D., Martina, R.A.: Digital escape rooms as innovative pedagogical tools in education: a systematic literature review. Sustainability (Switz.) **13** (2021). https://doi.org/10.3390/su13084587
13. María, G., González, T.: Benefits of Integrating Educational Tech Tools in Virtual Learning Environments, El Salvador (2023)
14. Ryan Syifa, Q., Syaom Barliana, M., Rahmanullah, F.: Prediction of facilities and infrastructure needs digital architecture technology in the era of industry revolution 4.0 and readiness of vocational high school program. In: Advances in Social Science, Education and Humanities Research, vol. 520 (2021)
15. Sociedad, U.Y., Diagnóstico Y Tratamiento De Enfermedades Y Plagas En Plantas Ornamentales, P. EL, Lucía Sandoval Pillajo, A., Antonio Checa Cabrera, M., Azucena Díaz Vásquez, R., Lenin Acosta Espinoza, J.: EXPERT SYSTEM FOR THE DIAGNOSIS AND TREATMENT OF DISEA-SES AND PESTS IN ORNAMENTAL PLANTS SISTEMA EXPERTO. Revista Científica de la Universidadde Cienfuegos. 1, (2021)
16. Hayadi, B.H., Bastian, A., Rukun, K., Jalinus, N., Lizar, Y., Guci, A.: Expert system in the application of learning models with forward chaining method (2018)
17. Julianti, M.R., Nurmaesah, N., Prayogo, W.: Expert system for diagnosing early symptoms of COVID-19 using the certainty factor method. JURNAL SISFOTEK GLOBAL **12**, 24 (2022). https://doi.org/10.38101/sisfotek.v12i1.475
18. Pérez Aguada, D., Milanés Luque, M., Mar Cornelio, O., Orellana García, A., La Habana, X.: Design of rule-based system to support population decision making in the procurement of medicines. Revista Cubana de Informática Médica **15** (2023)
19. Kodak, T., Bergmann, S.: Autism spectrum disorder: characteristics, associated behaviors, and early intervention. Pediatr Clin North Am. **67** (2020)
20. Cruz Arcani, P.G.: PROGRAMA DE ORIENTACIÓN VOCACIONAL DIRIGIDO A ESTUDIANTES DE SEXTO DE SECUNDARIA DE LA UNIDAD EDUCATIVA AMÉRICA I DE LA CIUDAD DE EL ALTO (2021)
21. Adalberto, K., Rojas, M.: Evaluación del desempeño de tres algoritmos de inferencia bayesiana, implementados como sistema experto para la identificación de modos de falla en ejes (2019)
22. Van de Schoot, R., et al.: Bayesian statistics and modelling (2021). https://doi.org/10.1038/s43586-020-00003-0
23. Yang, M., et al.: Invariant learning via probability of sufficient and necessary causes: In: Conference on Neural Information Processing Systems 37 (2023)
24. Shingaki, R., Kuroki, M.: Probabilities of potential outcome types in experimental studies: identification and estimation based on proxy covariate information. In: The Thirty-Seventh AAAI Conference on Artificial Intelligence (2023)
25. Montes Rivera, M., Olvera-Gonzalez, E., Escalante-Garcia, N.: UPAFuzzySystems: a python library for control and simulation with fuzzy inference systems. Machines **11** (2023). https://doi.org/10.3390/machines11050572
26. Sociedad, U.Y., Oviedo Bayas, B., Zambrano-Vega, C., Bayas, O., Zambrano: Nuevo clasificador bayesiano simple para el análisis de Datos Educativos. Revista Científica de la Universidad de Cienfuegos **11**(2), 278–285 (2019)

27. Arias-Duart, A., Mariotti, E., Garcia-Gasulla, D., Alonso-Moral, J.M.: A confusion matrix for evaluating feature attribution methods. In: Conference on Computer Vision and Pattern Recognition Workshops (CVPRW) (2023). https://doi.org/10.1109/CVPRW59228.2023.00380
28. Cesar, A.O., Acuña-Rodríguez, M., Emma, P.O., Gatica, G., Córdova, A.: A CHASIDE test-based analysis for identifying adolescents characteristics impacting their vocational orientation: case of private schools in the city Barranquilla, Colombia. Procedia Comput. Sci. 654–659. (2024). https://doi.org/10.1016/j.procs.2023.12.168
29. Paz Pérez, D.E., Rodríguez-Alberto, M.: Interés y aptitud vocacional como predictores del rendimiento académico universitario. Revista de Psicologia y Ciencias del Comportamiento de la Unidad Académica de Ciencias Jurídicas y Sociales **14**, 70–87 (2023). https://doi.org/10.29059/rpcc.20231201-163

Detection of Basic Motorcycle Faults Using a Fuzzy Bayesian Expert System

David Alonso Carranza Escobar[✉]

Research and Postgraduate Studies Department, Universidad Politécnica de Aguascalientes, Aguascalientes, Mexico
mc230011@alumnos.upa.edu.mx

Abstract. The accelerated increase in motorcycles in the country has led to a gap in the knowledge and maintenance of these vehicles. This gap has generated a growing demand for repair and maintenance services and, in turn, has led to a lack of information and education among motorcycle owners about basic mechanics and essential maintenance procedures. Motorcyclists face difficulties in identifying and troubleshooting common problems due to an insufficient understanding of mechanical systems and a lack of accurate and accessible information on proper motorcycle maintenance, resulting in undetected mechanical failures, which may present themselves early as noises or alerts before becoming more serious problems. These undetected failures not only affect the performance and durability of the vehicle but can also put the rider's safety at risk. Issues such as brake wear, transmission system degradation, or suspension problems may not be visible until they become critical faults. Thus, the development of an expert system is necessary for new users to know the problems through basic alerts and based on a series of questions to know where the main failure occurs and the percentage of the same, avoiding more complex situations and improving safety and efficiency in the maintenance of the same. In this context, Python is used as a programming language, the UPAFuzzySystems library, and the expert library. Forward and backward chaining is used to develop diagnostic questions. Bayesian reasoning is used to obtain probabilities based on user responses, generating effective answers and solutions for users based on statistics. On the other hand, fuzzy logic is used to improve the accuracy and efficiency of the expert system by addressing the uncertainty and imprecision of user responses, allowing for a more detailed and accurate assessment of potential motorcycle mechanical problems. Thus, we present a fuzzy Bayesian expert system to contribute to the predictive maintenance of a motorcycle.

Keywords: Expert System · Forward · Backward · Motorcycle · Maintenance · Bayes · Fuzzy logic

1 Introduction

A motorcycle is defined as a two-wheeled vehicle, with a chassis, handlebars for steering, a drive unit powered by a combustion engine, management modules for brakes, damping systems and transmission which is responsible for shifting gears [1–3].

In recent years the use of motorcycles has increased significantly both statewide and nationally, according to INEGI the vehicles with the highest annual growth rate are motorcycles with 10.89% (2019–2020) [4], Due to this, many of the people who begin to make use of this means of transport are not familiar with the warning signs that may indicate the condition or mechanical problems of the motorcycle. This lack of knowledge about the various problems that the motorcycle may present can lead to gradual deterioration and in the long run seriously affect the overall performance of the motorcycle. It is therefore necessary for new riders to understand the signs and possible causes to ensure good maintenance and prevent costly and dangerous damage [5].

That is why this expert system has as its main objective to provide an effective tool for the detection and resolution of mechanical failures in motorcycles, providing drivers with specific recommendations according to the problem and the actions to be taken to make the motor vehicle work properly, in addition to providing a probability that a certain mechanism or component is not working properly in order to generate a better understanding of the main causes of the problems they face daily, from the ignition to the sounds generated by the engine, thus preventing further damage and reducing the percentage of accidents caused by mechanical problems.

In this way, the expert system seeks to ask a series of basic questions to make the system easy to use and accessible even for those users who have no mechanical experience, and through questions and answers predict the main problem that the user may face in various situations and present a solution based on their answers.

For the construction of the expert system, fuzzy logic is used to process imprecise information for data analysis [17]. In the application context, this tool allows to observe and evaluate ambiguous information, such as the maintenance history of a vehicle and its rating (e.g., good, poor or deficient). Based on this information, the system can estimate the possible state of a specific component and allow the user to know its current state, suggesting possible changes according to the history.

In conjunction with fuzzy logic, the expert system also makes use of the Bayes theorem for the evaluation of probabilities because it facilitates the updating of estimates of the occurrence of an event or failure of a mechanical problem, as new evidence is incorporated. Bayes theorem focuses on the initial or priori probability (initial hypothesis) and the conditional probability (the likelihood of the evidence under a specific hypothesis) [13, 14]. Thus, by using this theorem, the accuracy of the expert system itself is improved, with the specific probability of knowing the mechanical problem.

2 Contributions

This work seeks to contribute to motorcycle riders, with and without previous experience, by providing a complete framework for understanding and diagnosing basic motorcycle mechanical problems. Through the implementation of an expert system that facilitates the identification of problems and provides specific recommendations, it seeks to generate a better understanding of these basic faults by increasing the knowledge of motorcycle mechanics, so that they can keep their vehicles working properly.

3 Theoretical Framework

Expert Systems

An expert system, known as a knowledge-based system, is a computer program designed to capture and utilize the knowledge of human experts in a topic, with the objective of providing specialized information to the user who uses the system. Its operation is based primarily on a human expert, using his knowledge composed of facts and rules, that, using a series of questions and taking into account the user's answers, decides which elements the system will use to continue asking questions until it reaches a concrete objective [5]. Being based on a human expert, the system recognizes which questions to ask based on the user's needs, showing a high level of intelligence [6–8].

There are two methods for inference in expert systems: forward chaining and backward chaining [2, 5, 9].

Forward Chaining

This is an artificial intelligence reasoning method which is based on objectives and the application of inference rules to existing data to extract additional data to reach an end point. It starts from the input information generating questions until reaching its conclusion, it is produced by data and it works correctly to obtain the facts through the use of IF-THEN rules [10, 11].

The functionality of the inference engine seeks to initialize with the collection of data in the knowledge base and the rules it contains, then analyzes the rules to determine if the conditions meet the facts, this process is generated with all the rules until you get a goal reached [7].

Backward Chaining

This method consists primarily in going backwards from the end point or the hypothesis, through the steps that lead to it, this process allows to establish logical steps that can be used to obtain the objective or hypothesis in a faster way, this method is easier for the user, but requires a good coding to work correctly. Also, unlike the operation of forward chaining in backward chaining the inference engine has the instruction to search in the knowledge base the rules that can obtain the expected solution or hypothesis, these rules must have the action, and if the engine analyzes that matches the condition of the database the rule concludes and proves the hypothesis [12].

Bayesian Reasoning

Bayesian reasoning is based on Bayes' theorem, an extremely important tool for calculating the probability of one event given another by aggregating various observations in the process. This theorem focused on expert systems allows to calculate the probability of a hypothesis or multiple hypotheses given evidence by integrating the expert's knowledge and the new information provided in a coherent way based on the IF-THEN rule [13].

This reasoning is based on updating initial beliefs (Priori) about a hypothesis by incorporating new evidence, which results in a conclusion or hypothesis (Posteriori). This method is not only based on counting uncertainty probabilistically, but also allows adjusting the estimates as new data is obtained.

In expert systems H represents the hypothesis and E the evidence that supports the hypothesis and makes use of the conditional probability formula in Eq. 1 shows the conditional probability formula and the meaning of the terms used [14].

$$p(H|E) = \frac{p(E|H) \times p(H)}{p(E|H) \times p(H) + p(E|\neg H) \times p(\neg H)} \quad (1)$$

where:

p(H) is the prior probability that the hypothesis H is true.
p(E|H) is the probability that hypothesis H is true when the evidence is also true E.
p(¬H) is the prior probability that hypothesis H is false.
p(E|¬H) is the probability of finding evidence when the hypothesis is false.

Another of the formulas needed for Bayesian analysis is Sufficient probability (LS) which represents the measure of the expert's belief in hypothesis H when evidence E is available, which is described by the following equation, Eq. 2 [14].

$$LS = \frac{p(E|H)}{p(E|\neg H)} \quad (2)$$

Similarly, it is necessary to obtain the likelihood of necessity (LN) which is a measure of the hypothesis when evidence E is absent as described in Eq. 3 [14].

$$LN = \frac{p(\neg E|H)}{p(\neg E|\neg H)} \quad (3)$$

Fuzzy Logic

Fuzzy logic, whose foundations were laid in 1970 by Lotfi A. Zadeh, has been an innovative tool that revolutionized our approach to problems involving imprecise information within a continuous spectrum from [0 to 1] [13, 15, 16].

This capability allows us to more accurately represent real-world scenarios in which categories are not rigidly defined, but show varying degrees of membership. These fundamental concepts have paved the way for substantial advances in control systems and decision-making processes in complex scenarios. In 1975, Ebrahim Mamdani introduced the Fuzzy Inference System (FIS), which has since become a widely adopted soft computing technique. This system provides a linguistic methodology for dealing with complex procedures by facilitating the translation of human experience into rule-based structures, which usually follow an If-Then form [17].

The evolution of fuzzy logic and fuzzy inference system has laid the foundation for contemporary fuzzy interface systems. These systems involve the fusion of input variables, whereby imprecise and fuzzy data from diverse origins are merged, and the subsequent defuzzification of output variables to obtain results that are interpretable and applicable in practical scenarios [18]. Fuzzy systems convert the knowledge repository into a mathematical framework, which exerts a substantial influence in various fields, ranging from engineering applications to decision making processes in uncertain environments; this process is represented by Fig. 1 in which the inference system process is shown.

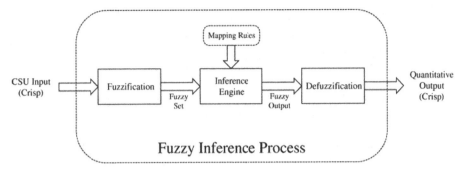

Fig. 1. Diagram of a fuzzy inference system recovered from [18].

Python
Similarly, to generate the analysis of the expert system and perform its verification, Python is used, which is among the most widely used programming languages. In addition,[19]. Python has a series of characteristics which standout such as: its adaptability, its open code which helps to promote the advancement and extension of programs and libraries, as well as its large number of libraries for the analysis of various topics, these characteristics stand out favorably because they make the analysis of the systems less complex [19].

UPAFuzzySystems
The UPAFuzzySystems library is a tool for the development of fuzzy inference systems, which allows researchers and users to perform control simulations in an efficient way. UPAFuzzySystems facilitates the derivation of transfer functions and models in state spaces as well as the derivation of information from universes which can be discrete or continuous. In addition, the library optimizes the formulation of rules within the inference system, favoring the creation of graphs. The library is compatible with fuzzy controllers, such as the Mamdani and Takagi-Sugeno models. It is important to highlight that the library is developed in Python and is open source, making it accessible to the scientific and development community [17].

Experta Library
Similarly, use is made of the Experta library which has a set of tools that facilitate the construction of expert systems to analyze, these libraries are programmed in order to meet the decision-making process of a specific topic, and has a great support in the documentation for those users who start from scratch the subject and thus learn more quickly through reading and examples, for the analysis and coding of expert systems.

But it currently has an update problem which makes it only work on Python 3.8.10 versions [20].

4 Methodology

To carry out the methodology of the system it is necessary to know the objective of the system, which is based on the application of maintenance for a motorcycle. It is also important to say that, if you do not have a good experience in the use of tools, it is better to take it to the agency or to the mechanic workshop to make the corresponding adjustments. The purpose of this system is to learn about the basic problems that can occur on a motorcycle. When generating the system and not getting a good answer, it may be a major problem and follow the suggestion of the expert system. Similarly, the system rules based on probabilities can help the user to know if a motorcycle component needs to be replaced or the motorcycle's performance needs to be checked.

Using the generated rules, Python version 3.8.10, the Experta library and the UPA-FuzzySystems library, Fig. 2 shows the flow chart corresponding to the operation of the system and how the questions are developed from the user's answers, thus graphically showing the decision-making process. In addition, Fig. 3 shows how fuzzy logic is used to determine the probability of needing a clutch replacement based on the age of the vehicle and its mileage.

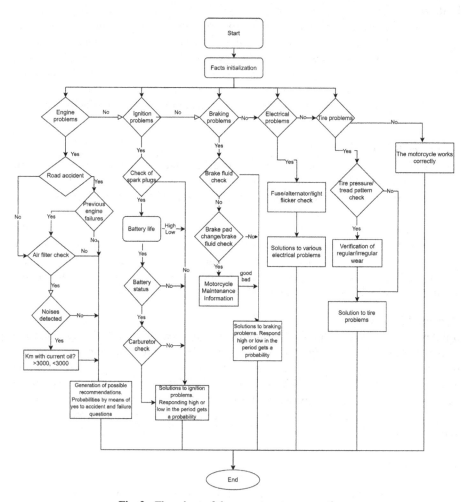

Fig. 2. Flowchart of the expert system operation.

This flowchart clearly shows the steps the system follows and how it develops as the questionnaire progresses and the user answers the questions. This allows new users or drivers to understand the faults of their motorcycle from basic signals, adding probabilistic answers based on the inputs provided by the user.

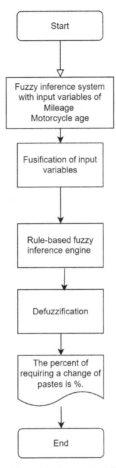

Fig. 3. Fuzzy inference system to determine the probability of clutch replacement based on motorcycle mileage and age.

After understanding the internal workings of the system, it is necessary to load the Experta library in version 3.8.10 and the UPAFuzzySystems in order to make use of its functions and to correctly and effectively generate the program based on the maintenance of a motorcycle as shown in Fig. 4.

```
from experta import *
import UPAFuzzySystems as UPAfs
```

Fig. 4. Expert library usage method.

```
class Ignition(Fact):
    """Ignition problems"""
    pass

class Braking(Fact):
    """Braking problems"""
    pass

class Electrical(Fact):
    """Electrical problems"""
    pass

class Engine(Fact):
    """Engine problems"""
    pass

class Tires(Fact):
    """Tire problems"""
    pass
```

Fig. 5. Definition of the main facts.

For the analysis of the facts, Fig. 5 shows the main facts defined in the system that are responsible for providing a knowledge base that the system will later use to diagnose the various specific problems that arise. Derived from the main facts are the secondary facts which generate a foundation for the expert system and thus know what the problems of the motorcycle are.

- Ignition: Represents problems related to the vehicle's ignition system, which can be checked by checking the spark plugs or the correct adjustment of the carburetors.
- Braking: Refers to problems in the braking system, related to brake fluid level or brake pads.
- Electrical: Refers to problems in the motorcycle's electrical system, such as battery, alternator or fuse failures.

- Engine: Focuses on problems associated with the engine, such as air filter checks, oil changes or tightening of hardware.
- Tires: Considers tire-related problems, such as irregular tire wear, tread checks or tire pressure.

There is a wide variety of problems that may be related to the operation of the motorcycle, this expert system is based solely on the early detection of some problems or irregularities that may present the same in a basic way, facilitating the detection and diagnosis for inexperienced people, and thus know the causes and how the problems of the vehicle are presented.

Similarly, Fig. 6 shows part of the code that was created for the optimal operation of the expert system. Part of the rules used for the engine problem can be seen, which starts with a backward chaining method to find out where the problem arises and followed by a forward chaining, a sweep of questions is made based on the selected problem, and finally some of the answers obtained from the decisions made by the user are shown.

```
@Rule(Error(Ex='yes'), NOT(Engine(p_engine=W())))
def question_p_engine(self):
    self.declare(Engine(p_engine=input("Do you have engine problems? (yes/no)")))

@Rule(Engine(p_engine='yes'), NOT(Accidents(p_accidents=W())))
def question_accidents(self):
    self.declare(Accidents(p_accidents=input("Has the motorcycle been in any accidents? (yes/no)")))

@Rule(AND(Engine(p_engine='yes'), Accidents(p_accidents='yes'), NOT(EngineFailures(p_failures=W()))))
def question_p_failures(self):
    self.engine_failures = EngineFailures(state=input("Has the motorcycle engine presented any failures? (yes/no)"))
    self.declare(self.engine_failures)
```

Fig. 6. Rules used to solve problems in the expert system.

Bayesian Rules

To carry out the Bayesian analysis, use is made of a database obtained from [21] in which various types of vehicles and some records such as damage report, maintenance history, ondometer record, service history, fuel efficiency, and brake, tire and battery conditions are presented, based on the above. These variables directly impact the mains variables. Likewise, the analysis and application of the Bayesian system, it is necessary to calculate the sufficient probability (LS) and the required probability (LN), which are established in Eqs. 2 and 3. These values are fundamental since they allow generating an update of the initial probability of an event while the evidence is present or absent, providing a more accurate evaluation of the problem contained in the engine and informing the user of the possibilities of engine failure.

It can be observed that in the specific case of the engine questions shown in Fig. 6 the LS value is 1.09, indicating that the presence of the evidence reinforces the hypothesis that the problem is related to the considered cause, increasing the diagnostic probability of the expert system. On the other hand, the LN value is 0.91, which is equivalent to a decrease in the belief of the hypothesis when the evidence is absent, suggesting that the problem is less likely to be related to that specific cause in the absence of the evidence.

The obtained values of LS and LN are substituted into the equations to update the posterior probabilities. The result is shown in Eqs. 4 and 5, which refer to the formulas for obtaining the posterior probability when evidence is present or absent.

$$LS = \frac{p(E|H)}{p(E|\neg H)} = \frac{0.51}{0.47} = 1.09 \qquad (4)$$

$$LN = \frac{p(\neg E|H)}{p(\neg E|\neg H)} = \frac{0.48}{0.52} = 0.91 \qquad (5)$$

Finally, after analyzing the values shown above, it is necessary to enter them into the code. The figure Fig. 7 shows the pragmatic representation applied in the expert

```
@Rule(EngineFailures(state='yes') | EngineFail-
ures(state='no'))
    def engine_failure_info(self):
        LS = 1.09
        LN = 0.91
        OS = 0.5
        OS = OS / (1 - OS)
        if self.engine_failures['state'] == 'yes':
            OS_T = OS * LS
            OS = OS_T / (1 + OS_T)
        if self.engine_failures['state'] == 'no':
            OS_N = OS * LN
            OS = OS_N / (1 + OS_N)
        self.failure_prob = FailureProbability(OSr=OS)
        self.declare(self.failure_prob)
        print('Due to engine problems, fuel efficiency may be
        affected with a probability of %f' % self.failure_prob['OSr'])
```

Fig. 7. Question based on probabilities

Detection of Basic Motorcycle Faults Using a Fuzzy Bayesian Expert System   173

system, which shows how the initial probabilities are adjusted according to the presence or absence of evidence. And based on the user's answers to the questions related to the variables of reported failures, accident history, fuel efficiency, a probability is obtained. That alerts the user by the probability of some engine failure.

Fuzzy Logic

To obtain the values of the fuzzy sets for each of the universes, the maximum value of the mileage of the database [21] was taken and divided into 3 parts up to the minimum value, thus obtaining the fuzzy sets for the universe shown in Fig. 8; in the same way, this process was repeated for the age of the vehicle and the clutch pastes, which are shown in Figs. 10 and 12.

And finally, the code and the results obtained from the implementation of the fuzzy systems are shown, in which the universes generated with their respective fuzzy sets and the graphs of each one can be observed, in which the mileage of the motorcycle

```
mileage_universe = np.arange(0, 100001, 1000)
Mileage1 = UPAfs.fuzzy_universe('Mileage', mileage_universe, 'continuous')
Mileage1.add_fuzzyset('Low', 'trimf', [0, 5000, 9000])
Mileage1.add_fuzzyset('Medium', 'trimf', [9000, 15000, 20000])
Mileage1.add_fuzzyset('High', 'trapmf', [20000, 50000, 100000, 100000])
Mileage1.view_fuzzy()
```

Fig. 8. Code to define the fuzzy sets of the universe called mileage with the UPAFuzzySystems library.

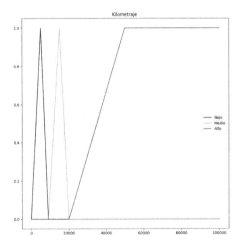

Fig. 9. Graph of the fuzzy sets of the universe 'Kilometer'.

and the years of use of the same one is taken into account, in Fig. 8 and Fig. 9 the universes generated with their fuzzy sets and the respective graphs for each one of them corresponding to the first universe or premise 1 are shown.

Following the first universe corresponding to the mileage of the motorcycle, Fig. 10 shows the pragmatic mode for the generation of the universe corresponding to the age of the vehicle with their respective fuzzy sets which are new, used and old, referring to the grouping and data of the years of the vehicle, Fig. 11 shows the graph generated from the coded fuzzy sets.

```
vehicle_age_universe = np.arange(0, 21, 1)
Vehicle_Age = UPAfs.fuzzy_universe('Vehicle Age', vehicle_age_universe, 'continuous')
Vehicle_Age.add_fuzzyset('New', 'trapmf', [0, 0, 1, 3])
Vehicle_Age.add_fuzzyset('Used', 'trimf', [3, 5, 7])
Vehicle_Age.add_fuzzyset('Old', 'trapmf', [7, 10, 15, 20])
Vehicle_Age.view_fuzzy()
```

Fig. 10. Pragmatic mode for the definition of the fuzzy sets of the universe corresponding to the age of the vehicle.

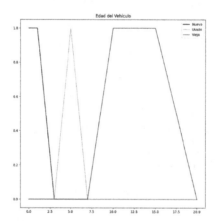

Fig. 11. Graph of the fuzzy sets of the 'Vehicle age' universe.

Finally, the universe corresponding to the system output, which refers to clutch pastes as a function of vehicle mileage and age, is shown in Fig. 12 and Fig. 13, and the coding of this universe and its graphical perspective.

```
clutch_pads_universe = np.arange(0, 101, 1)
Clutch_Pads = UPAfs.fuzzy_universe('Clutch Pads',
clutch_pads_universe, 'continuous')
Clutch_Pads.add_fuzzyset('Low', 'trapmf', [0, 0, 20, 30])
Clutch_Pads.add_fuzzyset('Medium', 'trimf', [25, 50, 75])
Clutch_Pads.add_fuzzyset('High', 'trapmf', [60, 80, 100, 100])
Clutch_Pads.view_fuzzy()
```

Fig. 12. Code to define the fuzzy sets of the output universe called clutch pastes with the UPAFuzzySystems library.

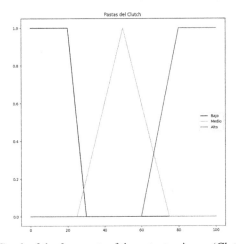

Fig. 13. Graph of the fuzzy sets of the output universe 'Clutch pastes'.

5 Results and Experimentation

- Identification of the lack of knowledge regarding basic mechanical failures in motorcycles.
- Analysis and cleaning of the maintenance database to improve the Bayesian reasoning of the system and the fuzzy logic system.
- Creation of fuzzy sets for variables such as mileage and vehicle age to handle uncertainty regarding clutch pastes.
- Calculation of probabilities described in the methodology to adjust the diagnosis according to user responses.

- Rule coding to generate accurate recommendations based on user responses.
- The system provides clear recommendations for the unexperienced user to understand the problems their motorcycle is having.

The results obtained from the investigation and programming of the expert system based on basic problems that a novice motorcyclist may encounter are useful at times when the user has no knowledge of motorcycle mechanics and needs to know whether the problem with his vehicle is serious or simply requires an adjustment to keep it working properly. Figure 14 shows the results obtained from the questions sent to the engine showing the suggestions offered by the system based on the user's requests, likewise, it shows the results obtained from one of these probabilistic questions i.e. applying the Bayesian part and finally, it shows the percentage of need to change clutch pastes based on the fuzzy inference system built with the UPAFuzzySystems library, demonstrating a correct application performance. It is important to mention that if you do not have experience with the use of tools it is better to go to a specialized center and likewise if the regulations concerning the engine determine that it is not a basic problem to go to the repair of the motorcycle, since the life of a driver may be at risk if you do not have prior knowledge of the fault or problem.

```
The recommendations are as follows:
According to the diagnosis, changing the clutch pads is required at a
rate of 14.499999999999996 %
Due to engine problems, fuel efficiency may be affected with a proba-
bility of 0.521531
No oil changes are required until a maximum of 3000 km have been
driven or the next service.
```

Fig. 14. Results obtained from the question related to the km traveled with the current oil.

Also, the Fig. 15 shows 3D graph shows the response of the fuzzy inference system to determine the need for clutch fluid replacement as a function of vehicle mileage and age. It is observed that the higher the mileage and the older the vehicle, the higher the probability of requiring a change of the clutch pastes, represented in the higher areas of the surface. On the other hand, in vehicles with low mileage and lower age, the need for change is lower, indicating that the clutch is still in good condition.

Surface Response: Clutch Pads Inference

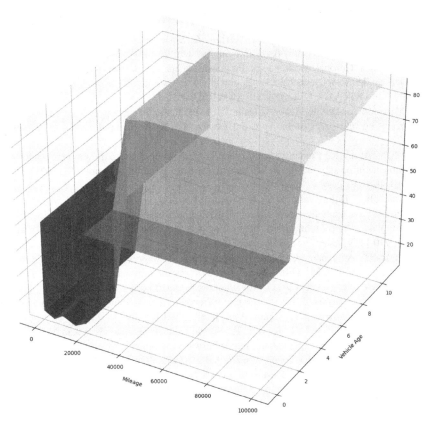

Fig. 15. Results of tridimensional Fuzzy Inference System.

6 Conclusion

As a conclusion, I consider that expert systems are useful when a person who does not know a topic wants to know it because it is supported by the knowledge of an expert, in the same way this type of systems can be applied to countless topics, adapting easily to specific fields and providing solutions to real problems. The implementation of these systems for the basic maintenance of a motorcycle, the implementation of these systems for basic maintenance of a motorcycle, not only facilitates that users with intermediate experience in motorcycles easily detect and solve the problem, but also due to the probabilities, they can know if any of the components described in the system may present any failure and this is useful to obtain various diagnoses and that somehow that experience is instilled in people and know how to recognize the problems without the system. Generating a greater efficiency in the analysis of the problem, and reducing above all the costs in basic repairs.

References

1. Universidad Pública de Navarra: Manual de montaje y mantenimiento de una motocicleta de competición (1999)
2. Arifin, F., Hasanah, N., Irmawati, D., Arifin, Z.: Smart system for diagnosing motorcycle damage using adaptive neuro-fuzzy inference system for future transportation (2020)
3. Noe Spencer Coronel Montenegro: Propuesta de mejora del servicio de mantenimiento preventivo de motocicletas (2023)
4. CMOV: PROGRAMA ESTATAL DE MOVILIDAD (2021)
5. Al-Mansour, A., Lee, K.W.W., Al-Qaili, A.H.: Prediction of pavement maintenance performance using an expert system. Appl. Sci. (Switz.) **12** (2022). https://doi.org/10.3390/app12104802
6. Ayvaz, S., Alpay, K.: Predictive maintenance system for production lines in manufacturing: a machine learning approach using IoT data in real-time. Expert Syst. Appl. **173** (2021). https://doi.org/10.1016/j.eswa.2021.114598
7. Naryanto, R.F., Delimayanti, M.K., Kriswanto, Musyono, A.D.N.I., Sukoco, I., Aditya, M.N.: Development of a mobile expert system for the diagnosis on motorcycle damage using forward chaining algorithm. Indonesian J. Electr. Eng. Comput. Sci. **27**, 1601–1609 (2022). https://doi.org/10.11591/ijeecs.v27.i3.pp1601-1609
8. Odunayo, D.: An expert system for diagnosing faults in motorcycle (2012)
9. Saibene, A., Assale, M., Giltri, M.: Expert systems: definitions, advantages and issues in medical field applications (2021). https://doi.org/10.1016/j.eswa.2021.114900
10. Anwar, M.R.: Analysis of expert system implementation in computer damage diagnosis with forward chaining method. Int. Trans. Artif. Intell. (ITALIC) (2023)
11. Seppewali, A., Mulyo, W.H., Riswan, R.: Sistem Pakar Diagnosa Kerusakan Motor Suzuki Smash Titan 115 Cc Menggunakan Metode Forward Chaining. Jurnal Teknologi Dan Sistem Informasi Bisnis **5**, 13–20 (2023). https://doi.org/10.47233/jteksis.v5i1.728
12. Budiman, T., Rohmat, S.A.: Design of a LAN Network Disorder Diagnostic System With a Knowledge-Based System Using the Forward Chaining Method (Case Study: Star Energy Geothermal Darajat II, Limited) (2022)
13. Hernández, M.P., Montes Rivera, M., Hernández, R.P., Escobar, R.M.: Brake maintenance diagnostic with fuzzy-bayesian expert system. In: Calvo, H., Martínez-Villaseñor, L., Ponce, H., Zatarain Cabada, R., Montes Rivera, M., Mezura-Montes, E. (eds.) MICAI 2023. LNCS, vol. 14502, pp. 77–101. Springer, Cham (2024). https://doi.org/10.1007/978-3-031-51940-6_8
14. Mantik, J., Yanti, N., Syahputri, A.: 2549–2555 Accredited (2021)
15. Castillo, O., Amador-Angulo, L., Castro, J.R., Garcia-Valdez, M.: A comparative study of type-1 fuzzy logic systems, interval type-2 fuzzy logic systems and generalized type-2 fuzzy logic systems in control problems. Inf. Sci. (NY) **354**, 257–274 (2016). https://doi.org/10.1016/j.ins.2016.03.026
16. Zadeh, L.A.: Fuzzy Sets (1965)
17. Montes Rivera, M., Olvera-Gonzalez, E., Escalante-Garcia, N.: UPAFuzzySystems: a python library for control and simulation with fuzzy inference systems. Machines **11** (2023). https://doi.org/10.3390/machines11050572
18. Ojha, V., Abraham, A., Snášel, V.: Heuristic design of fuzzy inference systems: a review of three decades of research. Eng. Appl. Artif. Intell. **85**, 845–864 (2019). https://doi.org/10.1016/j.engappai.2019.08.010
19. Python documentation: The Python Tutorial. https://docs.python.org/3/tutorial/index.html. Accessed 02 June 2024

20. Experta Documentation: Experta Library. https://experta.readthedocs.io/en/latest/. Accessed 31 May 2024
21. Vehicle Maintenance Data. Kaggle: Your Machine Learning and Data Science Community. https://www.kaggle.com/datasets/chavindudulaj/vehicle-maintenance-data. Accessed 21 July 2024

Intervention Model with Data Mining Techniques to Work with Dating Violence Victims Using a Social Support Network

Rogelio Rodríguez-Hernández[1]([✉]), Nemesio Castillo-Viveros[1], and Alberto Ochoa-Ortiz[2]

[1] Department of Social Sciences, Autonomous University of Ciudad Juárez, Ciudad Juárez, Mexico
`rogelio.rodriguez@uacj.mx`
[2] Doctorate in Technology, Autonomous University of Ciudad Juárez, Ciudad Juárez, Mexico

Abstract. Dating violence is a problem that affects a large proportion of young women in Mexico, so it is necessary to design and implement policies to support its victims. Although progress has been made in this area, technological and data mining solutions can help to eradicate the problem. Therefore, the purpose of this chapter is to propose an automated violence assessment process, through the design and implementation of a technological intervention model in RedSIAM, in order to provide users with appropriate care according to the type and level of violence detected. Subsequently, the risk factors for violence are evaluated and data mining techniques are used to propose the model. Through processing with clustering algorithms, violence profiles are created according to the characteristics of the users. Later, the resulting information is used to document, generate behavioral patterns and classify each user into one of the profiles already identified. This processing is done using classification algorithms such as EM (Expectation-Maximization) and Simple K-Means. Within the social network, each profile has access to different content depending on the group it belongs to. The intervention model allows us to have analyzable data to understand the patterns of violence among the victims, in order to help the social workers in their decision-making task to provide them with timely assistance.

Keywords: Dating violence · Intervention model · social support networks · Data mining

1 Dating Violence

The United Nations Declaration on the Elimination of Violence against Women (UN, 1993) states that violence against women constitutes a violation of human rights and freedoms and prevents women from fully or partially enjoying them. In Mexico, the General Law for Women's Access to a Life Free of Violence recognizes the existence of the problem and seeks to establish measures to prevent, punish and eradicate it (Federal

Government, 2007). Therefore, there must be policies aimed at preventing and eradicating gender-based violence. Although public recognition of the problem is a step forward, violence persists in Mexico and around the world. This is evidenced by data showing that a significant proportion of women experience violence in their daily lives. In this country, it is estimated that 70.1% of women have experienced some form of violence in their lives (National Institute of Statistics and Geography [INEGI], 2021). Now, one area where women often experience abuse is in intimate partner relationships, where available information shows that percentages of victimization in romantic relationships in Latin America and the Caribbean range from 21% to 38% (Sardinha et al., 2022), while in Mexico, percentages of 39.9% are observed in women over the age of 14 (INEGI, 2021). The prevalence of violence is forcing the world's governments to implement policies and actions to address the problem. These policies must take into account multidisciplinary collaboration. They must also take advantage of current scientific and technological advances. Therefore, the purpose of this chapter is to propose an automated violence assessment process, through the design and implementation of a model in RedSIAM, in order to provide adequate guidance to users, taking into account the types and levels of violence detected.

For the World Health Organization (WHO, n.d.), intimate partner violence is any behavior in an intimate relationship that causes physical, sexual or psychological harm and includes acts of physical aggression, sexual coercion, psychological abuse and controlling behavior. One type of relationship in which abuse occurs is dating, which has been defined as a dyadic relationship that involves joint interactions and activities with the implicit or explicit intention of continuing until one or both parties decide to end it or until another more committed relationship is established (Straus, 2004). Thus, dating relationships in adolescence and young adulthood are qualitatively different from those in adulthood (Castro & Casique, 2010). Regarding the extent of the problem, the most recent prevalence study among a representative sample of adolescents and young people aged 15–24 in Mexico indicates that 19.9% have experienced emotional violence, 2.7% physical violence and 2% sexual violence at least once in the previous year (Rodríguez-Hernández, et al., 2023). However, according to Poy (2008), dating violence tends to go unnoticed by institutions and young people, resulting in a lack of institutional and family support for victims. For reasons such as these, and because of the growing awareness of the suffering involved and the negative consequences for the victims, research efforts on the subject have increased in this country since the first decade of this century.

Dating violence does not occur randomly, but is influenced by a number of individual and social variables, including childhood abuse (Hébert et al., 2019), alcohol and illicit drug use (among victims) (Duval et al., 2020), early sexual experiences (Duval et al., 2020), lack of parental support and supervision (Hébert et al., 2019), bullying by friends (Garthe et al., 2017), poverty and living in dangerous neighborhoods (Gracia-Leiva et al., 2019), perceived unsafety at school (Earnest & Brady, 2016), and belonging to a cultural minority (Gracia-Leiva et al., 2019).

Therefore, dating violence is a complex phenomenon that requires consideration of its determinants to address it.

2 Interventions in Dating Violence Based on New Technologies

As mentioned above, dating violence requires interventions based on multidisciplinary approaches and supported by the latest technological advances. In this context, the question arises as to why the intervention should be combined with information and communication technology tools. In response, it can be said that for several years it has been stated that the changes observed today are leading us to new realities in time and space, within the so-called digital society (Villar, 2001, p.95). Thus, it is necessary to take advantage of the advanced development of ICTs.

However, traditional methods are still mainly used, both for assisting victims of violence and for analyzing the problem. Quantitative studies are limited to the use of non-automated statistics, excluding artificial intelligence techniques such as data mining, which offer important advantages in classifying the information collected. This makes it difficult to devise equations or automated methods capable of producing a numerical value that assesses the level of violence suffered by each victim, considering its different forms (physical, psychological and sexual).

One of the advances in ICTs that can be useful for intervention in dating violence is that observed in social media. According to Gómez (2013), social media is an effective solution for the integration of a group of people with common interests, but it is also often a source of problems due to its lack of control and anonymity. Although there are groups on social networks where users identify with the issue and decide to join them, they do not use a methodology of specific support, but only encourage a social etiquette of support for victims and interest in the issue, regardless of whether they are victims.

The use of data mining techniques to analyze, classify and group the characteristics of victims of violence into profiles would facilitate the timely assessment of violent cases and the appropriate attention and referral to specialists in the field. In addition, an equation that indicates within a numerical scale the level of violence received would allow prioritizing attention within the victims belonging to the identified profiles, continuing the concept of focusing on the most urgent cases. This work proposes the use of data mining techniques, essentially clustering and classification algorithms, to detect behavioral patterns. The idea of using web platforms and resources is not new. A 'hackathon' event held in Washington D.C. in 2013 set the challenge of creating technological solutions to the problem of domestic violence. It brought together programmers from different parts of the world to create prototype solutions through collaborative work or applications for smartphones. Among the proposals is a website with a series of questionnaires that allow users to self-assess their level of risk and, based on the results, provide recommendations, help and guidance. Another "hackathon" project with participants from Panama proposes to use data to generate statistics and project them in the form of a map to visualize the situation in different regions. The data can be downloaded for future analysis. Information on legal procedures, institutions involved, news and other information can be included to help victims and institutions act and protect women's lives.

Based on the results of these studies, the search for answers to the causes of violent behavior is intensified. The importance of providing a space for continuous and automated intervention to periodically monitor the resulting changes in dating violence is highlighted.

3 Virtual Social Media and RedSIAM

The concept of social network has adopted different approaches over time, with the Internet being a watershed to distinguish them, and "refers to a community in which individuals are connected in some way, through friends, values, work relationships or ideas" (Calvo & Rojas, 2009, p.68). But today it also refers to "the web platform where people connect with each other" (Oliva, 2012, p.3). According to Wasserman & Faust (1996), social networks facilitate the formation of relational ties between pairs of social actors (entities) and groups (finite sets of social actors). This is conducive to interaction between members of the same profile of violence, so that they share their experiences and are therefore interested in staying within it to receive attention and be listened to, both by their own peers and by their counsellors. In recent years, computational data analysis techniques have been developed and applied to identify clusters and to justify what characteristics they share (especially in targeted advertising and market profiles), and it is also necessary to identify which variables are involved in dating violence.

According to Leiva (2009), there are three main groups of virtual social networks: specialization (horizontal and vertical), life (personal and professional) and a final group where there is a mixture of all the above (hybrids). RedSIAM can be considered a vertical network, as users may share similar interests and experiences related to violence, in addition to the characteristics that make them suitable to participate in the study.

According to Margain et al. (2013), the RedSIAM project was created to help women who are physically, emotionally and/or sexually abused. Support in this project is provided through a virtual social network and is executed through a mobile device capable of activating an emergency alert via voice recognition. This alert can be sent to family members, police, and members of the same social network. The location of the victim at the time of the attack can be obtained through the GPS location system, thus facilitating timely support. For RedSIAM, when the woman is being attacked (moment of violence), the role of mobile computing will be to extract the "wake-up call"; this function is identified in a hard layer in which the voice recognition service is used, and this data is stored on a server. For mobile device use, four important elements are identified in the hard layer: voice, device, language, and microphone-speaker. And three processes: capture application supported by mobile computing, voice recognition service and storage of captured data. The violence tendency map module is associated with the analysis of the problem, which is addressed in this chapter with a proposed intervention model within the social network.

Elgg® is an open-source platform for building social websites. Costello (2012) clarifies that because it is open source, the application can be freely used for whatever you want, with the freedom to modify and redistribute it. RedSIAM was developed under Elgg® because it is a framework designed for social websites, with the basic features needed for the project, in addition to being scalable and modular. According to Elgg® (2013), it uses a unified data model based on atomic data units called entities. This allows the addition of extension modules (EM) that can interact directly with the database, creating a more stable system and a better user experience because content created by different EMs can be integrated in a consistent way. With this approach, extensions are developed with greater flexibility and the ability to change the behaviour and appearance of Elgg® through views or event handling. The intervention model includes phases related to data acquisition, data processing and information presentation. The MoEs offer the possibility of modifying RedSIAM according to the needs of the model, being able to implement the necessary algorithms, the evaluation instrument and, if necessary, some connection with other analysis and database tools.

4 Social Data Mining

Social Data mining is generally defined as "the process of (automatically) discovering new and significant relationships (profiles), patterns and trends by examining large amounts of data" (Pérez & Santín, 2007, p. 2). Its aim is to extract knowledge from large amounts of information using advanced data analysis techniques. Today, data mining has evolved to the point where its implementation requires several steps. However, data mining is a non-trivial stage of knowledge discovery in data during the KDD (Knowledge Discovery in Database) process, where concrete algorithms are used that "generate valid, novel, potentially useful and understandable identification of patterns hidden in the data from pre-processed data" (Fayyad et al., 1996). The reason why data mining has been chosen as the cornerstone to determine the success of the present project is because, according to Duque (2014), its techniques allow finding apparently hidden relationships in the data and interesting patterns that are achieved through classification and prediction, which complements the statistical data provided periodically by government institutions and various research studies.

According to Garrido (2012), among the most representative characteristics of data mining are:

- It explores data stored for several years, often with the discovery of valuable and unexpected results. These data are concentrated in a warehouse, data mart or on an Internet or intranet server (with client-server architecture). In this research, such a feature cannot be exploited, as the data obtained are newly created, but a data warehouse may be beneficial if the study is longitudinal.
- Data mining algorithms can be easily combined to perform efficient analysis and processing of information (data). In addition, the data miner (in this case, psychologists or social workers) is often a user with little or no programming skills but has software that uses probing tools to get quick answers that overcome the limitations of software development or data management. It produces five types of information: associations, sequences, classifications, groupings, and forecasts.

Regarding the phases of a data mining project, Gonzalez et al. (2006) point out that the steps to follow for the realization of a data mining project are always the same, regardless of the specific knowledge extraction technique used: Data filtering, variable selection, knowledge extraction, and interpretation and evaluation. Through data mining, descriptive and predictive patterns can be found (Han et al., 2012). Descriptive patterns represent characteristic properties of the data, while predictive patterns infer events or characteristics with possible future occurrence. Hernández et al. (2004) point out that predictive models aim to estimate future or unknown values of the variables of interest, called target or dependent variables, using other variables or fields of the database, called independent or predictor variables. In contrast, descriptive models identify patterns that explain or summarize the results, i.e., they are used to explore the properties of the information under study, not to predict new data.

5 Methods and Materials

The present research is composed of several phases that together allow to develop and implement the intervention model for the social network RedSIAM. As a result of the final operation, it is intended to obtain an intervention model to support decision making with respect to the evaluation of the situation of violence of each user in the social network. Subsequently, these data will be used to classify each user (Fig. 1).

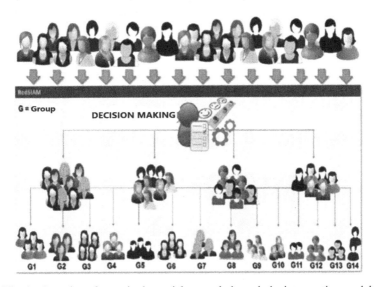

Fig. 1. Grouping of users in the social network through the intervention model.

5.1 Data Collection and Participants

To collect the data, the instrument "Dating Relationships in University Students" was used, which consists of 121 items to detect and measure physical, psychological and sexual violence in dating relationships. This instrument, designed by members of the Autonomous University of Ciudad Juárez (UACJ), was validated by the Committee of SNI Researchers of the Psychology Department of the Autonomous University of Nuevo Leon (UANL).

The inclusion criteria for the selection of participants were: 1) being female; 2) being between 17 and 29 years old; 3) being a university student; and 4) being in a relationship for more than 6 months.

5.2 Data Preprocessing and Transformation

The analysis technique used to perform the preprocessing was data mining, since it allows the generation of patterns using specific algorithms, but it is difficult to obtain good results with the data in its original state. Therefore, a pre-processing stage is necessary to increase the classification accuracy and improve the learning so that the process handles fewer parameters, which makes the process of detecting behavioral patterns more efficient. After the data cleaning process, 112 records were obtained from the UPA, 133 records from the UACJ, 39 records from the ITP, and 83 records from the UPSIN. The final database consists of 368 records that are unified with respect to valid response values.

The data transformation stage is necessary for the data mining tools to produce better results when working with the data. This process is not automatic; it is necessary to build a data mining exploitation model. The first step consists in building the data input representation scheme, in which the available information is known and the processes to be followed for data modeling are determined. Once this is done, the result obtained from each of the transformations performed is evaluated. In this way, the input data set is improved by eliminating noise and improving its quality (Fig. 2).

Fig. 2. Transversal study of our sample after apply a specific instrument online to obtain our dataset.

5.3 Elgg® Database Design and Statistical Analysis

In order to implement the intervention model in the RedSIAM social network, developed on the Elgg® platform, it is necessary to design a database that allows access to the data collected during the application of the instruments through a Web interface. The correct design of the database will facilitate the processing of the records as well as any future modifications. For this reason, the design principles of Microsoft Office (2010) were followed.

Age, height, weight, partner's age, and frequency of violence are the data to be considered in the statistical analysis. A correlation analysis is then conducted between all the variables, selecting the pairs whose correlation coefficient is greater than 0.8. For descriptive statistical analysis, mean, standard error, median of grouped data, kurtosis, skewness and correlation coefficient are used.

5.4 Weka Input Files and Attribute Selection

Although the Weka tool accepts various file types such as ".csv" to obtain input data and perform queries, Weka's own format is ".arff". This format, according to Garcia in his Weka manual, has three basic parts: attributes, relationships, and data. Attribute or feature selection consists of a series of methods to find the dataset that has the attributes with the most weight in the classification process, so that accurate results are obtained. It also makes it possible to reduce the size of the data so that only the most influential features in the problem are used, and even to avoid the problem of overfitting. Overfitting occurs when the number of features is large compared to the number of instances, and it is necessary to avoid it in order to reduce noise and obtain accurate results. The Weka program allows the evaluation of the variables through its "Explorer" interface. To do this, it is necessary to load the file with the ".arff" extension. Then go to the "Select Attributes" window and select the attribute to be analyzed. Finally, select the attribute evaluator and the search method to start the analysis.

5.5 Knowledge Extraction with Data Mining and Evaluation of Results

For the knowledge extraction stage to produce good results, it is necessary that the preprocessing stage has been performed correctly, seeking the highest accuracy percentages. Depending on the algorithms and methods used for its execution, the quality of the results obtained will depend on the quality of the results obtained. Two clustering algorithms are used to compare the results: K-Means and EM.

The evaluation step is important to know and compare the performance of the classification and clustering algorithms. One way to evaluate is to divide the data set into two parts, one for training and one for testing. The training data set is used to build the data mining model, while the test data set is used to evaluate its accuracy. It is worth mentioning that the percentage corresponding to each of the sets is a parameter that requires the split percentage. Another option is to use a test set that is independent of the initial set, since it is made up of new data.

Finally, cross-validation consists in dividing the instances (depending on the folding parameter) so that in each evaluation one of them is taken as test data and the rest as training data. In this case, the calculated error corresponds to the average of the errors of each run.

In classification, the confusion matrix indicates the number of correctly and incorrectly classified instances, where the hits are those corresponding to the diagonal.

The error rate is obtained by dividing the sum of false positives and false negatives by the total number of test samples, as shown in Eq. 1.

$$(FP + FN)/(TP + TN + FP + FN) \qquad (1)$$

6 Results

6.1 Redesign of the Database for Elgg®

The relational model was redesigned to integrate into the database the entities and relationships corresponding to the Dating Violence Perception Assessment Instrument, considering the existence of users and responses, as shown in Fig. 3. In addition, a type of question was specified, with a weight corresponding to the one assigned in the equation of violence. The relational model was generated using MySQL™ Workbench software, considering the data types and their specifications in each entity.

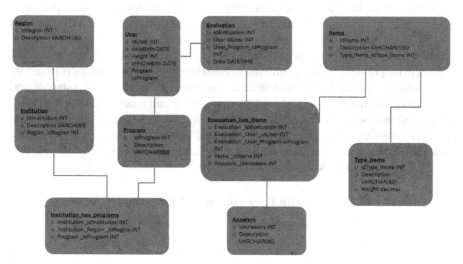

Fig. 3. Relational database model for instrument application designed in MySQL™ Workbench.

6.2 Statistical Analysis and Selection of Variables

The descriptive statistical analysis includes only the quantitative variables of the assessment instrument, as only these can be subjected to measures of central tendency, measures of distribution and measures of dispersion. The results show that the average age of the participants was 19 years. They were 1.62 meters (m) tall and weighed 58.7 kilograms (kg). The average age of the couple was also 21 years. As far as the ranges obtained are concerned, the age range is from 17 to 28 years, the height range is from 1.48 to 1.89 m, the weight range is from 36 to 160 kg and the age range of the couple is from 17 to 46 years.

The correlation analysis performed between all the variables has been reduced to show the pairs of variables whose correlation coefficients are weakly positive (+0.5) and strongly positive (1).

The variable selection stage makes it possible to reduce the size of the data so that only the most influential characteristics in the problem are worked with. In this case, the problem as such involves the set of more than 100 variables, so we worked to find their combinations in different assumptions.

The Weka program allows the evaluation of variables through its "Explorer" interface. To do this, it is necessary to load the file with the .arff extension. Then go to the "Select Attributes" window and select the variable we want to analyze. The "CfsSubsetEval" search method and the "Greedy Stepwise" search method are selected for the in-depth evaluation. Finally, the analysis starts. In this case, the new cluster attribute was tested (Table 1).

Table 1. Selected attributes for clusters.

	Attribute		Attribute
10(100%)	changed-way-of-being	10(100%)	insults-you
9(90%)	loss-contact	9(90%)	he-pushes-you
10(100%)	critics	10(100%)	you-argue-a lot
10(100%)	I-don't-respect	8(80%)	you-destroy-things
10(100%)	you-blame	10(100%)	you-fight-a lot
10(100%)	checking-without-consent	10(100%)	you-don't-consider-risks
10(100%)	addictive-substance	9(90%)	argues-a lot
10(100%)	tobacco	10(100%)	attracts-attention
10(100%)	alcohol	10(100%)	destroys-things
8(80%)	frequency-consumption	10(100%)	breaks-rules
10(100%)	modifies-relation	10(100%)	fights-a lot
10(100%)	affectionate-altercation-violent	8(80%)	attacks-physically
10(100%)	Need-being-loved	10(100%)	screams-much
10(100%)	aggression-in-parents-relationship	10(100%)	he's-stubborn
10(100%)	you-shut-him-up	10(100%)	his-mood-changes
10(100%)	you-yell-at-him	10(100%)	doesn't-consider-risks
9(90%)	you-insult-him	10(100%)	mocks
10(100%)	you-push-him	10(100%)	gets-angry
9(90%)	ignores-opinions	10(100%)	threatens-hurt
10(100%)	yells-at-you		

6.3 Extracting Information Using Data Mining

The first step is to group the users according to the characteristics used in the assessment tool. The first algorithm used to process the data was EM. It generated four profiles or groups. Cluster or group one represents 4% of the total instances. Its members are characterized by not answering most of the questions, coupled with the fact that most of their answers express that they do not experience a certain (negative) situation regarding their behavior or within their relationship. Although the group is small, most of its members belong to the Polytechnic University of Sinaloa.

Next, cluster zero represents 52% of the instances. Most of its members belong to the Polytechnic University of Aguascalientes and the Autonomous University of Ciudad Juárez, although the highest percentage of students is found in the Technological Institute of Parral (62%). Their behavior is characterized by low consumption of addictive substances by the user and moderate consumption of addictive substances by the partner. The couple tends to be affectionate after a violent argument, and the woman tends to accept this out of fear of being alone or out of a need to feel loved. Users live with their parents, so they experience more violence in their parents' relationships, which leads them to normalize it and consider it healthy. It is more common for them to attack each other psychologically within the relationship, as well as to show physical violence that can be passed off as play (scratching, hair pulling, biting, giving hickeys, etc.). It should be noted that the age of the members of the couple is older than in the other groups.

Cluster two represents 11% of the cases. Its members mostly belong to the Autonomous University of Ciudad Juárez (19%). In terms of their behavior, the age of both members of the couple is younger than that of the other groups. Their consumption of addictive substances is more moderate in terms of variety and frequency, especially alcohol and tobacco. They are the group with the most members who have started their sexual life, had sexual relations against their will and given in to sexual requests. Couples use blackmail and show affection after violent arguments because women fear being alone and need to feel loved. They attack each other psychologically and engage in mild to severe physical violence. Cluster three represents 33% of the cases. The majority of its members belong to the Autonomous University of Ciudad Juárez, although the highest percentage of students from the Polytechnic University of Aguascalientes is concentrated in this group (37.6%). As far as their behavior is concerned, women tend to be younger than the other groups, both consume more addictive substances, especially alcohol, tobacco, and marijuana, and do so more frequently. The number of women who use these substances is even greater than that of men. They also make up the largest proportion of couples who have had sexual relations against their will. In addition to the psychological assaults on both members of the couple, the threat of physical violence is beginning to appear, going beyond what is considered playful (scratching, hair pulling, biting, hickeys, etc.).

The second algorithm used was Simple KMeans. First, cluster three represents 43.4% of the cases and corresponds to cluster one of the EM algorithms, that is, simulating the state of a traffic light, it would correspond to the green color, in which the attention required is preventive in terms of sexual violence and risky physical violence, as well as corrective behavior of psychological violence on both sides. Then, cluster one represents 35.1% of the total number of cases and corresponds to cluster zero of the EM algorithm, i.e., it would correspond to the yellow category of the traffic light, which, although it continues to call for preventive actions, requires the adoption of strategies compatible with the consumption of addictive substances that increase the risks of sexual violence.

Cluster zero represents 19.8% of the total and corresponds to cluster two of the EM algorithm, i.e., the orange color of the traffic light, due to the use of addictive substances, compliance with sexual demands, blackmail and manipulation, and moderate psychological violence. It also corresponds to a moderate level of psychological and physical violence. Finally, cluster two represents 1.7% of the cases and corresponds to cluster two of the EM algorithm, i.e., it would correspond to the red color of the traffic light, as the physical violence has reached dangerous extremes and requires immediate professional attention. Other's algorithms were used to the four samples.

6.4 Design of the Equation to Assess the Level of Violence

The violence equation is based entirely on the characteristics of the assessment tool and, as a result of its solution, provides a numerical value that aims to assess the level of violence detected in each user. To construct the equation, the questions of the assessment tool are first classified according to five criteria: psychological violence received, psychological violence perpetrated, physical violence received, physical violence perpetrated and sexual violence.

6.5 Model of Technological Intervention

The model of technological intervention consists of five basic phases:

1. Recruitment phase (Fig. 3).
2. Data collection phase (Fig. 3).
3. Data Processing Phase (Fig. 4).
4. Results Interpretation Phase (Fig. 5).
5. Intervention phase.

It should be noted that the intervention model is a tool to facilitate decision making, which is why it aims to work hand in hand with psychologists or social workers from educational institutions. The recruitment phase begins with a rapid assessment in university groups. In other words, a small talk with the student. In the case of women, this would serve to determine whether they are a candidate to join RedSIAM, according to the characteristics mentioned above. Or to send a hyperlink to a web form where the necessary data can be collected, such as gender, age, marital status, length of relationship (if any) and e-mail. The above in order to establish communication with students who fall within the analysis profile proposed in this research.

To register in the RedSIAM social network, you must first create a username and password. To guarantee the student's privacy, she will be assigned a user code that is independent of her name or her institutional email. It will be up to the student to choose the one he/she prefers and feels most comfortable with. The data collection phase will then begin, in which the user will be asked to fill in the web form corresponding to the dating relationship assessment tool, whose answers will be recorded in a database with the structure mentioned above. We resume our different analysis in Table 2.

The third stage of data processing is the stage where data mining techniques are used to classify each user into one of the identified violence profiles. This is done based on the characteristics obtained through the assessment tool. Similarly, the weights of the responses are considered to obtain the numerical result of the equations for the level of violence received and the level of violence exerted.

Table 2. Comparative Table of Algorithms and Detection Percentages

University	Number of Students	Algorithm Used	Control over Activities (%)	Emotional Manipulation (%)	Verbal Abuse (%)	Physical Abuse (%)	Social Isolation (%)	Excessive Jealousy (%)	Threats (%)
Autonomous University of Ciudad Juárez (UACJ)	112	Random Forest	85	78	69	88	75	82	79
Polytechnic University of Aguascalientes (UPA)	85	K-Nearest Neighbors	80	85	74	82	70	84	76
Technological Institute of Parral	93	Support Vector Machine	87	81	72	85	76	86	80
Polytechnic University of Sinaloa	97	Decision Trees	83	76	71	79	73	81	75
4 Universities	258	Naive Bayes	82	79	70	84	72	83	77
4 Universities	252	Gradient Boosting	86	82	73	87	77	85	78
4 Universities	270	Neural Networks	88	83	75	90	78	87	81

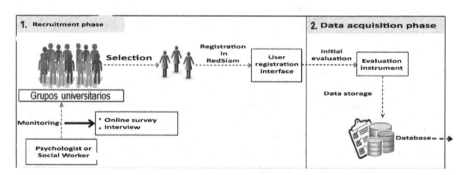

Fig. 4. Phases one and two of the technological intervention model.

The next step is to interpret the results. This includes the visualization of the user's violence profile, the numerical result of the level of physical, psychological, and sexual violence, the level of general violence received, perpetrated and the level corresponding to the Violentometer. Similarly, the psychologist or social worker has access to a general report on all users, the average level of violence in each type (physical, psychological, and sexual) and the number of users in each violence profile. These data will facilitate the decision-making process to choose the intervention strategies appropriate to the user's situation (Fig. 6).

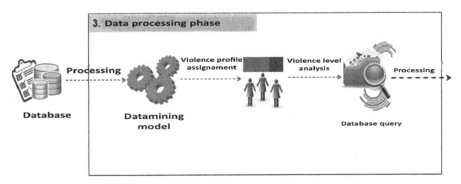

Fig. 5. Phase Three of the Technological Intervention Model.

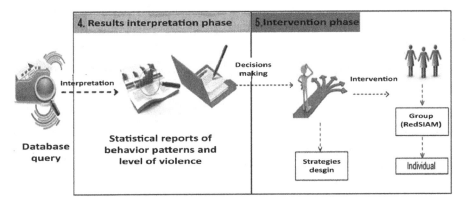

Fig. 6. Phases Four and Five of the Technological Intervention Model.

The last phase corresponds to the intervention phase, in which the psychologist or social worker must use the results of the model to support the choice of an action plan to intervene in a group or individually.

Among the strategies that can be adopted are

1. External referral.
2. Face-to-face counseling.
3. Personal interviews.
4. Workshops.
5. Social network support.

Strategies are recommended according to the violence profile. Individually, external channeling is suggested for the emergency profile, face-to-face counseling for the risk profile, support through the social network for the prevention profile, and face-to-face talks and workshops for groups that want to go. The recommendation for psychologists or social workers in relation to group one (indifferent or green) is to encourage group and personal interaction with these users so that their confidence increases and they can be honest in self-assessing their relationship.

For its part, for group zero (preventive or yellow), it is recommended to carry out workshops, conferences and interactions with users through the RedSIAM social network to make them realize the importance of respect in relationships, that psychological violence is also harmful in courtship and, above all, that both members can feed it by the simple fact of accepting it. Another important point is to make them realize that violence tends to be normalized, so if they begin to practice and accept minor acts of physical violence, they could get worse in the future.

As for group three (risk or orange), it is recommended to offer counselling and face-to-face talks on the effects of addictive substances on dating relationships, as well as workshops on how to improve communication in relationships. On the other hand, referral to an external service or regular counselling to help the woman reject these practices should be offered if the indicators of sexual violence are high. Finally, for the second group (emergency or red color), it is suggested to make greater use of external channels and personal counselling so that each case can be dealt with in a personalized manner. It is worth mentioning that, for each group, the accompaniment of the social network is extremely useful, especially to carry out exercises in which experiences, opinions, doubts and answers need to be shared.

7 Conclusions and Future Research

The focus of this work is the application of data mining techniques to the analysis of the problem of intimate partner violence in university dating relationships, through a technological intervention model that assigns a violence profile to each woman. The intervention model can act as a tool for automating the process of assessing the situation of violence in university dating relationships and as a support for the decision-making that social workers or psychologists must make in each intervention. In terms of data mining techniques, the evaluation and search methods for the selection of variables allow us to focus on the attributes that contribute most to grouping the users of the RedSIAM social network into violence profiles. The results of the EM algorithm were more suitable for maintaining four profiles of violence: neutral, preventive, risk, and emergency. It identifies the characteristic patterns of behavior of each profile and the relationships between the different attributes. And the decision tree generated by the J48 algorithm makes it possible to classify new users of the social network into the previously defined profiles. Using the equations of violence perpetrated and violence received, it is possible to quantify the level of violence on a scale from 0 to 1. This is based on the Violence Perception Assessment Instrument and compared with the Violentometer scale to show its equivalent result. In addition, when interpreting the results, it is useful to have a breakdown of the level of violence by type (physical, psychological and sexual). The problem of intimate partner violence requires further research in all its forms. As a next step, it is proposed to carry out a longitudinal study on the problem of intimate partner violence at university level through the social network RedSIAM. Its intervention model will provide the necessary information to evaluate whether this tool has improved the participants' behavior, their awareness of the problem or their way of dealing with it in a positive way. Evaluations can be carried out at the end of each term (four or six months) to follow up the case and prepare action strategies for counsellors. It would also be necessary to allocate a dedicated server to host the network, so that users can access it both inside and outside the institutions. In the long term, it is proposed to make the necessary adaptations to the data collection tool to implement the intervention model in care centers outside the university environment, such as women's institutes. In this way, on the basis of the profiles of violence identified, it will be possible to integrate new profiles in order to determine how the intervention at university level contributes to reducing the risk of living in a situation of violence during the life of a couple in general, as well as the characteristics common to each profile in its different contexts (university, non-university, marriage, cohabitation, courtship, etc.). Regarding the use of data mining techniques to contribute to the solution of social problems such as violence, other algorithms that Weka does not offer by default could be tested. This would compare their advantages and limitations, their performance, and the ease of interpreting the learning process. This research has laid a foundation for using data mining techniques to address intimate partner violence (IPV) in university dating relationships through a technological intervention model. However, there are several areas where future research can be expanded and deepened. The following are proposed areas for future studies:

Longitudinal Studies on Intervention Outcomes
One of the most promising areas for future research is to conduct longitudinal studies to monitor the long-term effectiveness of the RedSIAM intervention model in reducing IPV incidents. These studies would follow participants over time, collecting data at regular intervals (e.g., each semester or academic year). This approach will provide insights into how attitudes, behaviors, and coping mechanisms evolve over time with exposure to the intervention.

Research Question: Does long-term use of RedSIAM lead to sustained reductions in violent behavior and improved relationship health in college dating contexts?

Issues Measures:

- Frequency of violent incidents over time.
- Changes in violence perception and behavior.
- Psychological well-being of participants.

Proposed Methodology:

- Surveys and interviews at baseline, midterm, and endpoint of the study.
- Control group comparison with those not exposed to the intervention model.

Expansion of Profiles Beyond University Contexts
Another significant area for future exploration is adapting the violence profiles identified by RedSIAM to apply beyond the university context. This could involve tailoring the intervention model for use in community settings, such as women's shelters, non-governmental organizations, or even corporate environments. By expanding the profiles and intervention mechanisms, the system could better address violence in a variety of intimate relationships, including cohabitation, marriage, and non-university dating relationships.

Research Question: How can the violence profiles identified in university settings be adapted to community-based interventions for a wider demographic of IPV survivors?

Issues Measures:

- Effectiveness of interventions across different relationship contexts (e.g., cohabiting vs. married vs. dating relationships).
- Rate of IPV recurrence in non-university contexts.

Proposed Methodology:

- Cross-sectional studies comparing the impact of RedSIAM's violence profiles in university vs. non-university environments.
- Modifying the data collection tool to accommodate broader demographics.

Implementation of Machine Learning Algorithms Beyond Weka

In addition to improving the accuracy of violence profiles, future research could explore the use of more advanced machine learning algorithms beyond those provided by Weka. Testing algorithms such as **Random Forests**, **Gradient Boosting**, or even deep learning techniques could offer improved classification accuracy, better handling of complex interactions between variables, and more nuanced insights into the patterns of violence.

Research Question: Which machine learning algorithms provide the most accurate and interpretable results in identifying violence profiles in university dating relationships?

Issues Measures:

- Classification accuracy of violence profiles.
- Computational efficiency and interpretability of different algorithms.

Proposed Methodology:

- Comparison of algorithm performance on a dataset of university dating relationships.
- Evaluation of algorithm complexity, ease of interpretation, and potential for real-time application.

Socio-Economic Impact of Technological Interventions

Another key area to study is the socio-economic impact of deploying IPV intervention models like RedSIAM. By tracking the outcomes of these interventions, researchers could analyze their cost-effectiveness, impact on local economies, and potential for creating new jobs (e.g., in counseling services or technology development).

Research Question: What are the socio-economic impacts of deploying a technological intervention model for IPV in university settings?

Issues Measures:

- Cost savings in healthcare and counseling services.
- Job creation linked to technological deployment.
- Economic opportunities for the local community (e.g., partnerships with local therapy providers, equine therapy centers).

Cross-Cultural Validation of Violence Profiles
Given the cultural variability in how intimate partner violence is perceived and addressed, future research should explore how well the violence profiles defined in the RedSIAM model apply across different cultural contexts. A comparative study could help adapt the intervention model for use in various countries or regions, adjusting for local social norms and legal frameworks.

Research Question: How do the identified violence profiles vary across different cultural contexts, and how can the RedSIAM model be adapted to suit these differences?

Issues Measures:

- Comparison of violence profiles across different countries or regions.
- Cultural factors influencing the perception and reporting of IPV (Table 3).

Table 3. Proposed Areas for Future Research

Research Area	Key Question	Expected Outcome
Longitudinal Study on IPV Interventions	Does RedSIAM lead to sustained behavior change?	Better understanding of long-term intervention efficacy
Expansion of Profiles Beyond University Contexts	How can profiles be adapted for non-university IPV cases?	Broader applicability of intervention model
Testing Advanced Machine Learning Algorithms	Which algorithms offer better classification for violence profiles?	Improved classification accuracy and real-time application
Socio-Economic Impact	What is the economic benefit of technological IPV interventions?	Evidence of cost savings and economic growth in therapy-related sectors
Cross-Cultural Validation	How do violence profiles vary across cultures, and how can models be adapted?	Adaptation of profiles to various cultural contexts

References

Castro, R., Casique, I.: Violencia en el noviazgo entre los jóvenes mexicanos. In: Casique, I. (ed.) Betting on Adolescent Empowerment: Relationships with Sexual and Reproductive Health and Dating Violence]. National Autonomous University of Mexico, México (2010)

Costello, C.: Elgg® 1.8 Social Networking. Packt Publishing Ltd (2012)

Duque, N.: Autonomous University of Colombia. http://www.virtual.unal.edu.co/cursos/sedes/manizales/4060029b/und_8/index.html. Accessed 28 May 2024

Earnest, A.A., Brady, S.S.: Dating violence victimization among high school students in Minnesota: associations with family violence, unsafe schools, and resources for support. J. Interpers. Violence **31**(3), 383–406 (2016)

Elgg®. Elgg®. https://elgg.org/. Accessed 15 June 2024

Fayyad, U., Piatetsky-Shapiro, G., Smyth, P.: From data mining to knowledge discovery in databases. Am. Assoc. Artif. Intell. 37–54 (1996)

Garrido, M.: Pattern search in databases and its application in PYMEs. Xalapa-Enríquez, México (2012)

Garthe, R.C., Sullivan, T.N., McDaniel, M.A.: A metaanalytic review of peer risk factors and adolescent dating violence. Psychol. Violence **7**(1), 45–57 (2017)

González, E., Pérez, Z., Espinosa, I.: Extraction of patterns and rules in the academic process of the University of Computer Science using data mining techniques. Havana, Cuba (2006)

Gracia-Leiva, M., Puente-Martínez, A., Ubillos-Landa, S., PáezRovira, D.L.: Violencia en el noviazgo (VN): Una revisión de metaanálisis [Dating Violence (DV): a review of meta-analyses]. Anales de Psicología/Ann. Psychol. **35**(2), 300–313 (2019)

Han, J., Kamber, M., Pei, J.: Data Mining: Concepts and Techniques. Elsevier (2012)

Hébert, M., et al.: A meta-analysis of risk and protective factors for dating violence victimization: the role of family and peer interpersonal context. Trauma Violence Abuse **20**(4), 574590 (2019)

Leiva, J.: Baratz. Knowledge management. http://www.baratz.es/portals/0/noticias/Redes%20Sociales_J.Leiva_Baratz.pdf. Accessed 21 June 2024

Margain, L., Hernández, G., De Luna, C., Ochoa, A.: Mobile computing in the emotional support social network (RedSiam) to reduce intimate partner violence. Santiago, Santiago, Chile (2013)

National Institute of Statistics and Geography. National Survey on the Dynamics of Household Relationships ENDIREH 2021. Main results. https://www.inegi.org.mx/contenidos/programas/endireh/2021/doc/endireh2021_presentacion_ejecutiva.pdf. Accessed 01 July 2024

Oliva, C.: Social networking and young people: Internet privacy challenged. Aposta 1–16 (2012)

Poy, L.: La Jornada (2008). https://www.jornada.com.mx/2008/11/12/index.php?section=sociedad&article=044n1soc

Rodríguez Hernández, R., Castillo-Viveros, N., Esquivel-Santoveña, E.E.: Prevalencia y correlatos de la violencia en el noviazgo en las adolescentes y jóvenes mexicanas [Prevalence and correlates of dating violence among girls and adolescents in Mexico]. Psychol. Soc. Educ. **15**(1), 68–75 (2023)

Sardinha, L., Maheu-Giroux, M., Stöckl, H., Meyer, S.R., García-Moreno, C.: Global, regional, and national prevalence estimates of physical or sexual, or both, intimate partner violence against women in 2018. Lancet **399**(10327), 803–813 (2022)

Straus, M.A.: Prevalence of violence against dating partners by male and female university students worldwide. Violence Women **10**(7), 790–811 (2004)

United Nations. Declaration on the elimination of violence against women. https://www.acnur.org/fileadmin/Documentos/BDL/2002/1286.pdf?file=fileadmin/Documentos/BDL/2002/1286. Accessed 21 June 2024

Villar, L.: University, educational evaluation and curricular innovation. In: Pantoja, A., Aranda, T. (eds.) Universidad de Sevilla, Instituto de Ciencias de la Educación, España, pp. 95–128 (2001)

World Health Organization. Definition of intimate partner violence. https://apps.who.int/violence-info/intimate-partner-violence/. Accessed 21 June 2024

Enhanced Pest Detection Using Quaternion-Based Image Segmentation in Yellow Sticky Trap Samples for Precision Agriculture

Esquivel-Félix Ramiro[1]($\boxtimes$)[iD], Solís-Sánchez Luis Octavio[1][iD], Ochoa-Zezzatti Alberto[2][iD], Castañeda-Miranda Celina Lizeth[1][iD], and Guerrero-Osuna Héctor Alonso[1][iD]

[1] Doctorado en Ingeniería y Tecnología Aplicada, Universidad Autónoma de Zacatecas, Zacatecas, Mexico
resfera@gmail.com, {lsolis,celina,hectorguerreroo}@uaz.edu.mx
[2] Doctorado en Tecnología, Universidad Autónoma de Ciudad Juárez, Ciudad Juárez, Mexico
alberto.ochoa@uacj.mx

Abstract. Pest detection remains a significant challenge in precision agriculture, particularly in accurately identifying and classifying insects, which are pivotal for deploying effective pest management strategies. Optimizing farming practices while ensuring environmental sustainability demands robust Integrated Pest Management (IPM) systems. These systems encompass detection, identification, targeted application of management techniques, and meticulous recordkeeping. The success of IPM hinges on the timely execution of pest detection tasks, emphasizing continuous monitoring and comprehensive data collection. Despite the critical importance of frequent monitoring, the current insect detection and surveillance methodologies are often labor-intensive, time-consuming, and prone to inefficiencies, leading to potential crop damage due to the prioritization of other pressing agricultural tasks. Typically, inspections are conducted several times a week; however, increasing the frequency of daily checks would significantly enhance the accuracy of pest management decisions. Yet, the practical implementation of daily inspections is hindered by challenges related to cost, labor, and feasibility. This paper addresses these challenges by focusing on the complexities of quaternion-based image segmentation for pest detection. The integration of machine vision techniques into mobile embedded systems presents a promising solution to enhance the efficiency and effectiveness of pest management practices in agriculture, ultimately contributing to more sustainable and productive farming outcomes. Looking towards future advancements, the application of Few Shot Learning (FSL) algorithms offers a transformative potential for pest detection. FSL algorithms, designed to learn from a minimal number of annotated examples, can significantly improve the precision and adaptability of pest detection systems. By leveraging FSL, agricultural systems can overcome the limitations of data scarcity, providing accurate pest identification even with limited labeled examples. This capability is precious in dynamic agricultural environments where new pest species and disease symptoms frequently emerge.

Keywords: Quaternions · image processing · color imaging · image segmentation · pest management · precision farming

1 Introduction

Pest detection represents a significant challenge in modern agriculture, particularly detecting and classifying insects, which are critical for effective pest management strategies [1]. Efficient pest management is a cornerstone in optimizing agricultural practices, enhancing the quality of crop production, and upholding environmentally sustainable practices [2]. The increasing demand for food production in a changing climate further amplifies the need for precise and timely pest control methods. Integrated Pest Management (IPM) systems have been widely adopted as a comprehensive approach to pest control. These systems involve a structured process that includes four critical stages: detection, identification, appropriate management application, and meticulous documentation of the methods used [3]. The success of IPM heavily depends on the accuracy and timeliness of pest detection activities. Continuous monitoring and detailed recording of pest behavior are crucial components that strengthen the overall effectiveness of IPM systems. Despite the importance of these activities, insect detection and monitoring remain labor-intensive, time-consuming, and prone to human error [4]. The manual nature of these tasks often leads to their neglect, as farmers prioritize other pressing agricultural activities. Inspections are typically conducted only a few times per week, which may not be sufficient to capture the dynamic nature of pest infestations. Ideally, daily inspections would be conducted to provide a more comprehensive understanding of pest populations and to inform the selection of the most effective control strategies [5]. However, the practicalities of daily monitoring—given its complexity, cost, and labor requirements—make it a challenging task to implement consistently. To address these challenges, the development of an automated, autonomous vision system for pest detection presents a promising alternative. Such a system would enable the daily recording and analysis of pest populations, offering continuous monitoring without the extensive labor input required by traditional methods [6]. Recent advancements in machine vision, robotics, artificial intelligence, and autonomous vehicles suggest that these technologies could play a pivotal role in enhancing the precision and efficiency of pest monitoring activities within IPM frameworks [7, 8]. Research in this domain has explored various approaches to crop monitoring and pest detection. For instance, [9] developed a technique to differentiate between crops and weeds, while other studies have focused on identifying color variations in fruits to detect contaminants and pests [10]. Muppala and Guruviah [11] have proposed a machine vision system designed to inspect crops, drawing comparisons to the visual capabilities of human inspectors in an effort to boost agricultural yields. Moreover, computational image analysis systems [12], when integrated with autonomous vehicles [13], demonstrate significant potential for the detection and identification of shapes, particles, and insects across diverse crop types. These systems offer a more objective and consistent approach to pest detection, reducing the likelihood of human error and improving the reliability of pest management decisions. At the Autonomous University of Zacatecas (UAZ), the Laboratory for

Research and Development in Applied Information Technologies (LIDTIA) within the Postgraduate Degree in Engineering and Applied Technology (PITA) has been conducting pioneering research in this area. Their studies focus on the use of adhesive tapes to capture insects, followed by image acquisition and processing using advanced mathematical techniques. Specifically, the research involves segmentation and vectorized stereographic projection of color using quaternions [14]. The application of quaternion-based image segmentation for pest detection in yellow sticky traps offers significant potential to enhance the efficiency and accuracy of traditional monitoring methods in IPM systems. By integrating this technique into standalone vision and image analysis systems for mobile embedded systems, it is possible to revolutionize crop pest detection, making it more efficient, reliable, and scalable across different agricultural settings.

2 Background a Related Work

There is a large body of information related to image processing techniques applied to improve pest detection and classification methods, among the most widely used are those that use sticky traps to capture insect samples in crops [15], studies show us the possibility of detecting and recognizing insects in greenhouses using tools such as deep learning [16], it is possible to recognize whitefly, aphids and other diseases [17], insect samples are collected in controlled habitats and then taken to laboratory analysis and diagnosed by entomologists [18], the color contrast in the yellow sticky traps generates color contrast which is favorable for the identification of the insects, this facilitates the artificial vision systems in the station of the morphological characteristics of the pests, the images represent a large amount of information to acquire large amounts of data for the processing in the acquisition of the image where it is represented mathematically [19] in the form $I(x,y) = f(x,y)$ in the format of the following equation:

$$Imagen = f(x,y) = f_{nxm} = f(v_{pixel})_{ij}$$

$$= \begin{bmatrix} f(v[rgb]_{pixel})_{n,m} & \cdots & f(v[rgb]_{pixel})_{n,m+j} \\ \vdots & \ddots & \vdots \\ f(v[rgb]_{pixel})_{n+i,m} & \cdots & f(v[rgb]_{pixel})_{n+i,2+j} \end{bmatrix} \quad (1)$$

$$= \begin{bmatrix} [rgb] & \cdots & [rgb] \\ \vdots & \ddots & \vdots \\ [rgb] & \cdots & [rgb] \end{bmatrix}$$

2.1 Quaternion

Quaternion or quaternion numbers are an extension and generalization of the complex numbers, represented as a sum of the imaginary units i, j, k, and a scalar. This subject was first treated by Hamilton, and William Rowan [20] and the way Quaternion is represented as follows.

$$q = q_r + q_i i + q_j j + q_k k \quad (2)$$

where i, j, k the imaginary units and qr, qi, qj, qk, are real numbers, the

$$i^2 = j^2 = k^2 = -1$$
$$ij = -ji = k$$
$$jk = -kj = i \quad (3)$$
$$ki = ik = j$$

For a Quaternion the modulus and conjugate are:

$$q = \sqrt{qq'} = \sqrt{q_r + q_i + q_j + q_k} \quad (4)$$

2.2 Color Images Based on Quaternion

Representation of color images in quaternion [21] is like a function of two variables representing an ordered pair f(x, y)RGB, each pixel can be represented as a quaternion [22] which would form the representation of a color pixel:

$$f(x, y) = f_R(x, y)i + f_G(x, y)j + f_B(x, y)k \quad (5)$$

where fR(x, y)i corresponds to the color red, fG(x, y)j to the color green, and fB(x, y)k to the color blue as color components of a pixel [23], an image is then represented by Eq. (5) as a hypercomplex matrix [24].

2.3 The Quaternion Fourier Transform

The process of compressing an image can be carried out employing the Quaternion Fourier Transform algorithm [24]:

$$H(j\omega, kv) = \int_{-\infty}^{\infty} \int_{-\infty}^{\infty} e^{-j\omega} h(t, \tau) e^{-kv\tau} dt d\tau \quad (6)$$

Following the above algorithm, the inverse Fourier transform of quaternion [25] is a subsequent deconvolution process of the direct transform to recover the characteristics of the original image:

$$h(t, \tau) = \int_{-\infty}^{\infty} \int_{-\infty}^{\infty} e^{j\omega} H(j\omega, kv) e^{kv\tau} dt d\tau \quad (7)$$

To maintain image fidelity the modulation coefficient for RGB channels or i,j,k color components of the Discret Quaternion Fourier Transform (DQFT) is set to the quantization index modulation $\mu = ((i + j + k))\sqrt{3}$ or $\mu = ((0i - 2j + 8k))\sqrt{68}$ which is the one that will be used for testing in this paper [21]. Since machine vision is deployed on digital computers it is necessary to implement a direct discrete Quaternion Fourier Transform compression:

$$F(u, v) = \frac{1}{\sqrt{MN}} \sum_{m=0}^{M-1} \sum_{n=0}^{N-1} e^{-\mu 2\pi (\frac{mu}{M} + \frac{mv}{N})} f(m, n) \quad (8)$$

The recovery of the original image is a special case of the complex transform function, which is a quaternion inverse discrete quaternion FFT compression:

$$f(m, n) = \frac{1}{\sqrt{MN}} \sum_{u=0}^{M-1} \sum_{v=0}^{N-1} e^{\mu 2\pi \left(\frac{mu}{M} + \frac{mv}{N}\right)} F(u, v) \quad (9)$$

where f(m,n) is a color image of dimensions M × N represented in Eq. (5), μ is the hypercomplex unit which can be defined as a linear combination of the components i, j, k such that $\mu = \alpha i + \beta j + \gamma k$ where $\alpha, \beta, \gamma \in R$ and $\|\mu\| = 1$.

2.4 Color Edge Detection Sobel Filter

Color edge tracing with Sobel filter gradient adjustment of an image f(x, y) at a point(x, y), is defined as a two-dimensional vector. Vertical edge detection:

$$G_x = \begin{bmatrix} 1 & 0 & -1 \\ 2 & 0 & -2 \\ 1 & 0 & -1 \end{bmatrix} \quad (10)$$

Horizontal edge detection:

$$G_y = \begin{bmatrix} -1 & -2 & -1 \\ 0 & 0 & 0 \\ 1 & 2 & 1 \end{bmatrix} \quad (11)$$

The gradient of magnitude estimation Vertical and horizontal Sobel edge correlations:

$$G[f(x, y)] = \begin{bmatrix} G_x \\ G_y \end{bmatrix} = \begin{bmatrix} \frac{\partial}{\partial x} = f(x+1, y) - f(x-1, y) \\ \frac{\partial}{\partial y} = f(x, y+1) - f(x, y-1) \end{bmatrix} \quad (12)$$

$$|G| = \sqrt{G_x^2 + G_y^2} = |G_x| + |G_y| \quad (13)$$

Contour angle adjustment:

$$\varphi(x, y) = \tan^{-1} \frac{G_y}{G_x} \quad (14)$$

3 Materials and Methods

For a research study that includes a comprehensive and detailed analysis of image processing in pest detection, we have chosen to use a generic image of a yellow sticky trap from a free database of the Wageningen University & Research [28], which contains small white spots that simulate the presence of insects. This type of trap is widely used in agriculture for insect monitoring and trapping, allowing early detection of pests that could affect crops.

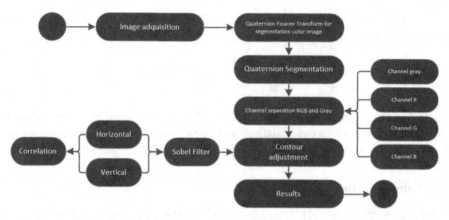

Fig. 1. General initial processing algorithm for contour matching.

In general terms, the methodology used in pest detection work is based on the extraction of regions of interest within an image Fig. 1. This extraction is crucial, as it allows the analysis to focus on the most relevant areas, facilitating the identification and classification of specific insects or pests. The first step in this process is usually image preprocessing, which aims to improve the quality of the image so that subsequent processing algorithms can operate more efficiently and accurately. However, in this study a particular methodological choice was made: no denoising was performed during preprocessing. This choice allows us to observe the direct effects of the proposed algorithms, without the influence of additional filters that could alter the original image characteristics.

One of the fundamental techniques applied in this study is quaternion-based segmentation. Quaternions are an extension of complex numbers that allow color images to be represented more comprehensively, keeping the information of the color channels (red, green, and blue) together with an additional channel that may contain greyscale or other information. In mathematical terms, a color image can be represented by a quaternion $q_{r_i}(x, y)$, $q_{g_j}(x, y)$, $q_{b_k}(x, y)$, where each component of the quaternion is associated with one of the color channels.

Edge detection in color images is a major challenge in computer vision, and to address it, we use a Sobel filter modified to operate on quaternions. The traditional Sobel filter is used to detect edges in grayscale images by applying a gradient operator in horizontal and vertical directions. In this study, we adapt the Sobel filter to work in the quaternion domain, which allows us to detect edges in colour images without the need to convert them to greyscale.

The methodology used for quaternion image compression consists of obtaining four color channels. Although the aim of the steps is similar to that of traditional methods, it was decided to omit conventional Gaussian smoothing to observe the direct effects of the proposed algorithms. This is because we are faced with the task of working with the color features in a broad way, without the separation of the RGB channels or the conversion to grayscale.

For the implementation of this methodology, we used Python 3.10 together with the Scipy libraries and their sub-libraries Signal and Fftpack, as well as the OpenCV

4.7 framework. These tools allowed us to develop functions for both direct and inverse compression of images into quaternions, thus generating an image with four channels. The first step was to convert the original RGB image to a quaternion matrix, which represents the information of each pixel in the red, green, blue and greyscale channels. This representation preserves the wealth of information contained in the original image, which is essential for the detailed analysis of the color characteristics.

Once the quaternion image is obtained, we proceed with the preprocessing using the Sobel filter technique for color edge detection. This process is performed horizontally and vertically, generating two color edge maps that are subsequently combined to form a complete edge map. These maps are essential for identifying color transitions in the image, which in turn allows for highlighting morphological features of regions of interest, such as insect outlines.

It is important to note that, in digital image pre-processing, it is common to use Gaussian smoothing filters to reduce noise and improve image quality before segmentation. However, in this study, it was decided to omit this step and directly apply the image transformation to quaternions. Also, it was decided not to convert the image to greyscale, since the main objective of the preprocessing in this work is to preserve and enhance the color characteristics of the image, taking full advantage of the capabilities of the quaternion method.

The final result of this process is an image with color contours, which can be used for the implementation of closed contour counting techniques and the morphological identification of closed areas. These techniques are fundamental for insect classification and pest detection, as they allow the identification of the characteristic shapes and patterns of the species present in the sticky trap.

4 Results

The initial test image used in this study is a photograph called '1281.jpg' Fig. 2 which was selected from the Wageningen University Research database [29] in RGB color format of 5184×3456 pixels, cropping between the points (79, 564, 1940, 789) was extracted to form an image of size 789×1940 pixels, which serves as the basis for further analysis. The image is stored in a 3×3 tensor matrix variable, which effectively represents the combination of RGB pixels in the whole image. This matrix structure is essential to maintain the integrity of the color information, which is essential to accurately identify areas where adult whiteflies are present. The image captures several regions of interest (ROIs) where white-flies are concentrated. These ROIs are vital for downstream processing steps, as they provide the focus points for edge detection and color segmentation tasks. The RGB channels are essential for preserving the chromatic characteristics of the whiteflies, which differentiate them from the background and other image elements.

4.1 RGB Channel Separation and Grayscale Transformation

To better understand the color distribution and to facilitate further analysis, an RGB color channel separation test was conducted using traditional segmentation methods, such as

Fig. 2. Test image 789 × 1940 pixels RGB yellow sticky trap section. (Color figure online)

the color space method [26]. In this approach, the RGB color image I is defined with a resolution of $M * N * 3$ where M and N denote the rows and columns of the image, respectively. The image is then decomposed into its constituent color channels: $R(i,j)$, $G(i,j)$, and $B(i,j)$, where i and j represent the horizontal and vertical axes of each pixel Fig. 3.

Fig. 3. Separation of color channels a) Red channel, b) Green channel, c) Blue channel (Color figure online)

Following the separation, a transformation is applied to extract the grayscale image using the luminosity method [27], also known as color weighting by weights. This method modifies the RGB channels by applying specific weights to each channel, effectively converting the color image into a grayscale representation. The grayscale image provides a simplified version of the original image Fig. 4, emphasizing the intensity variations that are critical for edge detection and further processing.

4.2 Fourier Transform and Quaternion-Based Image Compression

To delve deeper into the image's structural characteristics, a Fourier transform was applied to the original image, with the compression being handled through quaternion mathematics. The Fourier transform is pivotal in understanding the frequency distribution of colors within the image, which can reveal patterns and regularities that are not immediately apparent in the spatial domain. The application of the Fourier spectrum to the image compression process provides insights into the color frequency distribution 5, offering a unique perspective on how different color frequencies contribute to the overall image composition. This step is particularly useful for tasks that require the

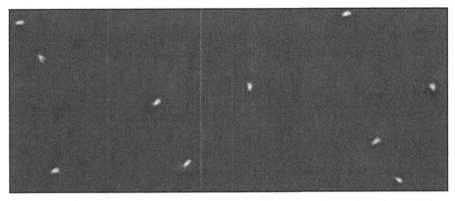

Fig. 4. Gray channel extraction by weighted RGB channel method also known as luminance method

identification of periodic patterns or the isolation of specific color features within the image. Quaternion-based image compression represents a significant advancement over traditional methods. By transforming the image into a quaternion representation, we can store it in a 4-channel matrix Table 1, where the i, j, and k components correspond to the RGB colors Fig. 6, and the real component corresponds to the grayscale information. This approach not only preserves the color information in a more compact form but also enhances the ability to manipulate and analyze the image in subsequent processing steps (Fig. 5 and Table 1).

Fig. 5. Original Image and Quaternion-based Compression Image

Fig. 6. Image on frequency domain Fourier spectrum

4.3 Image Segmentation Using Quaternion Transformation

The quaternion transformation and subsequent compression of the image lead to a notable enhancement in the clarity of the objects within the image. By outlining the contours of

Table 1. Quaternion based color transformation which in 4-channel

	Image shape	Channels
Simple Greyscale	789 × 1940	1
Simple RGB	789 × 1940	3
RGB and Grey Hipercomplex	789 × 1940	4

the areas of interest, this method effectively highlights the characteristics of the objects, making them more distinguishable from the background. This enhancement is particularly beneficial when applying edge detection filters, as it allows for the extraction of the most representative features of the objects, which are often accentuated by color. When the image is transformed into quaternions, it results in the mixing of four layers of color. Although this mixing alters the appearance of the colors, the underlying information remains intact. This characteristic of quaternion transformation is crucial, as it ensures that while the visual representation may change, the essential data required for analysis is preserved (Figs. 7, 8 and 9).

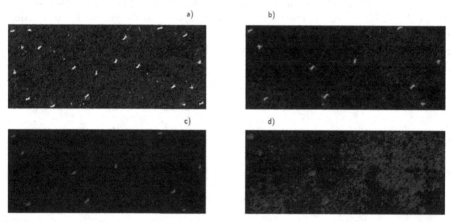

Fig. 7. Output images a) real component grey color, b) component j green color, c) component i red color, d) component k blue color (Color figure online)

4.4 Convolution for Sobel Filter Contour Detection

Finally, the Sobel filter is applied to the Quaternion Fourier Transform (QFT) image in the frequency domain to detect contours. The Sobel filter, with its vertical and horizontal derivative kernel operators, is particularly effective in identifying edges within an image. By applying this filter to the QFT image, we can detect the contours of the objects within the image with high precision. The combination of quaternion-based image compression and Sobel filter contour detection represents a robust methodology for analyzing complex images with multiple color channels. This approach not only enhances the visibility of

Fig. 8. Traditional segmentation, compared Quaternion-based segmentation

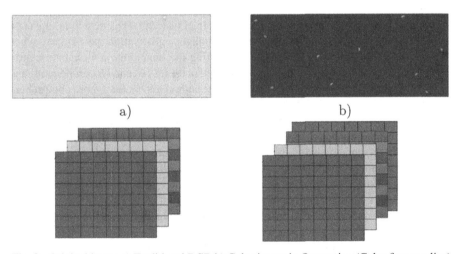

Fig. 9. Original image a) Traditional RGB b) Color image in Quaternion (Color figure online)

the regions of interest but also facilitates the subsequent classification and identification of objects within the image (Fig. 10).

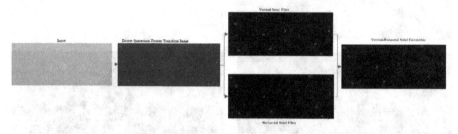

Fig. 10. Quaternion segmentation a color edge detection with Sobel filter correlated

4.5 Summary of Findings

The results demonstrate that the quaternion-based approach, coupled with traditional and advanced image processing techniques, provides a comprehensive framework for analyzing RGB images with high accuracy. The ability to maintain color integrity while enhancing edge detection and segmentation processes highlights the potential of this methodology in various applications, particularly in agricultural monitoring and precision farming, where accurate detection of pests like whiteflies is critical for maintaining crop health.

5 Discussion

The results obtained from processing an image using Quaternion Discrete Fourier Transform (QDFT) reveal significant advantages in image segmentation, particularly when dealing with complex color spaces. By segmenting the image into a 4-channel matrix (4D image matrix), where the colors are embedded in the iR, jG, and kB components, and the real component holds a grayscale image, this method allows for a more comprehensive analysis. This structure inherently highlights the contours of objects in the image, such as the insects present on yellow sticky traps. The ability to work within both the RGB and grayscale color spaces simultaneously within a single four-dimensional matrix is particularly advantageous for contour detection and subsequent analysis. This edge-enhancing capability is particularly useful for algorithms focused on contour marking. In this study, we utilized the Sobel algorithm, which is adept at detecting and highlighting the edges of areas and intersecting objects. The Sobel filter, applied in both horizontal and vertical sweeps, effectively closes the contours by correlating the resulting images from these sweeps. The outcome is a refined image where the contours of the objects—like insects—are delineated, almost as if marked by a watermark. This phenomenon can be understood through the lens of hypercomplex number theory, where the separation of an image by its RGB and grayscale components enhances edge detection and contour marking, offering a significant improvement over the traditional method. Traditional image processing typically operates within a three-dimensional (3D) framework, where color components are separated into RGB channels, and grayscale components are processed independently. This separation introduces limitations, particularly when these components must be combined for tasks such as object recognition—in this case, the detection of pests and insects. The conventional approach requires managing two separate processes: one involving a 3D RGB matrix and another for the grayscale channel, which is generated through an additional processing step. However, with QDFT, these processes are streamlined into a single step that segments all four channels (RGB and grayscale) concurrently, offering a more efficient and integrated approach (Fig. 11).

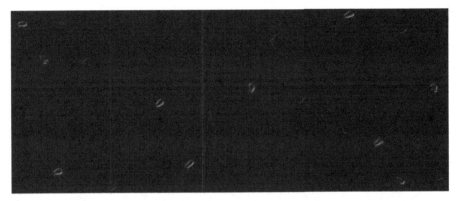

Fig. 11. Color edge detection and highlighter with correlated Sobel Filter (Color figure online)

5.1 Potential Improvement with Few-Shot Algorithm

While the QDFT method provides a robust framework for image segmentation and edge detection, there is potential for further enhancement using Few-Shot Learning algorithms. Few-Shot Learning (FSL) is particularly beneficial in scenarios where the available data samples are limited, as it enables the model to generalize from a small number of training examples. In the context of agricultural monitoring, especially in field conditions such as those in Zacatecas, where the number of samples may be fewer than expected due to environmental variability or limited pest incidence, FSL can be a game-changer. Implementing a Few-Shot Learning algorithm could significantly enhance the model's ability to accurately detect and classify pests even with limited data. By leveraging the existing QDFT framework, the FSL algorithm can be trained on a minimal set of annotated images, learning to recognize patterns and features that are crucial for identifying pests like whiteflies. This approach could lead to more precise detection, reducing the need for extensive data collection and annotation, which can be time-consuming and resource-intensive. Moreover, the integration of Few-Shot Learning could improve the adaptability of the system across different regions and pest types. For instance, if a specific type of pest is prevalent in Zacatecas but not in other regions, the model can quickly adapt to this new context with minimal additional training. This flexibility is particularly valuable in agricultural environments, where pest populations can fluctuate rapidly, and the ability to respond to these changes with high accuracy is crucial for effective crop management. In addition, the use of QDFT for image segmentation and contour detection provides a strong foundation for pest detection in agricultural settings. However, by incorporating Few-Shot Learning, the system can be further refined to handle cases with limited data, making it more versatile and effective in real-world applications, especially in regions like Zacatecas where sample sizes might be constrained. This combined approach holds the potential to significantly advance the state of the art in precision agriculture, offering more reliable and scalable solutions for pest management.

6 Conclusions

Findings, quaternion-based color segmentation enables the acquisition of new information, broaden the possibility of extracting the details of the features of the region of interest by highlighting the contours of the objects to analyze the information resulting from the quaternion-based color segmentation process, allows us to work with a new dimension space in a single four-dimensional image matrix with RGB color components and grayscale and opens the possibility of selecting other image processing alternatives to then implement object recognition processes, working with the application of autonomous vision and image analysis systems, this represents a promising perspective in the field of crop pest detection, offering the possibility to improve the efficiency and performance of traditional monitoring methods for pest detection, recognition and classification. Such advances in these areas of machine vision, robotics, and artificial intelligence provide new opportunities for the development of more accurate and automated systems for Integrated Pest Management systems. The integration of Few Shot Algorithm techniques into the domain of agricultural pest detection presents a transformative advancement for enhancing the precision and efficiency of identifying plant diseases caused by pests. The utilization of Few Shot Learning (FSL) models offers significant improvements in how we approach and analyze pest-related data, particularly in scenarios where labeled training data is scarce or difficult to obtain.

Enhanced Accuracy and Precision in Pest Detection: Few Shot Learning algorithms have demonstrated remarkable capabilities in learning from a limited number of examples. In the context of agricultural pest detection, this means that even with a small dataset of annotated images or instances of plant diseases, the system can generalize and make accurate predictions about new, unseen cases. This is especially valuable in pest detection, where the diversity of pests and symptoms can lead to complex and varied manifestations of plant diseases. By leveraging FSL, we can achieve higher accuracy and precision in identifying specific pest-related indicators, thereby improving the overall effectiveness of pest management strategies.

Reduction in the Need for Extensive Data Collection: Traditional methods of training machine learning models for pest detection often require large volumes of labeled data to achieve reliable results. However, Few Shot Learning alleviates this requirement by enabling models to learn and adapt from only a few examples. This reduction in data collection efforts not only saves time and resources but also addresses challenges related to the availability of high-quality annotated datasets. As a result, agricultural practitioners can implement advanced pest detection systems without the need for exhaustive data-gathering processes.

In addition, Few Shot Learning represents a significant advancement in the field of agricultural pest detection, offering improved accuracy, reduced data requirements, and enhanced object recognition. Its integration with advanced vision systems and image processing techniques paves the way for more effective and efficient pest management solutions. As research and development in this area continue to progress, we can expect to see further innovations that will revolutionize pest detection and contribute to more sustainable and productive agricultural practices.

6.1 Future Research Directions

1. **Advancement of Few Shot Learning Algorithms**: While current FSL algorithms have shown promise, there is a need for continued improvement in their robustness and performance. Future research should focus on:

   - *Algorithm Optimization*: Developing more advanced FSL algorithms that can handle diverse and complex pest detection scenarios. This includes improving algorithms to better generalize from limited data and enhance their adaptability to various pest species and plant diseases.
   - *Hybrid Models*: Investigating hybrid approaches that combine FSL with other machine learning techniques, such as transfer learning or self-supervised learning. These models could leverage additional sources of information or pre-trained models to improve performance in scenarios with very few annotated examples.
   - *Model Interpretability*: Enhancing the interpretability of FSL models to provide more insights into the decision-making process. This could involve developing methods to visualize and understand the features that FSL models use for classification and detection.

2. **Integration with Multi-modal Data:** The integration of FSL with multimodal data sources can significantly enhance pest detection systems. Future research should explore:

   - *Combining Visual and Non-Visual Data*: Integrating FSL with data from various sensors, such as thermal cameras, hyperspectral imaging, and environmental sensors. This multi-modal approach can provide a more comprehensive view of pest-related conditions and improve detection accuracy.
   - *Contextual Information*: Incorporating contextual information, such as weather conditions, soil health, and crop variety, into FSL models. This can help in understanding the environmental factors affecting pest prevalence and developing more precise detection and management strategies. In addition, the future research agenda for Few Shot Learning in agricultural pest detection encompasses a range of areas, from algorithm development and multi-modal integration to real-world implementation and ethical considerations. By addressing these research directions, we can enhance the effectiveness and applicability of FSL-based systems, ultimately leading to more efficient and sustainable pest management practices in agriculture.

Acknowledgment. The authors would like to thank the National Council for the Humanities Science and Technology (CONAHCYT), Autonomous University of Zacatecas (UAZ) and the Zacatecas State Technological University (UTZAC) for their support in the development of this research.

References

1. Boissard, P., Martin, V., Moisan, S.: A cognitive vision approach to early pest detection in greenhouse crops. Comput. Electron. Agric. **62**(2), 81–93 (2008). https://doi.org/10.1016/j.compag.2007.11.009
2. Furtick, W.R.: Implementing pest management programs: an international perspective. In: Apple, J.L., Smith, R.F. (eds.) Integrated Pest Management, pp. 29–38. Springer, Boston (1976). https://doi.org/10.1007/978-1-4615-7269-5_3
3. Glass, E.H.: Pest management: principles and philosophy. In: Apple, J.L., Smith, R.F. (eds.) Integrated Pest Management, pp. 39–50. Springer, Boston (1976). https://doi.org/10.1007/978-1-4615-7269-5_4
4. Ehler, L.E.: Integrated pest management (IPM): definition, historical development and implementation, and the other IPM. Pest Manag. Sci. **62**(9), 787–789 (2006). https://doi.org/10.1002/ps.1247
5. Talaviya, T., Shah, D., Patel, N., Yagnik, H., Shah, M.: Implementation of artificial intelligence in agriculture for optimization of irrigation and application of pesticides and herbicides. Artif. Intell. Agric. **4**, 58–73 (2020). https://doi.org/10.1016/j.aiia.2020.04.002
6. Kapach, K., Barnea, E., Mairon, R., Edan, Y., Ben-Shahar, O.: Computer vision for fruit harvesting robots – state of the art and challenges ahead. Int. J. Comput. Vis. Robot. **3**(1–2), 4–34 (2012). https://doi.org/10.1504/IJCVR.2012.046419
7. Dara, S.K.: The new integrated pest management paradigm for the modern age. J. Integr. Pest Manag. **10**(1), 12 (2019). https://doi.org/10.1093/jipm/pmz010
8. Di Martini, D.R., et al.: Machine learning applied to UAV imagery in precision agriculture and forest monitoring in Brazililian Savanah. In: 2019 IEEE International Geoscience and Remote Sensing Symposium, IGARSS 2019, pp. 9364–9367 (2019). https://doi.org/10.1109/IGARSS.2019.8900246
9. Bossu, J., Gée, Ch., Jones, G., Truchetet, F.: Wavelet transform to discriminate between crop and weed in perspective agronomic images. Comput. Electron. Agric. **65**(1), 133–143 (2009). https://doi.org/10.1016/j.compag.2008.08.004
10. Pereira, C.S., Morais, R., Reis, M.J.C.S.: Recent advances in image processing techniques for automated harvesting purposes: a review. In: 2017 Intelligent Systems Conference (IntelliSys), pp. 566–575 (2017). https://doi.org/10.1109/IntelliSys.2017.8324352
11. Muppala, C., Guruviah, V.: Machine vision detection of pests, diseases, and weeds: a review. J. Phytol. 9–19 (2020). https://doi.org/10.25081/jp.2020.v12.6145
12. Chelladurai, V., Karuppiah, K., Jayas, D.S., Fields, P.G., White, N.D.G.: Detection of callosobruchus maculatus (F.) infestation in soybean using soft X-ray and NIR hyperspectral imaging techniques. J. Stored Prod. Res. **57**, 43–48 (2014). https://doi.org/10.1016/j.jspr.2013.12.005
13. Khan, S., Tufail, M., Khan, M.T., Khan, Z.A., Iqbal, J., Alam, M.: A novel semisupervised framework for UAV based crop/weed classification. Plos One **16**(5), e0251008 (2021). https://doi.org/10.1371/journal.pone.0251008
14. Rodríguez, L.A.R., Castañeda-Miranda, C.L., Lució, M.M., Solís-Sánchez, L.O., Castañeda-Miranda, R.: Quarternion color image processing as an alternative to classical grayscale conversion approaches for pest detection using yellow sticky traps. Math. Comput. Simul. **182**, 646–660 (2021). https://doi.org/10.1016/j.matcom.2020.11.022

15. Moreno-Lucio, M., et al.: Extraction of pest insect characteristics present in a mirasol pepper (Capsicum annuum L.) crop by digital image processing. Appl. Sci. **11**(23), Article no 23 (2021). https://doi.org/10.3390/app112311166
16. Rustia, D.J.A., et al.: Automatic greenhouse insect pest detection and recognition based on a cascaded deep learning classification method. J. Appl. Entomol. **145**(3), 206–222 (2021). https://doi.org/10.1111/jen.12834
17. Solis-Sánchez, L.O., et al.: Scale invariant feature approach for insect monitoring. Comput. Electron. Agric. **75**(1), 92–99 (2011). https://doi.org/10.1016/j.compag.2010.10.001
18. "Agriculture | Free Full-Text | Codling Moth Monitoring with Camera-Equipped Automated Traps: A Review. https://www.mdpi.com/2077-0472/12/10/1721. accedido 18 de junio de 2023
19. Grigoryan, A.M., Agaian, S.S., Panetta, K.: Quantum-inspired edge detection algorithms implementation using new dynamic visual data representation and short-length convolution computation. arXiv, 31 de octubre de 2022. https://doi.org/10.48550/arXiv.2210.17490
20. Hamilton, W.R.: Elements of Quaternions. Longmans, Green, & Company (1866)
21. Chen, B., Coatrieux, G., Chen, G., Sun, X., Coatrieux, J.L., Shu, H.: Full 4-D quaternion discrete Fourier transform based watermarking for color images. Digit. Signal Process. **28**, 106–119 (2014). https://doi.org/10.1016/j.dsp.2014.02.010
22. Ye, H.-S., Zhou, N.-R., Gong, L.-H.: Multi-image compression-encryption scheme based on quaternion discrete fractional Hartley transform and improved pixel adaptive diffusion. Signal Process. **175**, 107652 (2020). https://doi.org/10.1016/j.sigpro.2020.107652
23. Pei, S.-C., Chang, J.-H., Ding, J.-J.: Quaternion matrix singular value decomposition and its applications for color image processing. In: Proceedings 2003 International Conference on Image Processing (Cat. No.03CH37429), p. I-805 (2003). https://doi.org/10.1109/ICIP.2003.1247084
24. Wang, X., Wang, C., Yang, H., Niu, P.: A robust blind color image watermarking in quaternion Fourier transform domain. J. Syst. Softw. **86**(2), 255–277 (2013). https://doi.org/10.1016/j.jss.2012.08.015
25. Ell, T.A., Sangwine, S.J.: Hypercomplex Fourier transforms of color images. IEEE Trans. Image Process. **16**(1), 22–35 (2007). https://doi.org/10.1109/TIP.2006.884955
26. Zhang, C., Yang, W., Liu, Z., Li, D., Chen, Y., Li, Z.: Color image segmentation in RGB color space based on color saliency. In: Li, D., Chen, Y. (eds.) Computer and Computing Technologies in Agriculture VII. IFIP Advances in Information and Communication Technology, pp. 348–357. Springer, Heidelberg (2014). https://doi.org/10.1007/978-3-642-54344-9_41
27. Kumar, R., Kumar Bhandari, A.: Luminosity and contrast enhancement of retinal vessel images using weighted average histogram. Biomed. Signal Process. Control **71**, 103089 (2022). https://doi.org/10.1016/j.bspc.2021.103089
28. Nieuwenhuizen, A.T., et al.: Raw data from Yellow Sticky Traps with insects for training of deep learning Convolutional Neural Network for object detection. Wageningen University & Research (2019). https://doi.org/10.4121/uuid:8b8ba63a-1010-4de7-a7fb-6f9e3baf128e
29. Deserno, M., Briassouli, A.: Faster R-CNN and efficientnet for accurate insect identification in a relabeled yellow sticky traps dataset. In: 2021 IEEE International Workshop on Metrology for Agriculture and Forestry (MetroAgriFor), pp. 209–214 (2021). https://doi.org/10.1109/MetroAgriFor52389.2021.9628708
30. Liu, S., Chen, W., Zhang, Z.: Few-shot learning for plant disease detection in agriculture: a review. Comput. Electron. Agric. **197**, 106902 (2022)
31. Wang, H., Zhang, C., Zhang, C.: Few-shot learning with convolutional neural networks for plant disease recognition. Int. J. Comput. Vision **129**(7), 1856–1872 (2021)
32. Hsu, Y., Wang, Y., Li, Y.: Application of few-shot learning for pest detection in crops using deep learning techniques. J. Agric. Food Chem. **71**(2), 456–464 (2023)

Optimization of a Treatment Plant Through the Incorporation of New Waste Separation Components: A TOPSIS-Focused Multi-criteria Analysis Approach

Carlos Iván Ramón Diego[1], José Ismael Ojeda Campaña[2(✉)],
Virginia Berenice Niebla Zatarain[2], and Alberto Ochoa-Zezzatti[3]

[1] Centro de Alta Dirección en Ingeniería y Tecnologías, Universidad Anáhuac México, Naucalpan, Mexico
[2] Tecnológico Nacional de México/ITES de Los Cabos, Cabo San Lucas, Mexico
JIsmael.OC@loscabos.tecnm.mx
[3] Autonomous University of Ciudad Juarez, Ciudad, Mexico

Abstract. Efficient waste management is a transcendental challenge in the contemporary world, where waste treatment plants are essential actors in this process. In this paper, the optimization of a treatment plant through the strategic integration of innovative waste separation components is addressed in detail. This approach relies on a rigorous multi-criteria analysis and the application of the TOPSIS methodology to provide a comprehensive assessment. The research focuses on evaluating the effectiveness of these tools to enhance operational efficiency, mitigate adverse environmental impacts, and simultaneously maximize recovery of valuable resources. From exploring new perspectives to implementing practical solutions, this study seeks to contribute to the forefront of waste management, supporting informed and sustainable decision making in the design and operation of treatment plants.

Keywords: Waste Management · Treatment Plants · Multicriteria Analysis · TOPSIS · Optimization · Operational Efficiency · Environmental Sustainability · Resource Recovery

1 Introduction

Sustainable waste management is essential to preserve the environment and natural resources. In this context, the optimization of waste treatment plants becomes a priority to effectively address current challenges. The incorporation of new waste separation components presents itself as a promising strategy to improve the efficiency and sustainability of these facilities.

Today, water stress and scarcity of natural resources due to population growth and climate change require efforts to reduce the impact to the environment (Tchobanoglous, G., 2019). Humans today have created large volumes of polluted water commonly referred to as wastewater, which may contain domestic, commercial, industrial and agricultural wastes, or even a mixture of these (Hoyland G., 2013).

Many industrial sectors generate wastewater that, in excessive quantities, can cause serious environmental damage (Charazinska S., et al., 2019). Industrial wastewater is aqueous waste containing various toxic and hazardous compounds dissolved or suspended in water that when discharged directly can have negative effects on the environment (Başak, S., et al., 2023). The most commonly reported compounds may include nitrogen, ammonium, chlorides, sulfates and various types of heavy metals (Moody C.M., et al., 2017).

The various organic, inorganic, chemical and microbial components of wastewater fluids must be treated before they are discharged into the environment, for this purpose and in order to comply with government regulations on environmental protection as well as to improve efficiency, sustainability and ensure a safe working environment, wastewater treatment plants (WWTPs) play a critical role in protecting public health, protecting the environment and sustainable management of water resources (Gafri, O., 2021). Their importance in improving the quality of life and well-being of communities and in promoting a cleaner and healthier environment is undeniable (Başak, S., et al., 2023).

In the particular case of industries, significant amounts of wastewater are generated in the course of their operations and there is evidence that water used in manufacturing typically ends up almost entirely as wastewater, often with little or no treatment (Garrone, P., et al., 2018). However, some industries with the aim of reducing environmental impact employ treatment plants to remove harmful materials and pathogens found in wastewater fluids. (Gafri, O., 2021).

2 Methodology

This study aims to examine the efficacy of the technique known as Technique for Order of Preference by Similarity to Ideal Solution (TOPSIS) in a wastewater treatment plant. The methodology developed to achieve this goal involves the strategic integration of waste separation components, allowing for a comprehensive and holistic assessment to be conducted. This section outlines the methodology in detail.

2.1 Context and Current Challenges in Treatment Plants

The increasing amount of waste generated and the complexity of the materials present in waste streams have challenged the capacity of traditional treatment plants. The need for more efficient management becomes evident, considering the environmental impacts and the growing demand for resource recovery.

Water reuse in industry is an important practice that seeks to conserve and manage water resources in a sustainable manner. This involves taking measures to treat and recycle water used in industrial processes, thereby reducing the demand for clean water and minimizing the discharge of contaminated wastewater into the environment. Water reuse in industry is a crucial part of corporate responsibility and environmental sustainability. It helps to conserve a scarce resource and reduce the environmental impact of industrial activity, while bringing economic benefits to companies that effectively implement it. Reuse of treated wastewater is one of the most promising water management strategies, serving the dual purpose of increasing water resources and reducing the environmental impacts of untreated wastewater disposal. (Pereira, et al., 2002; Massoud, et al., 2018).

Industrial wastewater reuse has gained popularity in recent years worldwide. This may be due to many factors, such as groundwater depletion, high water abstraction costs, etc. Many countries have shifted the focus towards reuse of treated wastewater in water-intensive industries, mainly textiles, pulp and paper, leather, tanneries, thermal industries. Power plants, etc. (Lautze, J., et al., 2014). Typical industrial applications include the reuse of treated water as feed water in boilers, cooling towers and as process water for washing purposes, depending on the type of industry. In the United States and Mexico there are successful cases of treated water reuse as boiler feed water in thermal power plants. Due to the high-water consumption by industries, the use of treated wastewater for industrial applications is one of the most sustainable options, which can be further strengthened with strict water quality guidelines.

2.2 Multicriteria Analysis in the Optimization of Treatment Plants

Multi-criteria analysis provides a robust methodology for evaluating and comparing different optimization approaches. By considering criteria such as operational efficiency, economic feasibility and environmental benefits, informed decisions can be made on the integration of new separation components.

2.3 TOPSIS Methodology

The Technique for Order Preference by Similarity to Ideal Solution (TOPSIS) is presented as a valuable tool in multi-criteria decision making. By weighting criteria and calculating distances to ideal and non-ideal solutions, TOPSIS identifies the best optimization option for the treatment plant.

The TOPSIS method is performed using the multi-criteria point of view, based on the ranking according to the preferences assigned to each alternative, based on the criteria (Garcia, J., et al., 2006).

TOPSIS is a technique that considers the distance to the ideal solution and the distance to the anti-ideal solution. Mathematical algorithms help decision makers and engineers to compare and rank a set of alternative decisions (Schnitzler, H., et al., 2001). In these methods, the ranking of alternatives is based on real situations that give cost-benefit solutions of great importance to a company (Ochoa, A., et al., 2017).

Both ideal and anti-ideal solutions are fictitious solutions. The ideal solution is a solution for which all attribute values correspond to the optimal values of each attribute contained in the alternatives; the anti-ideal solution is the solution for which all attribute values correspond to the least desired values of each attribute contained in the alternatives. In this way TOPSIS provides a solution that is not only the closest to a hypothetically best solution, but also the farthest from the hypothetically worst. (Real, V., et al., 2011). The step of the TOPSIS method is shown in Fig. 1.

The TOPSIS (Technique for Order of Preference by Similarity to Ideal Solution) analysis method offers a number of advantages, including the following:

- Evaluation based on multiple criteria: TOPSIS is capable of evaluating alternatives based on multiple criteria, thus rendering it particularly useful in situations where a multitude of factors must be considered in decision-making processes.
- Incorporation of decision-maker preferences: The relative importance of different criteria can be defined by the decision maker, allowing the reflection of their specific preferences and priorities.

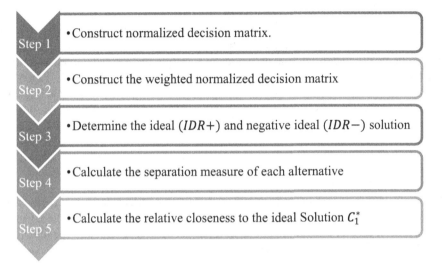

Fig. 1. The step of the TOPSIS method

In order to evaluate the m alternatives $(A_i, i = 1, 2, \ldots, m)$ based on the criteria $(C_j, j = 1, 2, \ldots, n)$, a Multi-Criteria Decision-Making matrix (MCDM) is constructed as shown in Table 1.

The notation x_{ij} is used to represent the valuation of alternative A_i with respect to criterion C_j and $W = [w_1, w_2, \ldots, w_n]$ is the vector of the weights that are associated with the criteria.

Table 1. Multi-Criteria Decision-Making matrix (MCDM).

Alternatives	Criteria (Wights	C_1 w_1	C_2 w_2	C_3 w_3	...	C_n w_n)
A_1		x_{11}	x_{12}	x_{13}	...	x_{1n}
A_2		x_{21}	x_{22}	x_{23}	...	x_{2n}
A_3		x_{31}	x_{32}	x_{33}		x_{3n}
$\vdots$						
A_n		x_{m1}	x_{m2}	x_{m3}	...	x_{mn}

The first TOPSIS step is to build the normalized decision matrix (R), which allows comparisons across criteria, as shown in Eq. 1.

$$r_{ij} = \frac{x_y}{\sqrt{\sum_{i=1}^{m} x_{ij}^2}} \; for \; i = 1, 2, 3, \ldots, m \; and \; j = 1, 2, 3, \ldots, n$$

2.3.1 Saturation Tests

When the MEA solution reaches its saturation point, it loses the ability to absorb more gas molecules, indicating that it has reached its maximum retention capacity. Knowing the saturation time is essential to optimize the adsorption process and to ensure a flow rate for maximum MEA adsorption capacity. During the experimental tests, two MEA concentrations, one at 10% and the other at 15%, were evaluated to analyze their influence on the adsorption capacity and the saturation time. The gas flow is continuous at the CO_2 concentration set under the operating conditions, while the liquid to be saturated is a quantity of sorbent solution loaded in the sorption column.

2.3.2 Continuous Testing

The continuous tests performed with 10% and 15% w/w MEA solutions were designed to operate the sorption column continuously to achieve a steady state condition. Consequently, achieving and maintaining a controlled steady state guarantees the maximum sorption capacity of the sorbent solution.

2.4 Software Development

A tool was developed to assist the user in interpreting and making decisions based on temperatures acquired through an Arduino circuit by automating the calculations and generating the graphs in real time, the waiting time was reduced, allowing early detection of potential errors in the experimental test. Table 2 describes the applications and programming language used in the development of this software. The choice of these tools is essential to ensure the efficiency and functionality of the data processed by the software.

Table 2. Description of the tools used and programming language for software development.

Tools	Description	Version
Programming Language	Python	3.8.10
Pandas Library	Library for handling and analyzing tabular data	1.3.0
Matplotlib Library	Library for generating charts and visualizations	3.4.2
User Interface	QT5 Designer	5
Arduino IDE	Integrated Development Environment for programming Arduino	1.8.13
Excel	Microsoft spreadsheet application for data export	2019, 365, etc.

2.4.1 Electrical Diagram

With the aim of increasing the efficiency of manual processes related to data acquisition and processing, the design of the electrical circuit includes an Arduino Uno connected to four DS18B20 temperature sensors, a 4.7 kΩ providing the correct amount of pull-up current for 1-wire system, and HC-05 Bluetooth module. This setup allows communication and data transfer from the sensors to the Arduino. The component layout of the circuit is shown in Fig. 3, with this configuration a more efficient and accurate temperature acquisition is achieved.

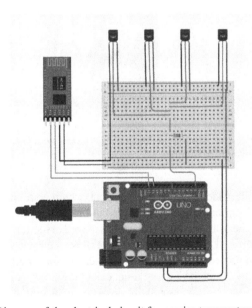

Fig. 2. Diagram of the electrical circuit for precise temperature acquisition.

2.4.2 Acquisition Code

The Arduino code was developed using the "DallasTemperature" library with DS18B20 sensors to acquire temperature data and send temperature readings to a mobile device via a Bluetooth module, allowing their visualization through a serial application that receives data from the Bluetooth connection.

2.4.3 Acquisition Interface of the Absorption System

Figure 5 shows the main interface used for the absorption process experiments, with data acquired from each sensor in the absorption column. The interface is divided into specific sections: in section (a) is where the Arduino port is configured to acquire data, the start button to begin temperature acquisition and allows to set the configuration and duration of the experiment is (b), on (c) has a tooltip with information about process experiments. Section (d) is responsible for calculating the saturation time based on the parameters entered in the form. The time of each test is displayed in section (e), and the acquired temperatures during the specified time in seconds are visualized in section (f).

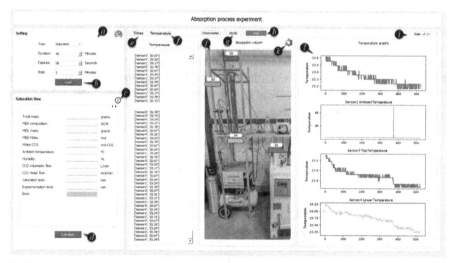

Fig. 3. Main GUI Interface for Absorption Process Experiments with Data Acquired from Each Sensor Located in the Absorption Column.

The interface also includes a timer in section (g), which continues to run even after data acquisition has been completed. (h) it's a button to stop the timer and thus the ongoing experiment. The actual computer date is displayed in section (i), providing a time stamp of the tests performed. Section (j) includes an illustrative picture of the absorption column, with colors representing the location of each sensor. The button (k) generates a graph representing the indicated capture time, calculating the average of the acquired temperatures. Finally, section (l) graphically displays the acquired temperatures, with each sensor represented by a different color for easy identification.

Figure 6 shows a graphical user interface (GUI) displaying the average temperatures over the waiting time, based on data obtained from the four sensors.

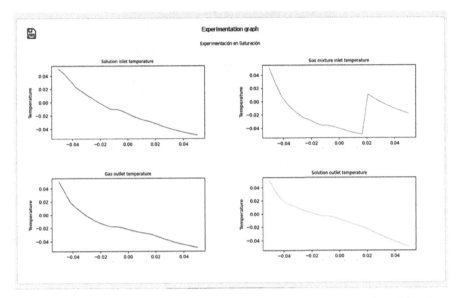

Fig. 4. Graphical User Interface (GUI) showing average temperatures over the waiting time.

2.4.4 Data Acquisition System Calibration

Accurate calibration of temperature sensors is considered a critical aspect in obtaining reliable data. To achieve accurate measurements, a comprehensive calibration process was implemented that allowed the adjustment of sensor readings using a thermal bath regulated and controlled by a cryostat.

The temperature displayed on the cryostat was verified using a reference RTD thermometer. During this procedure, the use of a hygrometer was considered essential to measure ambient temperature and humidity, to perform the thermal bath, it was necessary to define a temperature range in which the sensors would operate within the absorption column.

In this case, the range defined was from 0 °C to 65 °C, this range determines the intervals at which temperature measurements are taken for subsequent averaging and curve fitting. Figure 7 shows the sensors placed in the cryostat to perform the thermal bath and a reference RTD thermometer is used to verify the temperature and ensure that the tests are performed within the specified temperatures.

A comparative calibration approach was implemented to establish a direct correspondence between the sensor measurements and the reference values. This approach allowed the application of an individual fitting curve for each sensor, as shown in Fig. 8.

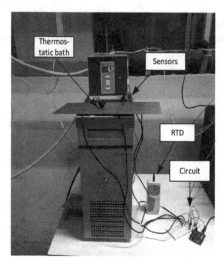

Fig. 5. Configuration of the thermal bath and verification of the cryostat temperature using a reference sensor before starting the temperature sensor tests.

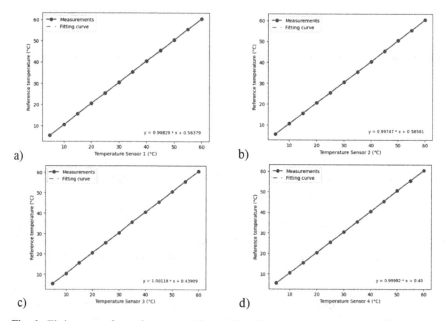

Fig. 6. Fitting curve for each sensor: a) Sensor 1, b) Sensor 2, c) Sensor 3, and d) Sensor 4.

Figure 9 shows the visual interface of the software developed to acquire and adjust the temperatures of the sensors in the Arduino circuit. The interface includes the configuration of data acquisition (a), specific tests for each sensor with visualization of averages and errors (b), (c) display of overall averages (d), the option to configure the Arduino port (e), count readings (f), start and stop temperature acquisition (g), (h), and individual temperature acquisition for each sensor (i), (j), (k), (l). It also includes functions for redirecting (m) and clearing the configuration of performed tests (n).

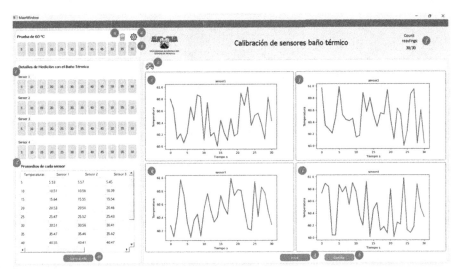

Fig. 7. Finalized main GUI with the indicated tests for each temperature sensor.

3 Experimentations and Results

Through this project's development, data acquisition from temperature sensors situated in the absorption column was optimized utilizing software created as a tool. The software allows the user to identify experimentation success in real-time and analyze temperature behavior based on the results gathered.

The primary goal is to effectively capture CO_2 while undergoing the absorption process, utilizing Monoethanolamine (MEA) as the absorbent. The study accounts for various controllable variables, including solution flow, MEA concentration, and packed section height, as well as uncontrollable variables, such as solution and gas inlet temperature. This resulted in the collection of temperature measurements for both the solution and gas outlet across multiple experimental conditions. These conditions encompassed saturation assessments and ongoing experiments.

3.1 Temperature

The paper presents absorption test results under steady-state and saturation conditions, examining the temperature behavior of gas outlet and solution under different experimental settings.

The software-collected data shows the results of experimental tests conducted at 10% MEA in both continuous and saturation stages. The saturation test indicates when the system has reached its maximum gas absorption capacity, while the continuous test involves a temperature increase to reach a stable equilibrium point. However, equilibrium cannot be fully observed due to the short duration of the experimental test. Refer to Fig. 10 for an illustration of this trend.

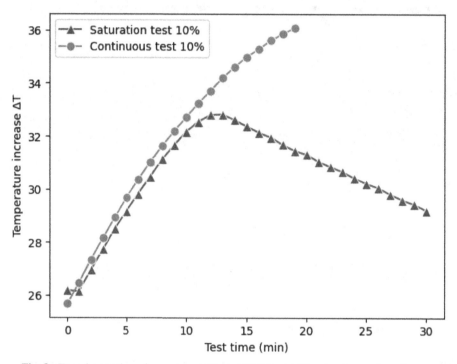

Fig. 8. Experimentation of experimental tests of MEA at 10% saturation and continuous.

An additional experiment was conducted using a 15% MEA solution. The obtained temperatures during the experiment are displayed in Fig. 11, illustrating the establishment of a steady state once equilibrium was reached. Subsequently, the continuous state indicated temperature stabilization, due to ongoing adsorption of MEA molecules. This explains the temperature maintenance. In saturation experiments, the temperature rises initially, reaches a peak, and then declines.

3.2 Data Analysis

In order to identify correlations between variables, an exploratory analysis was performed on the experimental data. Figure 12 illustrates a gradual temperature increase over time in Solution Outlet Temperature, Gas Mixture Inlet Temperature, and Gas Outlet Temperature.

Furthermore, an increasing Solution Outlet Temperature was found to be correlated with the above-mentioned temperatures. However, an inverse relationship was noted

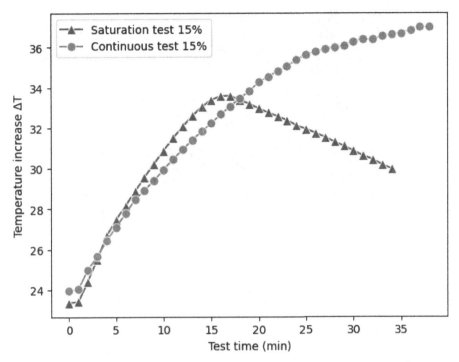

Fig. 9. Experimentation of experimental tests of MEA at 15% saturation and continuous.

between Solution outlet temperature and solution inlet temperature, indicating that as one increased, the other decreased.

In Fig. 13, the relationships are more pronounced, with CO2 gas temperature increasing over time, pointing to a faster absorption rate. Furthermore, a stronger negative correlation was observed between the concentrated solution temperature and the solution temperature, signifying higher sensitivity at greater concentrations.

Concentrating the MEA solution was found to be a more effective approach. The data shows a positive relationship between time and CO_2 gas temperature, indicating faster absorption at higher concentrations. Furthermore, the negative correlation observed between the temperature of the concentrated solution and the solution suggests more efficient heat transfer from the gas to the solution. Additionally, the text will follow conventional academic structure and format, with clear and objective language and precise word choice.

In both continuous experiments utilizing 10% and 15% MEA solutions, a significant negative correlation was discovered between time and the temperatures of the variables being analyzed. This finding suggests a trend toward stabilization or reduction in temperatures as the experiment progressed.

Figure 14 displays important negative correlations between the concentrated solution's temperature, the ambient and gas temperatures, illustrating the substance's noteworthy influence on the surrounding temperatures. Notably, we observed strong and moderate negative correlations.

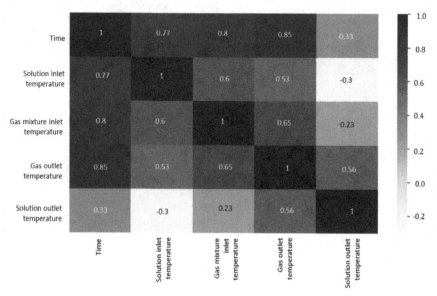

Fig. 10. Correlation of Variables in 10% MEA Saturation Test

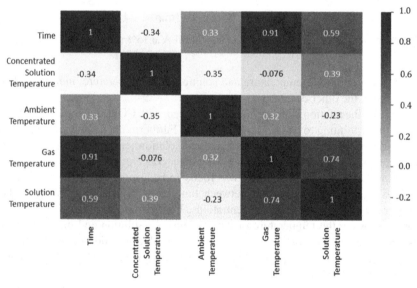

Fig. 11. Correlation of Variables in 15% MEA Saturation Test

In Fig. 15, the trend of temperature stabilization over time persisted. Finally, there were strong negative correlations observed between 'gas outlet temperatures', and 'solution outlet temperatures', indicating a simultaneous decrease in these temperatures at higher concentrations.

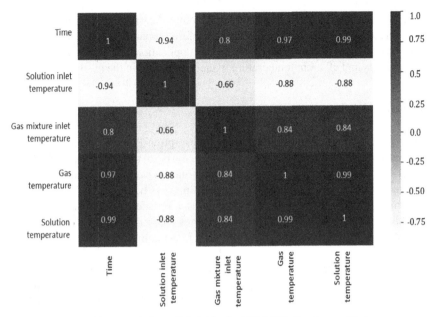

Fig. 12. Correlation of Variables in 10% MEA Continuous Test

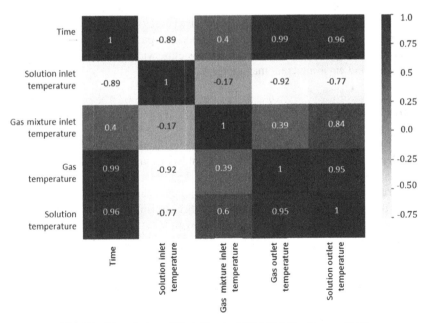

Fig. 13. "Correlation of Variables in 15% MEA Continuous Test"

Both experiments emphasize the importance of maintaining a controlled steady state to maximize efficiency in the continuous absorption of CO_2 with MEA solutions. The higher concentration (15%) exhibited more pronounced correlations, suggesting greater thermal sensitivity and more efficient heat transfer compared to the 10%. These findings are crucial for industrial and environmental applications, ensuring an optimal CO_2 capture process and efficient temperature management within the absorption system.

4 Future Work

Integration of Artificial Intelligence for CO_2 Absorption Process Optimization, based on the results obtained from the exploratory analysis of saturation experiments with 10% and 15% MEA solutions, alongside the continuous state, significant opportunities to enhance the efficiency of the CO_2 absorption process have been identified. Despite identifying crucial correlations between variables, a broader dataset is required to pinpoint additional key points. So far, the analysis has focused on linear relationships and observable patterns from existing data. However, the complexity of the CO_2 absorption process and the dynamic interaction of various variables necessitate more sophisticated approaches, such as constructing Machine Learning models, neural networks, deep learning algorithms, or support vector machine algorithms capable of capturing complex patterns and nonlinear behaviors in the data. These models could predict system behavior under various conditions, enabling real-time adaptation to maximize CO_2 absorption efficiency.

Optimization algorithms can explore a vast and complex parameter space far beyond the capabilities of traditional exploratory analysis. These tools will not only enhance the understanding of the CO_2 absorption process but also pave the way for more efficient and environmentally friendly carbon capture systems, with significant implications for sustainability and environmental management.

5 Conclusions

The experimental process was optimized through real-time data collection and analysis. The implementation of specialized software enabled accurate data collection and graph generation, showing the error margins and precision of each sensor.

Calibrated temperature sensors integrated into the packed absorption column and a graphical interface for data acquisition enabled the acquisition of temperatures across all 4 process streams. Experiments were conducted with MEA contractions at 10% and 15% by weight to mitigate the corrosion issues that arise with more concentrated solutions. Temperatures were acquired for MEA saturation tests and continuous tests using the developed software.

The software's ability to rapidly provide results and facilitate data-driven decision-making significantly contributed to enhancing the CO_2 capture process. Through the collected data's exploratory analysis, we identified correlations between variables and their association with temperature. Additional experiments are necessary to validate the observations; nonetheless, the results provided will facilitate future implementations of Artificial Intelligence techniques. These techniques offer a more advanced and flexible approach to optimizing the CO_2 capture process.

References

Intergovernmental Panel on Climate Change (IPCC). (2023, marzo 20). Urgent climate action can secure a liveable future for all. Switzerland. Retrieved julio 7, 2023, from https://www.ipcc.ch/2023/03/20/press-release-ar6-synthesis-report/

Alper, E., Orhan, O.Y.: Utilización de CO_2: Desarrollos en procesos de conversión. Petróleo **3**(1), 109–126 (2017)

Álvarez Pelegry, E.: La captura y almacenamiento de CO_2: una solución eficiente para luchar contra el cambio climático. Real Instituto Elcano, 1–21. Retrieved Julio 12, 2023 (2010).https://www.realinstitutoelcano.org/documento-de-trabajo/la-captura-y-almacenamiento-de-co2-una-solucion-eficiente-para-luchar-contra-el-cambio-climatico/

Arango, M.: Model risk assessment projects in thermal power generation. Revistas Espacios, **37**(9), 26 (2016). https://www.revistaespacios.com/a16v37n09/16370926.html

Ayala Blanco, E., Martínez Ortega, F.: Tecnologías de captura de CO en procesos de postcombustión de gas natural. MET&FLU (14), 22–33 (2019). https://www.cdtdegas.com/images/Descargas/Nuestra_revista/MetFlu14/7_CapturaCO2.pdf

Chao, C., Deng, Y., Dewil, R., Baeyens, J., Fan, X.: Post-combustion carbon capture. Renew. Sustain. Energy Rev. **138**(C), 110–490 (2021). https://doi.org/10.1016/j.rser.2020.110490

Chauvy, R., Meunier, N., Thomas, D., De Weireld, G.: Selección de productos emergentes de utilización de CO_2 para implementación a corto y mediano plazo. Energía aplicada(236), 662–680 (2019)

Díaz Cordero, G.: El Cambio Climático. Ciencia y sociedad, XXXVII (2), 227–240 (2012). http://www.redalyc.org/articulo.oa?id=87024179004

Environment LIFE Programme. (n.d.). LIFE-CO2-INT-BIO - CO2 emissions reduction by industrial integration and value chains creation. European Commission, Área prioritaria de Mitigación al Cambio Climático. Madrid, España: Ministerio para la Transición Ecológica y el Reto Demográfico. Retrieved Julio 12, 2023, fromhttp://www.lifeco2intbio.eu/images/docs/Captura_CO2.pdf

Iberdrola. (2020). Países más afectados por el cambio climático. Retrieved Julio 8, 2023, from Iberdrola Web site:https://www.iberdrola.com/sostenibilidad/paises-mas-afectados-cambio-climatico

Intergovernmental Panel on Climate Change (IPCC). (2023). AR6 Synthesis Report Climate Change 2023. IPCC. Retrieved Julio 8, 2023, fromhttps://www.ipcc.ch/report/ar6/syr/

Intergovernmental Panel on Climate Change (IPCC). Climate Change 2023: Synthesis Report. Contribution of Working Groups I, II and III to the Sixth. Geneva, Switzerland: H. Lee and J. Romero (2023). https://doi.org/10.59327/IPCC/AR6-9789291691647

Laguna Monroy, I.: La generación de energía eléctrica y el ambiente. Gaceta Ecológica(65), 53–62 (2002). Retrieved Julio 2023, fromhttps://www.redalyc.org/pdf/539/53906504.pdf

Olcina Cantos, J.: Cambio climático y riesgos climáticos en España. Investigaciones Geográficas. (49), 197–220. Retrieved from (2010)https://www.cervantesvirtual.com/nd/ark:/59851/bmcwq0k5

Organización Mundial del Comercio. (2009). *El Comercio y el Cambio Climático*. Suiza: Secretaría de la OMC. Retrieved julio 7, 2023, from https://www.uncclearn.org/wp-content/uploads/library/wto01_spn_0.pdf

Romeo, L., Bolea, I.: Overview Post-combustion CO2 capture. Boletín del Grupo Español del Carbón(35), 8–11. Retrieved from (2015).https://www.gecarbon.org/boletines/articulos/BoletinGEC_035_art2.pdf

Rosca Jusmet, J.: La política climática y los combustibles fósiles: una perspectiva desde la oferta. Revista de Economía Crítica(34), 9–25 (2022). https://revistaeconomiacritica.org/index.php/rec/article/view/649

Salaet Fernández, S., Roca Jusmet, J.: Agotamiento de los combustibles fósiles y emisiones de CO2: Algunos posibles escenarios futuros de emisiones. Revista Galega de Economía, 19(1), 1–19 (2019). Retrieved Julio 8, 2023, fromhttps://www.redalyc.org/comocitar.oa?id=39113124001

Samaniego, J., Schneider, H.: Cuarto informe sobre financiamiento para el cambio climático en América Latina y el Caribe, 2013–2016. Comisión Económica para América Latina y el Caribe (CEPAL) (2019).https://www.cepal.org/es/publicaciones/44487-cuarto-informe-financiamiento-cambio-climatico-america-latina-caribe-2013-2016

United Nations. *Causes and Effects of Climate Change.* Retrieved Julio 17, 2023, (2023). https://www.un.org/en/climatechange/science/causes-effects-climate-change

Zhu, Q.: Desarrollos en tecnologías de aprovechamiento de CO2. Energía limpia. Energía limpia **3**(2), 85–100 (2019)

Multicriteria Analysis Applied to the Selection of Shopping Centers for a Family-Owned Restaurant Business in Smart Cities: A Case Study in Ciudad Juárez

José Roberto Escamilla de Santiago, Alberto Ochoa-Zezzatti(✉), and Aida-Yarira Reyes-Escalante

Universidad Autónoma de Ciudad Juárez, Av. Universidad, Av. Heroico Colegio Militar y, Chamizal, 32300 Juárez, Chih, México
{jose.escamilla,alberto.ochoa,aida.reyes}@uacj.mx

Abstract. The location of an SME is a crucial factor that can determine its ability to stand out in the market; for restaurants, it becomes one of the most important aspects for success. This study employs the TOPSIS Multicriteria Model to identify the optimal location for a restaurant within a shopping center. The methodology used is quantitative in nature and focuses on analyzing seven existing shopping centers in the community of Ciudad Juárez. A multicriteria approach was adopted, evaluating 30 relevant aspects for choosing the location. Mahalanobis and Minkowski similarity functions were applied for the analysis, along with Dummy variables and a Sankey diagram to visualize competitive strategies. Additionally, market segmentation was conducted using the Kriging method, integrating geographic and demographic factors. The study results highlight the importance of location, identifying Plaza Juárez Mall as the best option. In conclusion, this study demonstrates the effectiveness of multicriteria techniques for data driven decision making.

Keywords: Competitiveness · Restaurant SMEs · Family-owned businesses · Multicriteria approach

1 Family Businesses in Smart Cities

In the digital era, Smart Cities [1] have become a focal point for urban development, integrating advanced technologies to improve the quality of life for their residents. A crucial aspect of these cities is the commercial sector, particularly shopping centers, which serve as hubs of economic and social activity [2]. With the emergence of Generation Z—a cohort that has grown up in a highly interconnected world market strategy must adapt to their unique expectations and behaviors.

Among the market offerings, small and medium sized enterprises (SMEs) play a vital role in economic growth, innovation, and social cohesion, making their support and development a priority for economic policies. Within this group are family-owned businesses. The concept of "familiness" [3] has been defined as a unique resource derived from the family structure [4].

The evolution of family business studies has reflected an increasingly sophisticated and multidimensional approach, with special attention to the internal dynamics and factors influencing long term sustainability and success. Much of the knowledge generated on family businesses is based on the three circle model (family, ownership, and business), which emphasizes the importance of succession planning and continuity in the marketplace [5]. This approach has highlighted the need for careful management of intergenerational relationships. However, with the turn of the millennium, new perspectives have emerged that broaden the understanding of the strategic advantages inherent in family businesses, such as technological approaches, sustainability, competitiveness, and globalization.

In the face of competitiveness challenges, the World Trade Organization (WTO) has stated that its international objective is to assist businesses in the goods and services sectors, as well as exporters and importers, in carrying out their activities [6]. In this context, family businesses gain value through their influence in development zones, job creation, economic growth, contribution to GDP, social responsibility, and improvements in the built urban environment [7].

The factors that have been identified as crucial for the success of family businesses include innovation, sustainability, and economic performance [8, 9]. Additionally, technological and organizational changes in various forms offer another opportunity to generate advantages over competitors, including the choice of location (OECD) [10].

The constant need to adapt and renew capabilities within family businesses is driven by changes in the environment, new needs, and the demands of new generations, which compel these businesses to seek organizational flexibility to adapt. This underscores the importance of active management within family firms [11].

More recently, research has begun to address heterogeneity within family businesses, as well as internal and external differences that can significantly influence business strategy and performance [12]. In this context, location becomes an important factor, as family businesses can be situated in a variety of spaces, such as urban, suburban, and rural areas, residential zones, commercial premises, and shopping centers.

Shopping centers have become a popular strategy, as they allow family businesses to operate in high traffic areas while benefiting from the modernity and ongoing updates of the surrounding environment [13].

1.1 Competitiveness and Location

Leading experts in business competitiveness [6, 14, 15] and [16] agree that a company's location is a crucial factor when evaluating its competitiveness. The impact of location can be observed through various key aspects (see Table 1).

1.2 Strategies for Determining a Commercial Location

The location of a food business is a critical decision that can significantly influence its success. There are several market methodologies used to determine the best location for

Table 1. Key Aspects of Commercial Location.

Key Aspect	Description
Market Access	Close to the target market, with easy access to premises, road transport, public transportation, and pedestrian access
Access to Resources	Proximity to raw materials, energy sources, human resources, and technology is essential
Infrastructure	City's infrastructure, company infrastructure, mobility, logistics, market access, and connectivity
Labor Force	Access to a skilled workforce with the necessary expertise
Salaries and Benefits	Location influences the cost of living, wages, and employee benefits
Innovation and Technology	Proximity to shopping centers, universities, technology parks, and industrial hubs
Local Competition	Presence of competitors. Differentiation strategies and innovation
Quality of Life	Areas with high educational levels, good healthcare services, and recreational options
Community Integration	Engagement with the local environment, contributing to the community's social and environmental needs

Source: Own elaboration

this type of business. Table 2 outlines various methods employed to identify commercial locations.

Research has been conducted to identify optimal investment locations for restaurants, highlighting the following findings: The pandemic crisis presented unique challenges, and many businesses faced operational difficulties. Some had to suspend activities, others shut down entirely, while those that survived did so successfully. The strategies that led to this success were the specific development strategies applied by family-owned SMEs [17]. Another related study explored decision making using the AHP technique combined with an extended version of the TOPSIS method. The results indicated the level of transparency or bureaucratic constraints compared to market conditions or the country's economic aspects [18]. Research conducted in 2023 using Fuzzy TOPSIS in the context of facility location selection provided clearer and more precise professional judgments regarding location, aiding decision makers at Rokomari.com and comparable organizations in choosing sites in a well informed and efficient manner [19]. In the case of Ciudad Juárez, two studies are related to the use of multicriteria decision making: TOPSIS and AHP for the location of a park and a new airport [20] and [21].

2 Methodology

Ciudad Juárez is a border community in northern Mexico with a high demand for dining services, which are spread throughout the urban area. There are seven major shopping centers that locals frequently visit for indoor shopping: Las Misiones, Galerías Tec, Plaza

Table 2. Commercial Location Selection Methods

Elements	Requirement	Characteristics
Demand and Demographic Analysis	Sufficient demand	Adequate range of products
Competition Analysis	Identify areas with low competition or where competition does not adequately meet demand	Highlight services
Accessibility	Various means of accessibility, both by foot and by car	Inclusive accessibility
Financial Feasibility	Location costs	Financial analysis of advantages and disadvantages
Scoring or Rating Method	Comparison between potential locations	Highlight advantages
Proximity and Clustering Analysis	Benefits of being near places that attract large crowds or being part of a business cluster	Joint benefit analysis
Geomarketing Studies	Optimize location choice by integrating data	Patterns and opportunities
Cost-Benefit Method	Favorable return on investment	Comparative analysis

Source: Own elaboration

Sendero, Plaza Juárez Mall, Plaza Sendero Las Torres, Plaza Campestre, and Plaza del Sol. These shopping centers are located at various points across the city, providing access to residents in different areas. It is worth noting that there are numerous restaurants throughout the city, ranging from small establishments to family run mini restaurants. The traditional entrepreneurial spirit in the city is based on established and proven business models, such as retail stores, restaurants, or service workshops. This type of entrepreneurship forms the backbone of the Mexican economy, as most businesses, which provide essential employment and services, fall into this category. As of June 2023, there were approximately 68,000 fast-food outlets in Mexico [22], with Ciudad Juárez accounting for 3,983 restaurants according to the 2022 census [23]. Based on this context, this study was conducted to identify the optimal location for a family-owned restaurant within the urban area of Ciudad Juárez.

The techniques used as part of the study were as follows (see Fig. 1):

a. **Creation of the main criteria for selecting a shopping center.** A survey was conducted in 2023 among 384 young adults over 18 years old. The survey aimed to gather the following information: geographic location of the main shopping centers (radius of influence), qualitative characteristics of the shopping centers (14 items), customer influx in the main shopping centers, and purchasing patterns (14 items).

b. **Similarity functions.** To ensure precise evaluation of the collected data, similarity functions such as Mahalanobis and Minkowski distances were applied. The Mahalanobis distance helps identify similarities by considering the correlation between variables, while the Minkowski distance allows for adjusting the relative importance of different dimensions in the analysis [24].
c. **Dummy variables.** Dummy variables were used to encode qualitative characteristics of the shopping centers, simplifying the data analysis [25].
d. **Sankey diagram.** The Sankey diagram was used to visualize the flow of customers and preferences across the different shopping centers, providing a clear representation of how consumers are distributed and how their preferences shift [26].
e. **Kriging model.** The Kriging model was applied to segment the market geospatially, optimizing strategies based on the geographic location of the shopping centers [27].
f. **TOPSIS method.** The TOPSIS method was employed to rank alternatives based on their proximity to an ideal solution, facilitating the selection of the most suitable shopping center for a restaurant business [28]. It was selected for its unique ability to handle nominal labels and transform subjective preferences into a quantifiable model. This methodology not only enables the integration of various qualitative and quantitative variables but also allows for an exhaustive evaluation of multiple alternatives based on established criteria. The flexibility of TOPSIS to adapt to scenarios where decisions require complex weighting makes it an ideal tool for strategic analyses, such as the selection of shopping centers in urban environments.

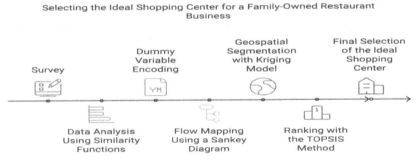

Fig. 1. Methodology process.

2.1 Multicriteria Analysis TOPSIS

Multicriteria analysis is an essential approach in the context of Smart Cities, where decision making must consider a wide range of factors. These factors include economic, social, technological, and environmental variables, which are interrelated in complex urban systems. The use of techniques such as Topsis allows for the simultaneous evaluation of multiple criteria, providing a ranking of the available options based on their proximity to the ideal solution. In the specific case of a third-generation restaurant business, multicriteria analysis facilitates the selection of the most suitable shopping center in Ciudad Juárez, considering both the geographic location and the consumer profile.

$$Si = \sum_{j=1}^{n} wj \cdot Cij \tag{1}$$

where:
Si Final score of shopping center i in the city.
wj Weight assigned to criterion j.
Cij Value of criterion j for shopping center i.

2.2 Selection of Shopping Centers and Customers

The selection of shopping centers in Ciudad Juárez is based on their relevance within the urban context, customer traffic, and their ability to attract Generation Z. The shopping centers chosen for this study are: Las Misiones, Galerías Tec, Plaza Sendero, Plaza Juárez Mall, Plaza Sendero Las Torres, Plaza Campestre, and Plaza del Sol. Each of these shopping centers presents unique characteristics regarding their location, type of stores, and demographic profile of visitors. Generation Z is a crucial demographic group for the success of a restaurant business, given their focus on unique experiences and their influence on consumption trends.

$$Pij = \frac{Dij}{\sum_{i=1}^{m} D_{ij}} \tag{2}$$

where:
Pij Probability of selecting shopping center i for customer j.
Dij Distance or dissimilarity between shopping center i and customer j.
m Total number of shopping centers.

2.3 Mahalanobis and Minkowski Similarity Functions

For a precise evaluation of shopping centers, similarity functions are used to measure the distances between different data points. The Mahalanobis distance is applied to identify similarities by considering the correlation between variables, which is useful for understanding the complex relationships among the evaluated criteria. On the other hand, the Minkowski distance, a generalization of Euclidean distance, allows for adjusting the relative importance of the different dimensions considered in the analysis, providing greater flexibility in the comparison of alternatives.

A) Mahalanobis

$$d_M(x, y) = \sqrt{(x-y)^T S^{-1}(x-y)} \tag{3}$$

where:
d_M Distance between points x and y.
S Covariance matrix.
p Parameter that defines the norm (p = 2 for Euclidean, p = 1 for Manhattan).
n Number of dimensions.

B) Minkowski

$$dM(x, y) = (\sum\nolimits_{k=1}^{n} \lceil X_k - y_k \lceil p)1/p \tag{4}$$

where:
d_M Distance between points x and y.
S Covariance matrix.
p Parameter that defines the norm (p = 2 for Euclidean, p = 1 for Manhattan).
n Number of dimensions.

2.4 Identification of Patterns Using Dummy Variables

Dummy variables are powerful tools for identifying patterns within categorical data sets. In this analysis, they are used to encode qualitative features of shopping centers, such as the presence of specific brands or additional services that attract Generation Z. These variables simplify modeling and facilitate the identification of trends and behavioral patterns among consumers.

$$Y = \beta_0 + \sum\nolimits_{k=1}^{p} \beta_k D_k + \in \tag{5}$$

where:
Y Dependent variable (expected result).
β_0 Intercept of the regression.
β_k Coefficients of the dummy variables DkD_kDk.
D_k Variables representing different categories.
$\in$ Error term.

2.5 Visualization of Strategies Using a Sankey Diagram

The Sankey diagram is used for visualizing flows and relationships between different elements of the analysis. In this case, the diagram illustrates the distribution of customers among the selected shopping centers and the transition of consumption preferences between various dining options. This visualization provides a clear understanding of how resources, such as time and money, are distributed among the different alternatives available to Generation Z customers.

$$F_{i \to j} = \sum\nolimits_{k=1}^{n} f_k \cdot R_{k, i \to j} \tag{6}$$

where:
$F_{i\to}$ Resource flow from i to j.
f_k Weight of resource k.
$R_{k,i\to j}$ Value of resource k that flows from i to j.
n Total number of criteria involved.

2.6 Association Rules and Market Segmentation Using a Kriging Model

Association rules are applied to identify frequent consumption patterns, which can be used to enhance marketing strategies and the personalization of dining offerings. The Kriging model, a geospatial interpolation method, enables market segmentation by considering the location of shopping centers and the demographic characteristics of surrounding areas. This approach offers a better understanding of how different population segments are spatially distributed and how business strategies can be optimized based on geographic location.

$$Z(x) = \mu + \sum_{i=1}^{n} \lambda_i Z(X_i)$$

where:
$Z(x)$ Estimated value at point xxx using Kriging.
μ Mean of observed values.
λ_i Weighting coefficients obtained from the covariance matrix.
$Z(X_i)$ Observed value at point x_i.
n Number of observed points.

The representation of the evaluation system for selecting the location of a restaurant in Ciudad Juárez integrates economic, social, technological, and environmental factors, key consumer profiles, consumer flow per shopping center, customer influx and attraction, location assessments, categorical site evaluations, and customer distribution by shopping center and consumption patterns (see Fig. 2).

3 Results

Based on the survey, which included a gender differentiation, 30 criteria were identified that influence the selection of a location for a family-owned restaurant. These criteria were used to construct the evaluation matrix using the TOPSIS method. The geographic location of the main shopping centers (radius of influence measured in meters), qualitative characteristics of the shopping centers (14 categories), customer influx in the main shopping centers (1 category), and purchasing patterns (14 categories) were considered.

The range of influence of shopping centers can be visualized, showing Plaza Campestre, Plaza Juárez Mall, and Las Misiones as having the greatest geographic influence, while Plaza del Sol, Plaza Sendero, and Galerías Tec have the least (see Fig. 3).

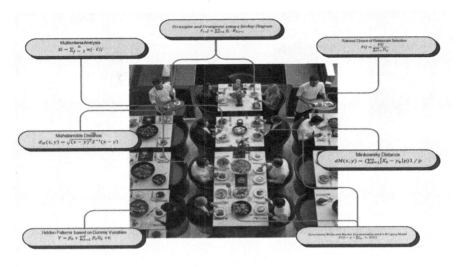

Fig. 2. Evaluation system for selecting a restaurant location.

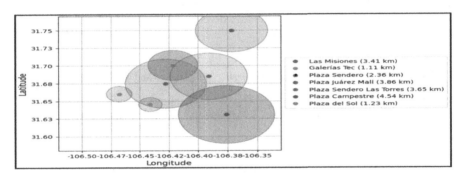

Fig. 3. Ranges of Influence of Shopping Centers in Ciudad Juárez

The results related to visits to shopping centers reveal differentiated trends based on gender: a) Shopping centers with higher female presence include Plaza Sendero, Plaza Juárez Mall, Plaza del Sol, and Galerías Tec. b) Shopping centers with higher male presence include: Las Misiones, Plaza Sendero Las Torres, and Plaza Campestre (see Fig. 4).

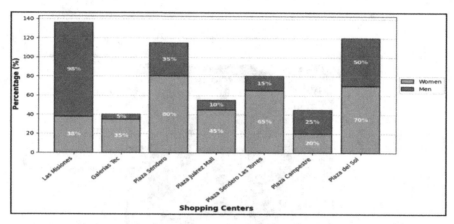

Fig. 4. Distribution of Visitors by Gender in Shopping Centers.

Using the Mahalanobis and Minkowski techniques, the highest index was obtained, reflected in the sum of the consumption values H and M, indicating that Plaza Sendero Las Torres scored the highest in the purchase index with a value of 16.10, followed by Galerías Tec with a value of 15.42, and Plaza del Sol in third place with a score of 14.05 (see Table 3).

Table 3. Shopping Center Purchase Index Using Mahalanobis and Minkowski Techniques

Shopping Center	Men Consumption	Women Consumption	Total Index (H + M)
Las Misiones	7.45	5.5	12.95
Galerías Tec	6.54	8.88	15.42
Plaza Sendero	5.5	6.68	12.18
Plaza Juárez Mall	7.97	4.89	12.86
Plaza Sendero Las Torres	8.69	7.41	16.1
Plaza Campestre	5.72	7.51	13.23
Plaza del Sol	5.66	8.39	14.05

By using dummy variables, it was detected that the intention to purchase when visiting shopping centers exhibits non-homogeneous behaviors during the visits. However, it was found that in all shopping centers, men show a higher purchasing indicator, as illustrated in Fig. 5.

The Sankey diagram visualizes future consumption behavior for men and women between 2024 and 2037. It identifies that men's consumption peaks at Galerías Tec Shopping Center starting in 2033 and consolidates in 2037. In the case of women, the peak is observed at Galerías Tec in 2030 and at Plaza Sendero (see Fig. 6).

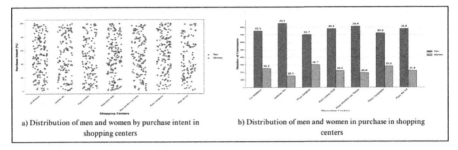

Fig. 5. Analysis of dummy variables in purchase intention.

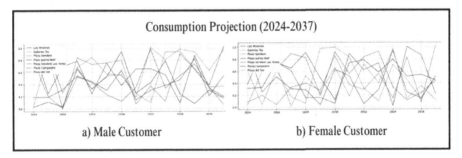

Fig. 6. Sankey Diagram by gender in mall centers from Juarez.

The results of the Kriging model reveal that the highest concentration of consumers is in the central northern part of the city (see Fig. 7). In these areas of concentration, it is evident that the shopping centers in the northern region attract the majority of consumers.

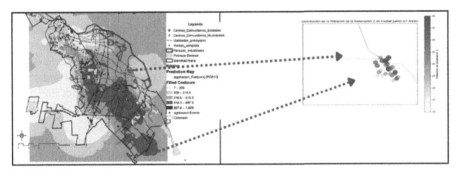

Fig. 7. Market segmentation through the Kriging model in the urban area of Ciudad Juárez.

The results of the multicriteria analysis using the TOPSIS method, with 30 dimensions 27 of which are categories involving services, commercial policies, shopping experience, social media recognition, purchase index, influence dimension, and future consumption index rank Plaza Juárez Mall as the top choice for both men and women.

Plaza Sendero Las Torres ranks second for men, while Plaza Campestre ranks second for women (see Fig. 8 and Table 4).

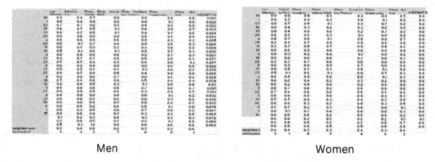

Men   Women

Fig. 8. Final TOPSIS matrix (Men and Women).

Table 4. Final TOPSIS results.

	Las Misiones	Galerías Tec	Plaza Sendero	Plaza Juárez Mall	Plaza Sendero Las Torres	Plaza Campestre	Plaza del Sol
Male	4	2	3	7	6	5	1
Female	3	1	5	7	4	6	2

4 Conclusions

The multicriteria analysis used to select a shopping center for a family-owned restaurant business in Ciudad Juárez reveals important patterns in the consumption behavior of men and women across different locations. The shopping centers with the highest total consumption indices are Plaza Sendero Las Torres and Galerías Tec (16.10 and 15.42, respectively), according to projections of goods and services consumption for the 2024–2037 period. This highlights a strong consumer preference for these areas, particularly when considering the balance of consumption between men and women.

The results of the multicriteria analysis show that the evaluated dimensions such as services, commercial policies, shopping experience, and social media recognition position Plaza Juárez Mall as the best location, based on all the measurement indicators.

The findings suggest that the best investment option for a restaurant is to locate it in Plaza Sendero Las Torres, based on the consumption index derived from the Mahalanobis and Minkowski techniques. This index, which reflects the investment's viability, yielded a score of 16.10, the highest in the analysis. However, if the investment is aimed at long term market stability, it is recommended to consider Plaza Juárez Mall. This recommendation is supported by the fact that Plaza Juárez Mall received the highest overall score in the analysis, which covered 30 criteria, not just the consumption index.

Although the latter is crucial for restaurant businesses to ensure a steady flow of customers, the TOPSIS model results highlight Plaza Juárez Mall as the most favorable option.

Overall, consumer behavior in the selected shopping centers has evolved over time, demonstrating the importance of adaptability and flexibility in business strategies for smart cities. The analysis indicates that in complex urban markets like Ciudad Juárez, a restaurant business can significantly optimize its performance by making informed decisions based on multicriteria data.

5 Future Research

Directions for Future Research

This research has demonstrated the effectiveness of multicriteria analysis in the strategic selection of a shopping center for a family-owned restaurant business in Ciudad Juárez, highlighting significant consumption patterns between men and women in various locations. Based on these findings, several opportunities arise to expand the research, with the aim of deepening the understanding of consumer behavior and optimizing business strategies in complex urban environments like Ciudad Juárez. Below are several future research directions that could contribute significantly to the field of study:

a. **Temporal and Geographic Expansion of the Study:**

It is essential to consider expanding both the temporal horizon and the geographic scope of the analysis. Although this study focused on consumption projections for the 2024–2037 period in Ciudad Juárez, extending the research to other periods or cities could offer a more comprehensive view of the stability and variability of consumption patterns. Additionally, comparing cities with diverse demographic and socioeconomic characteristics would allow for the identification of global and regional trends in consumer behavior, facilitating the adaptation of business strategies to different urban contexts.

b. **Detailed Analysis of Demographic and Socioeconomic Variables:**

It is crucial to delve deeper into the analysis of demographic and socioeconomic variables that influence consumer behavior in shopping centers. Incorporating factors such as age, educational level, income, and occupation would allow for more precise population segmentation, enabling the formulation of highly personalized marketing strategies. Understanding how these factors interact with gender to shape consumption preferences is critical for the development of more effective campaigns targeting specific market niches.

c. **Incorporation of New Technologies and Advanced Data Analytics:**

The advancement of technologies such as artificial intelligence, machine learning, and data mining presents unprecedented opportunities to enhance multicriteria analyses. In future research, it would be relevant to explore the application of these technologies to more accurately predict changes in consumption patterns and dynamically identify consumer preferences. Additionally, the integration of real time data, such as tracking user behavior through mobile devices and social networks, would provide valuable insights for business decision making.

d. **Evaluation of the Influence of Urban Environments and Smart Infrastructures:**
   Given that the study was conducted in a smart city context, a relevant future research line would be to explore how the urban environment and smart infrastructures influence consumer preferences. Analyzing the impact of factors such as mobility, connectivity, and accessibility to shopping centers on consumer decisions would provide a more comprehensive understanding of how smart cities are transforming shopping behavior, and how businesses can leverage these dynamics to optimize performance.

References

1. Singh, P., Lynch, F.M., Helfert, M.: Smart city development: positioning citizens in the service life cycle (citizens as primary customer). smart city dev. position. Citiz. Serv. Life Cycle Citiz. Prim. Cust. 112 (2020)
2. Reyes, A.Y., Rodríguez, F.T.S., Ibarra, S.I.G.: Los centros comerciales en una sociedad posmoderna: evidencia empírica en Ciudad Juárez," Adm. Organ., vol. 22, no. 42, Art. no. 42 (2029). https://doi.org/10.24275//uam/xoc/dcsh/rayo/2019v22n42/Reyes
3. Chrisman, J.J., Chua, J.H., Litz, R.: A unified systems perspective of family firm performance: an extension and integration. J. Bus. Ventur. **18**(4), 467–472 (2003). https://doi.org/10.1016/S0883-9026(03)00055-7
4. Habbershon, T.G., Williams, M., MacMillan, I.C.: A unified systems perspective of family firm performance. J. Bus. Ventur. **18**(4), 451–465 (2003). https://doi.org/10.1016/S0883-9026(03)00053-3
5. Sharma, P., Chrisman, J.J., Pablo, A.L., Chua, J.H.: Determinants of initial satisfaction with the succession process in family firms: a conceptual model. Entrep. Theory Pract. **25**(3), 17–36 (2001). https://doi.org/10.1177/104225870102500302
6. I.-A. D. Bank, Competitiveness: The Business of Growth : Economic and Social Progress in Latin America : 2001 Report. IDB (2001)
7. CEPAL. La competitividad y sus factores determinantes: un análisis sistémico para países en desarrollo | CEPAL. https://www.cepal.org/es/publicaciones/45005-la-competitividad-sus-factores-determinantes-un-analisis-sistemico-paises. Accessed 31 Aug 2024
8. World Economic Forum. Report World Economic Forum 2023. Foro Económico Mundial. https://es.weforum.org/publications/series/annual-report/. Accessed 34 Aug 2024
9. World Bank. Green Your Bus Ride : Clean Buses in Latin America. World Bank. https://documents.worldbank.org/en/publication/documents-reports/documentdetail. Accessed 27 Jun 2020
10. OECD. Índice de Políticas para PyMEs: América Latina y el Caribe (2024). https://www.oecd.org/es/data/dashboards/sme-policy-index-latin-america-and-the-caribbean-2024.html. Accessed 24 Aug 2024
11. Dima, A.M.: Innovation in family business. In: Proceedings of the International Conference on Economics and Social Sciences, Sciendo, pp. 786–802 (2020) https://doi.org/10.2478/9788395815072-078
12. Pittino, D., Chirico, F., Baù, M., Villasana, M., Naranjo-Priego, E.E., Barron, E.: Starting a family business as a career option: the role of the family household in Mexico. J. Fam. Bus. Strategy, 11(2), 1–9 (2020). https://doi.org/10.1016/j.jfbs.2020.100338
13. ZIGURAT. Evolución de las Ciudades Inteligentes (2024). https://www.e-zigurat.com/es/blog/evolucion-ciudades-inteligentes/. Accessed 31 Aug

14. M. E. Porter, Estrategia Competitiva: Técnicas para el análisis de los sectores industriales y de la competencia. Grupo Editorial Patria (2015)
15. Krugman, P.: Competitiveness: a dangerous obsession. Foreign Aff. **73**(2), 28 (1994). https://doi.org/10.2307/20045917
16. Hill, C.: International Business: Competing in the Global Marketplace (2023). https://www.mheducation.com/highered/product/international-business-competing-global-marketplace-hill/M9781260387544.html. Accessed 24 Aug 2024
17. Siuta-Tokarska, B., Juchniewicz, J., Kowalik, M., Thier, A., Gross-Gołacka, E.: Family SMEs in Poland and their strategies: the multi-criteria analysis in varied socio-economic circumstances of their development in context of Industry 4.0. Sustainability, **15**(19), 14140 (2023). https://doi.org/10.3390/su151914140
18. Porro, O., Pardo-Bosch, F., Agell, N., Sánchez, M.: Understanding location decisions of energy multinational enterprises within the european smart cities' context: an integrated AHP and extended fuzzy linguistic TOPSIS method. https://www.mdpi.com/1996-1073/13/10/2415. Accessed 07 Oct 2024
19. Rashid, M.: Multi-criteria decision making with fuzzy TOPSIS : a case study in Bangladesh for selection of facility location. laturi.oulu.fi. https://oulurepo.oulu.fi/handle/10024/42299. Accessed 07 Oct 2024
20. Reyes-Escalante, A.-Y., Ochoa-Zezzatti, A., Sandoval-Chávez, D.-A., Venegas-Ortiz, K.-S.: What is the best location of a smart airport in Juarez, Mexico? In: Ochoa-Zezzatti, A., Oliva, D., Juan Perez, A., (eds.) Technological and Industrial Applications Associated with Intelligent Logistics. LNITI, pp. 475–499. Springer, Cham (2021). https://doi.org/10.1007/978-3-030-68655-0_24
21. Reyes, A.Y.R., Sandoval, D.A., Vera, E.: A multi-criteria decision making for sustainable location of urban parks. Res. Comput. Sci. **149**(6), 16 (2020)
22. Statista. El portal de estadísticas. Statista. https://es.statista.com/. Accessed 22 May 2024
23. INEGI. Directorio de empresas y establecimientos. https://www.inegi.org.mx/temas/directorio/. Accessed 07 Oct 2024
24. Hou, Y., Chen, Z., Wu, M., Foo, C.-S., Li, X., Shubair, R.M.: Mahalanobis distance based adversarial network for anomaly detection. In: ICASSP 2020-2020 IEEE International Conference on Acoustics, Speech and Signal Processing (ICASSP), Barcelona, Spain: IEEE, May 2020, pp. 3192–3196 (2020). https://doi.org/10.1109/ICASSP40776.2020.9053206
25. Berkay. Market Basket Analysis Using Association Rules. Medium. https://iambideniz.medium.com/market-basket-analysis-using-association-rules-e049cdf260fe. Accessed 31 Aug 2024
26. Wang, Y., Zhu, Z., Wang, L., Sun, G., Liang, R.: Visualization and visual analysis of multimedia data in manufacturing: a survey. Vis. Inform. **6**(4), 12–21 (2022). https://doi.org/10.1016/j.visinf.2022.09.001
27. Riehmann, P., Hanfler, M., Froehlich, B.: Interactive Sankey diagrams, p. 240 (2005). https://doi.org/10.1109/INFVIS.2005.1532152
28. Cheng, N., et al.: Exploring light use efficiency models capacities in characterizing environmental impacts on paddy rice productivity. Int. J. Appl. Earth Obs. Geoinf. **117**, 103179 (2023). https://doi.org/10.1016/j.jag.2023.103179

Alarm Recommendation Intelligent System for Multilayer Ceramic Capacitor (MLCC) Electroplating Using Case-Based Reasoning and Natural Language Processing

Juan Pablo Canizales-Martinez[✉], Alberto Ochoa-Zezzatti, and Carmen Villar-Patiño

Universidad Anáhuac, 52760 Huixquilucan, Edomex, Mexico
{juan.canizales,alberto.ochoa,maria.villar}@anahuac.mx

Abstract. This study proposes an advanced alarm recommendation system tailored for a multilayer ceramic capacitor (MLCC) production line, harnessing the power of natural language processing (NLP) and Case-Based Reasoning (CBR) to enhance operational efficiency and minimize downtime. The primary goal is to provide relevant, prioritized alarm recommendations to operators, addressing the common issue of notification overload in traditional alarm systems, which often leads to user fatigue and critical alerts being overlooked. These oversights can cause significant operational disruptions.

By leveraging NLP techniques, the system analyzes alarm descriptions and historical data, identifying semantic patterns and relationships. This allows it to understand the severity and context of each alarm, ensuring that operators focus on the most critical issues. In doing so, unnecessary distractions are reduced, safety is improved, and overall productivity on the production line is enhanced. Additionally, the system offers contextualized recommendations by correlating alarms with key machine parameters, providing valuable insights into specific actions required to efficiently resolve the issue.

Moreover, CBR is integrated into the system to improve predictive maintenance processes. By analyzing past cases of equipment failures and repair actions, the system can offer optimized solutions for current maintenance needs, ultimately reducing downtime and preventing costly breakdowns. This approach enables operators to act proactively, addressing potential problems before they escalate into major failures.

The combination of NLP and CBR allows for a dynamic and intelligent response to fluctuating production conditions, and personalized recommendations tailored to operators' past interactions, preferences, and experience levels. This personalized approach further enhances decision-making, resulting in faster response times and more efficient workflows.

In summary, this alarm recommendation system not only tackles the issue of alarm fatigue but also introduces a more intelligent, context-aware, and personalized approach to managing alarms. The integration of CBR for predictive maintenance represents a significant advancement in ensuring the smooth operation of MLCC production lines, reducing downtime, optimizing human-machine interaction, and improving long-term production efficiency. In future research, it will be possible that through hyperparameter optimization using the system's recurrent neural network (RNN) algorithms, it will achieve a higher accuracy rate than the current one of around 86% to prevent fatal accidents in this type of factories in Industry 5.0.

Keywords: Alarm Recommendation System · Natural Language Processing (NLP) · Operational Efficiency · Downtime Minimization · Contextualized Recommendations · Personalized Recommendations · Industry 5.0

1 Introduction

The manufacturing industry is undergoing a digital transformation driven by technologies such as Artificial Intelligence (AI), Big Data, and the Internet of Things (IoT). In this context, predictive maintenance (PdM) is becoming a fundamental strategy to achieve operational excellence and minimize downtime [1]. Industry 5.0, characterized by human-machine collaboration, mass customization, and sustainability, presents new challenges in production management. Predictive maintenance, which aims to anticipate failures to prevent disruptions and minimize costs, becomes even more crucial in this context. To achieve greater efficiency and accuracy in predictive maintenance, an intelligent system is needed that can analyze failure data, identify patterns, and suggest personalized solutions. In this work, we present a case-based reasoning (CBR) system that integrates natural language processing (NLP) techniques to optimize predictive maintenance in Industry 5.0. Our objective is to develop a system that can analyze descriptions of failures on the production line and generate personalized recommendations to maintenance personnel. This system seeks to reduce downtime, minimize repair costs, and improve safety, contributing to more efficient and sustainable production.

There are other techniques and methods besides CBR for recommending explanation methods, such as collaborative filtering and machine learning techniques. However, CBR stands out for its ability to capture and utilize expert knowledge, which makes it particularly suitable for recommending explanation methods [20] (Fig. 1).

1.1 Towards an AI-Based Maintenance Recommendation System for MLCC Production Line

This research focuses on the development of a maintenance recommendation system for a Multilayer Ceramic Capacitor (MLCC) production line, integrating AI techniques such as Machine Learning (ML) and Natural Language Processing (NLP). MLCC production lines often face the challenge of unplanned production downtime due to equipment failures. These disruptions generate a significant cost for the company, impacting productivity and profitability [2]. While alert systems exist to detect anomalies, fault detection is not always timely, leading to unexpected downtime. The objective of this research is to develop a maintenance recommendation system that can anticipate and prevent failures on the production line. To achieve this, ML will be used to predict alerts and NLP to generate personalized solution recommendations based on failure reports. Electroplating of Multilayer Ceramic Capacitors (MLCCs) is a critical process in the manufacturing

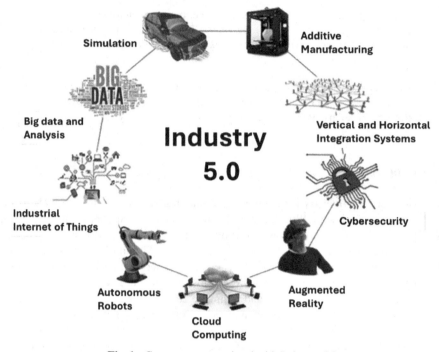

Fig. 1. Components associated with Industry 5.0

of these components, as the quality of the deposited metallic layer directly determines the electrical characteristics of the capacitor. Defects in the electroplating process can result in failures in the final product, leading to additional production costs and delivery delays (Figs. 2 and 3).

Fig. 2. Multilayer Ceramic Capacitors (MLCC)

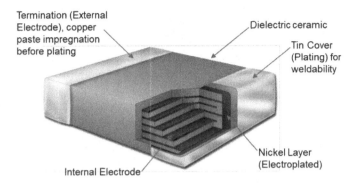

Fig. 3. Components that make up a Multilayer Ceramic Capacitor (MLCC)

Predictive maintenance plays a crucial role in improving efficiency and safety in production. By identifying and resolving faults in the electroplating process before they become major problems, downtime and associated costs can be minimized. However, identifying and resolving faults in electroplating can be complex, especially due to the large number of variables that can affect the process, such as temperature, concentration of electroplating baths, current, among others.

Machine maintenance is crucial to minimizing unforeseen interruptions in production. There are different types of maintenance: corrective (CM), preventive (PM), predictive (PdM), and prescriptive [3, 4].

Corrective maintenance (CM) involves repairing or replacing damaged components after a failure occurs. It is the most basic form of maintenance and is often reactive.

Preventive maintenance (PM), also known as time-based maintenance (TBM), was introduced in the 1950s. It is based on scheduling maintenance activities at predefined intervals to prevent potential failures. Its strategy is based on the "mean time between failures" (MTBF) and focuses on inspections, cleaning, lubrication, and parts replacement. [3, 4].

Predictive maintenance (PdM), also known as condition-based maintenance (CBM), focuses on determining the condition of equipment while it is operating. By monitoring variables such as temperature, noise, vibration, and lubrication, it is possible to predict when maintenance is needed. This allows for cost savings compared to preventive maintenance by avoiding unnecessary interventions.

Prescriptive maintenance is a step beyond predictive maintenance. It seeks to answer the question of how to control the occurrence of specific events and provides recommendations for maintenance decision-making. Its goal is to achieve a high degree of maturity in the maintenance system, enabling self-diagnosis and self-programming of maintenance (Fig. 4).

To overcome these challenges, this article presents an intelligent alert recommendation system that combines Case-Based Reasoning (CBR) and Natural Language Processing (NLP) to analyze failure descriptions and generate personalized recommendations for maintenance personnel. The CBR system stores historical cases of failures and their corresponding solutions, while NLP is used to compare new failure descriptions with past cases and generate tailored recommendations. Haptic interfaces are integrated

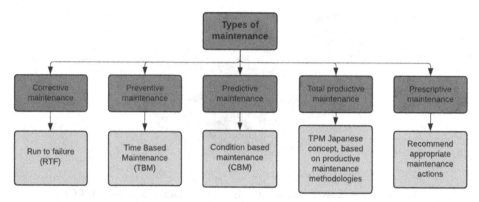

Fig. 4. Maintenance Types, based on [3, 4]

to provide feedback to maintenance personnel, improving their understanding of the recommendations and their overall experience.

The proposed recommendation system will help to:

Optimize maintenance planning: Accurately identify equipment maintenance needs, avoiding unnecessary interventions and optimizing resource allocation [5].

Reduce unplanned downtime: Anticipate failures and perform preventive maintenance before they occur, minimizing production disruptions and associated costs [6].

2 Review of the Literature

Case-based reasoning (CBR) is a machine learning technique that has been widely used in diverse applications, including predictive maintenance. CBR systems store historical cases of problems and their solutions, and then search for similar cases to solve new problems. Combining CBR with NLP techniques, such as text processing and sentiment analysis, has been successful in various areas, such as medical diagnosis and technical support. Predictive maintenance has been an active research area, and several methods have been developed to identify and address failures in production. These methods include sensor data analysis, machine learning, and vibration analysis techniques. [7], presents an interesting framework for prediction and decision-making in complex and unpredictable scenarios. Although the focus is not directly on predictive maintenance, the study highlights the importance of a well-defined case system for CBR and how the use of fuzzy sets and membership functions can improve prediction accuracy. The integration of qualitative and quantitative data through an attribute weighting approach and case similarity analysis provides an effective recommendation system for determining the risk and consequences of an event. While the context of the article is disaster management, the principles of CBR and attribute weighting could be applied to predictive maintenance for fault identification and maintenance planning. A paper developed by Yupeng Gao [7] presents a novel approach for mechanical equipment health management, combining (CBR) technology with a new intuitionistic fuzzy entropy method.

This approach seeks to improve fault prediction accuracy by incorporating both qualitative (expert experience) and quantitative (historical data) information into attribute weighting. The use of intuitionistic fuzzy sets allows for a more accurate representation of uncertainty and complexity in data analysis, which in turn facilitates the identification of failure patterns and more accurate maintenance action recommendations. This method, applied to mechanical equipment health management, could be adapted and used for the development of fault recommendation systems in the context of predictive maintenance, optimizing decision-making and resource management (Table 1).

Table 1. Comparative Table of Studies and their Strategies.

Paper	Year	Title	Description	Strategies/Algorithms/Methods
Jovani	2020	Machine Learning and Reasoning for Predictive Maintenance in Industry 4.0: Current State and Challenges	Examining academic advancements in fault prediction within the context of predictive maintenance (PdM) in Industry 4.0	This paper explores the use of machine learning (ML) and reasoning in predictive maintenance (PdM) within the context of Industry 4.0. We focus on decision support systems, utilizing concepts such as ontology, decision trees, random forests (RF), and various neural network architectures like artificial neural networks (ANN), support vector machines (SVM), k-nearest neighbors (k-NN), convolutional neural networks (CNN), recurrent neural networks (RNN), deep belief networks (DBN), deep Boltzmann machines (DBM), and generative adversarial networks (GAN). We also investigate the potential of deep reinforcement learning (DRL) and transfer learning in improving PdM strategies
Fink	2020	Potential, challenges, and future directions for deep learning in forecasting and health management applications	A comprehensive assessment of current developments, drivers, challenges, potential solutions, and future research needs in the field of deep learning applications (PHM) in Prognostics and Health Management (PHM)	Deep Learning (DL), Prognostics and Health Management (PHM), Generative Adversarial Networks (GAN), Transfer Learning, Deep Reinforcement Learning (DRL), Physics-Informed Machine Learning

(*continued*)

Table 1. (*continued*)

Paper	Year	Title	Description	Strategies/Algorithms/Methods
Lei	2020	Machine Learning Applications for Machine Fault Diagnosis: A Review and Roadmap	A Systematic Review Covering the Evolution of Machine Fault Detection from Past to Future, Highlighting Potential Research Trends for Researchers	Machine Learning (ML), Machine Fault Detection (IFD), Deep Learning (DL), Transfer Learning, Artificial Neural Networks (ANN), Support Vector Machines (SVM), k-Nearest Neighbors (k-NN), Convolutional Neural Networks (CNN), Recurrent Neural Networks (RNN), Deep Belief Networks (DBN), Deep Boltzmann Machines (DBM)
Zonta	2020	Predictive Maintenance in Industry 4.0: A Systematic Literature Review	Examining Publications on Predictive Maintenance in Industry 4.0	Predictive Maintenance (PdM), Industry 4.0, Artificial Neural Networks (ANN), Machine Learning Models, Stochastic Models, Physics-Based Models, Multi-model Approach
Montero	2020	Towards Multi-Model Approaches for Predictive Maintenance: A Systematic Literature Review on Diagnosis and Prognosis	Current Trends in Diagnosis and Prognosis, with a Focus on Multi-Model Approaches, Highlighting Current Research Challenges and Opportunities	Multi-model approaches, knowledge-based models, data-driven models, physics-based models, graphs, artificial neural networks (ANN), support vector machines (SVM), hidden Markov models, and genetic algorithms
Psarommatis	2023	Envisioning Maintenance 5.0: Insights from a Systematic Literature Review on Industry 4.0 and a Proposed Framework	Key Trends in Advanced Maintenance Techniques and the Consolidation of Traditional Maintenance Policies, Driven by the Growing Adoption of Industry 4.0 Technologies and the Need to Optimize Manufacturing System Performance and Reliability	Maintenance, Industry 4.0, Predictive Maintenance (PdM), Zero Defect Manufacturing (ZDM), Maintenance 5.0, Industry 5.0, Artificial Intelligence, Internet of Things (IoT), Neural Networks, Cyber-Physical Systems, KPIs, Human Factor, Sustainability, Maintenance Management Systems

(*continued*)

Table 1. (*continued*)

Paper	Year	Title	Description	Strategies/Algorithms/Methods
Gao	2022	A Data-Driven Predictive Maintenance System for Industry 4.0: State-of-the-Art and Research Opportunities	The Current State of Research in Data-Driven Predictive Maintenance (PdM) for Industry 4.0: From Data Collection to Decision-Making, Analyzing Key Challenges for Further Research	Predictive Maintenance (PdM), Industry 4.0, Artificial Neural Networks (ANN), Decision Support Systems, Machine Learning, Cyber-Physical Systems, Maintenance Management Systems
Canizales et al	2024	Alarm Recommendation Intelligent System for Multilayer Ceramic Capacitor (MLCC) Electroplating using Case-Based Reasoning and Natural Language Processing	This study proposes an advanced alarm recommendation system designed for a multilayer ceramic capacitor (MLCC) production line to enhance operational efficiency and minimize downtime	Leveraging the power of natural language processing (NLP) and Case-Based Reasoning (CBR)

3 Proposed Methodology

Predictive maintenance plays a crucial role in modern industry, especially in environments with complex systems where failures can have serious and costly consequences. With the proliferation of Internet of Things (IoT) sensors, companies are generating vast amounts of data that can be used to monitor equipment health and predict when failures will occur. While traditional machine learning (ML) approaches have been widely used to address this problem, they often face the challenge of limited availability of error data, which can lead to suboptimal results. Therefore, research in the field of case-based reasoning (CBR) has focused on addressing this issue through a knowledge-based approach that leverages experiences gathered regarding similar situations in the past. This article presents an intelligent recommendation system for predictive maintenance based on CBR. To address the limitations of ML, this framework combines the ability of CBR to leverage past experiences with that of rule-based systems to capture implicit expert knowledge.

To justify the use of CBR based on the following comparison:

Collaborative filtering: This technique is based on recommending articles similar to those that the user has preferred in the past. However, it does not consider expert knowledge, which can make the recommendations less precise.

Machine Learning: Machine learning techniques are capable of learning complex patterns from large data sets. However, they can be less transparent than CBR and more difficult to interpret.

CBR, on the other hand, is able to explicitly capture expert knowledge, making it more transparent and easier to interpret. Furthermore, CBR is a very flexible and adaptable technique, making it particularly suitable for recommending explanation methods in a dynamic environment where user preferences and explanation methods are constantly evolving. CBR is a powerful tool for recommending explanation methods. CBR recommendation systems are capable of learning from past experiences and using that experience to recommend the best explanation methods for new situations. CBR is also very flexible and adaptable to changes in the environment, making it an ideal technique for developing explanation method recommendation systems [20].

The intelligent alert recommendation system developed in this study comprises the following Pipeline Stages:

Data Collection and Preprocessing
Gather historical failure records from the MLCC electroplating production line. Clean and prepare textual descriptions of failures, solutions, and operational conditions.

Case Representation
Structure each failure record as a "case" with attributes (MLCC type, symptoms, operational conditions, solution, etc.). Use NLP techniques to vectorize textual attributes.

CBR System Development
Case-based reasoning (CBR) is a paradigm in artificial intelligence and cognitive science that models the reasoning process as primarily memory-based. Case-based reasoning systems solve new problems by retrieving stored "cases" that describe previous similar problem-solving episodes and adapting their solutions to meet new needs. Case-based reasoning has been formalized for computational reasoning purposes as a four-step process:

Retrieval: Given a target problem, retrieve from memory cases relevant to its resolution. A case consists of a problem, its solution, and usually annotations on how the solution was derived.

Re-use: Adapt the solution of the previous case to the target problem. This may involve tailoring the solution as needed to fit the new situation.

Revision: Once the previous solution has been adapted to the target situation, test the new solution in the real world (or in a simulation) and revise it if necessary.

Retention: After the solution has been successfully adapted to the target problem, store the resulting experience as a new case in memory.

Implementation and Evaluation
System Architecture: Design and build the system, including data storage, search algorithms, and interfaces for operators.

Evaluation: Use metrics like cosine similarity and actual downtime reduction to assess the system's accuracy and effectiveness (Figs. 5 and 6).

3.1 Data Collection and Preprocessing

Data is collected using historical records of failures in the MLCC electroplating process. These records include textual descriptions of the failures, the implemented solutions,

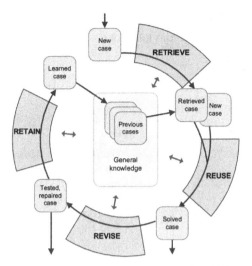

Fig. 5. CBR Diagram, based on Mathew [18].

$$Similarity(T,S) = \sum_{i=1}^{n} f(T_i, S_i) \times w_i$$

Where:
T is the target case
S is the source case
N is the number if features in each case
I is an individual feature form 1 to n
F is a similarity function for feature i in case T and S
w is the importance weighting of feature I

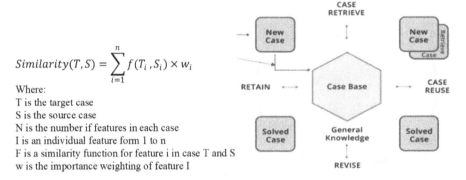

Fig. 6. Case-based reasoning lifecycle, Canizales et al.

and the operational conditions at the time of failure. The textual data is preprocessed to remove noise and prepare it for analysis. Tokenization is used to split the text into words or tokens, stop word removal is used to remove common words that do not contribute much meaning, lemmatization is used to reduce words to their base or root form, and normalization is used to convert all text to lowercase and remove special characters.

The dataset contains 1351 records of maintenance cases, each with the following variables:

Reporter: The individual who reported the failure on the production line.

Area: The specific production line where the failure occurred.

Machine: The specific component or equipment within the production line that experienced the failure.

Failure Description: A textual description detailing the nature of the failure, explaining what is happening.

Technician: The assigned technician responsible for resolving the failure.
Solution: A description of the actions taken to resolve the failure (Table 2).

Table 2. Maintenance cases dataset example.

Reporter	Area	Machine	Failure Description	Technician	Solution
Javier Iracheta	Autolinea 70	Touch screen	The carriers PM screen is very slow; it's not possible to report carriers properly	Antonio Rangel	Raspberry Pi reboot initiated. Machine is now operational
Carlos Gallardo	Autolinea 50	Scaner	The barcode scanner is not reading the barcode gun	Antonio Rangel	Make sure the keyboard is connected to the scanner for it to work properly
Edgar Herrera Lopez	Autolinea 60	Pc autolinea	It closed unexpectedly	Antonio Rangel	Application launch successful. Machine is now operational
Luis Rodriguez	Autolinea 100	Control	A pop-up window is appearing, blocking group downloads	Brian Jesus Puente	The equipment was inspected and found to be functioning properly
Javier Iracheta	Autolinea 90	Control	To change the number of remaining runs for a batch	Isaac Suarez Cortez	The touch screen has been reset

The failure report database contains records from the electroplating production line, detailing the failures that have occurred. The file serves as a problem reporting system, where production personnel record the failures they encounter, and maintenance personnel attend to them and document the solution (Fig. 7).

Each row of the file represents an individual failure report, with the following information:

ID: A unique identifier for each report.
Report Time: Exact date and time when the failure was recorded.

Alarm Recommendation Intelligent System 259

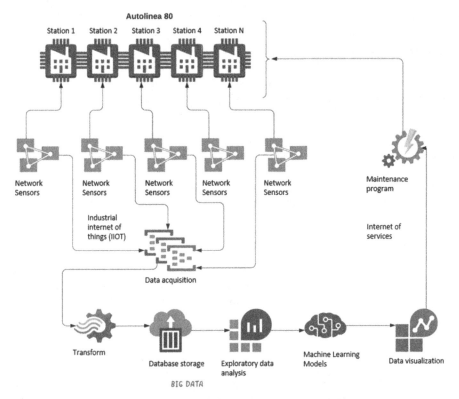

Fig. 7. Autolinea 80 Predictive Maintenance System Diagram

Reporter: Name of the operator who reported the failure.
Area: Specific area of the line where the failure occurred (e.g., Autolinea 70, Separators 3 in 1, Sample Pull, etc.)
Machine: Specific machinery or equipment where the failure occurred (e.g., TOUCH SCREEN, PLC, SCANER, etc.)
Description: Detailed description of the failure, including symptoms and observations.
Start Time: Time when the failure began.
End Time: Time when the failure ended.
Failure Category: General category of the failure (e.g., Electrical, Mechanical, Operating System, etc.)
colFalla2: Second failure category.
colSubSol1: Subcategory of the solution.
Solution: Description of the solution applied to resolve the failure.
Department Solved: Department that solved the failure (e.g., Software, Electronic Maintenance, Process Engineering, etc.)
Time Solved: Exact date and time when the failure was solved.
Response Time: Response time of the maintenance team from the time the failure was reported to the time it was started to be solved.
Downtime: Total time the machine was out of service due to the failure.

TPM: Indicates if the failure was handled based on the TPM (Total Productive Maintenance) system.
Apply Downtime: Indicates if a Downtime was generated due to the failure.
colCamara: Additional information related to the equipment camera.
colTamano: Additional information related to the equipment size.
colArchivo: Additional information related to the equipment file.
colReel: Additional information related to the equipment reel.

This file provides crucial information for production line management. The data can be analyzed to identify common causes of failures. It can be identified which types of failures occur most frequently, allowing preventive maintenance efforts to be focused.

The effectiveness of the maintenance team can be evaluated. Response time and downtime can be analyzed to identify areas where efficiency can be improved.

3.2 Case Representation

Each case is represented by a set of attributes that describe the failure and its solution. These attributes may include the MLCC type, the failure symptoms, the operational conditions at the time of failure, the solution applied, among others. Textual attributes are vectorized using NLP techniques, such as TF-IDF or Word2Vec, to convert them into feature vectors that can be processed by the system.

> Key Considerations for Case Representation in a CBR (Case-Based Reasoning) System Utilizing NLP (Natural Language Processing):
>
> **Text-to-Vector Conversion:**
> The CBR system needs to convert the textual descriptions of failures into a form that can be processed by the system. This is achieved using NLP techniques such as TF-IDF (Term Frequency-Inverse Document Frequency) or Word2Vec.
>
> **Case Attributes:**
> Each case in the CBR knowledge base is represented as a set of attributes that describe the failure and its solution. These attributes can be textual attributes: textual description of the failure, observed symptoms, and the solution applied.
>
> **Feature Selection:**
> Not all attributes are equally important for case comparison. It is necessary to select the most relevant features for retrieving similar cases. Feature selection techniques, such as mutual information feature selection, can be used to identify the most relevant attributes. This helps to improve the accuracy and efficiency of the CBR system by reducing noise and enhancing the quality of case retrieval.
>
> **Similarity Method:**
> To calculate the relationship between cases, a distance or similarity metric is used. The cosine method is one of the most widely used in NLP. This method measures the relationship between two vectors in a multidimensional space. The closer the vectors are, the more alike the failure descriptions will be.

3.3 CBR System Development

A search mechanism is implemented to retrieve similar cases from the case database. Cosine similarity is used to calculate the similarity between new failure descriptions and historical cases. An adaptation algorithm is implemented to adjust the solutions of retrieved cases to the new problem. A feedback system is used to evaluate the effectiveness of the recommendations and update the case database (Figs. 8 and 9),

4 Implementation and Evaluation of the Intelligent System

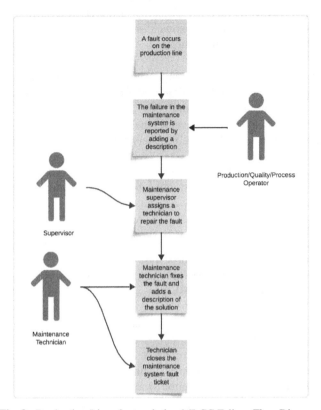

Fig. 8. Production Line electroplating MLCC Failure Flow Diagram.

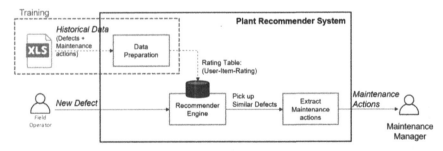

Fig. 9. Intelligent Maintenance Recommendation System Diagram based on [19].

The impact of failures on production can be measured. The total downtime of the line can be calculated to determine the cost of failures in production. A Python code was developed implementing a case-based reasoning (CBR) system to generate recommendations based on maintenance reports. This system utilizes Natural Language Processing (NLP) techniques to analyze and understand textual descriptions of failures and solutions. The process is divided into the following stages:

Data Preparation

Data Loading: The data is loaded from an Excel file using the Pandas library. This file contains two columns: "Failure Description" and "Solution," representing failure reports and solutions from the production system.

Data Filtering: Rows containing NaN values in the "Failure Description" or "Solution" columns are removed using the pandas dropna() function. This ensures that complete and valid data is being worked with.

Case Creation: A list of cases is created, where each case contains the failure description and its corresponding solution. This list will serve as the knowledge base for the CBR system.

Text Preprocessing

Library Imports: The necessary libraries for natural language processing are imported: nltk (Natural Language Toolkit) and stopwords.

Resource Download: The necessary nltk resources are downloaded for text preprocessing, such as the set of Spanish stopwords and the WordNet database for lemmatization.

Tokenization: The text of the failure description is divided into individual words (tokens) using the word_tokenize () function.

Lowercase Conversion: All tokens are converted to lowercase to avoid considering the same words with different capitalization as distinct.

Stopword Removal: Stopwords (common words like articles, prepositions, etc.) are removed from the text using the stopwords Words () function. These words do not contribute relevant semantic meaning to the analysis, and their elimination helps to simplify the process.

Lemmatization: Lemmatization of the words is performed, meaning each word is reduced to its root form or lemma. This allows grouping words with different grammatical forms (such as plurals or conjugated verbs) under the same term.

Similarity Calculation

Cosine_similarity () Function: This function calculates the similarity between two failure descriptions using the cosine method.

Evaluation of the CBR System

Evaluation would allow determining the accuracy and efficiency of the system to generate recommendations. In summary, a CBR system that utilizes NLP techniques to analyze failure descriptions and generate recommendations based on similar cases stored in a knowledge base. This system can be used to automate the problem-solving process and improve predictive maintenance efficiency. The CBR system can be used to identify patterns in failure reports and generate recommendations based on past experiences. The information obtained through the analysis of failure reports can be used to predict potential future problems and optimize maintenance strategies. This can help reduce downtime and repair costs, improving production system efficiency and reliability.

Limitations

The current system is simple and may have limitations for complex cases or those involving industry-specific vocabulary. It is important to continue developing and improving the system with new features, such as:

1. Adding new cases to the knowledge base.
2. Improving the text preprocessing process.
3. Evaluating the system's accuracy with a test dataset.
4. Integrating the system with other predictive maintenance systems.

The use of CBR and NLP for predictive maintenance is an area with great potential for development. It is a good starting point for creating more sophisticated and efficient systems that allow for the optimization of maintenance processes and the improvement of operational safety and profitability.

4.1 Explain the Value of Cosine Similarity

The cosine similarity metric measures the similarity between two vectors in a multidimensional space. This metric is used to compare the similarity between failure descriptions, represented as word frequency vectors. The closer the vectors are, the more similar the descriptions are. Cosine similarity ranges from 0 (completely different) to 1 (identical).

Steps to Calculate Cosine Similarity Metric

Text Preprocessing: Stop words are removed, and words are lemmatized.

Word Vectors: A word vector is created for each failure description, where each element of the vector represents the frequency of a specific word in the description. **Similarity Calculation:** The cosine similarity between the two-word vectors is calculated using the cosine similarity () function from the sklearn library.The cosine similarity formula is applied, which is the cosine of the angle between the two generated vectors.

To evaluate the performance of the CBR system, we can use metrics such as:

Cosine Similarity: This metric measures the similarity between the current failure description and the failures in the case base.

CBR System

cbr () Function: This function implements case-based reasoning to generate recommendations.

Search for the Most Similar Case: The function iterates through all cases stored in the list and calculates the cosine similarity between the current failure description and the failure description of each case. The similarity and corresponding case are saved in a list.

Identification of the Most Similar Case: The case with the highest cosine similarity to the current failure description is searched for.

Recommendation Return: The solution of the most similar case is returned as a recommendation (Figs. 10 and 11).

$$Similarity(T,S) = \sum_{i=1}^{n} f(T_i, S_i) \times w_i$$

Where:
T is the target case
S is the source case
N is the number if features in each case
I is an individual feature form 1 to n
F is a similarity function for feature i in case T and S
w is the importance weighting of feature I

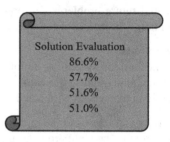

Fig. 10. Cosine Similarity Metric

	Caso	Similitud Coseno
0	LA PANTALLA DE CARRIERS PM VA MUY TRABADO, NO ...	0.866025
1	PANTALLA APAGADA PM CARRIERS	0.577350
2	PANTALLA NO DEJA REPORTAR CARRIERS	0.516398
3	PANTALLA PM CARRIERS FUERA INAVILITADA	0.516398
4	LA PANTALLA PARA REPORTAR CARRIERS NO ESTA FUN...	0.510310
5	PANTALLA INHIBIDA PARA REPORTAR LOS CARRIERS	0.471405
6	RESETEAR PANTALLA DELOS CARRIERS	0.433013
7	SE QUEDO TRABADO LA PANTALLA	0.387298
8	SE QUEDO TRABADO LA PANTALLA	0.387298
9	SE QUEDO TRABADO LA PANTALLA	0.387298

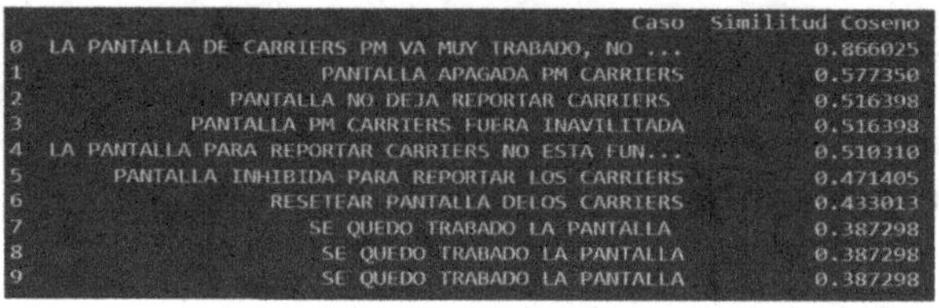

Fig. 11. Example of Cases and their Cosine Similarity Metric

Example of Use

Several examples are presented of how to use the CBR system to generate recommendations from new failure descriptions.

A new failure is introduced, the similarity with existing cases is calculated, and the recommendation of the most similar case is returned.

The following image shows examples of recommendations based on the CBR system programmed in Python (Fig. 12).

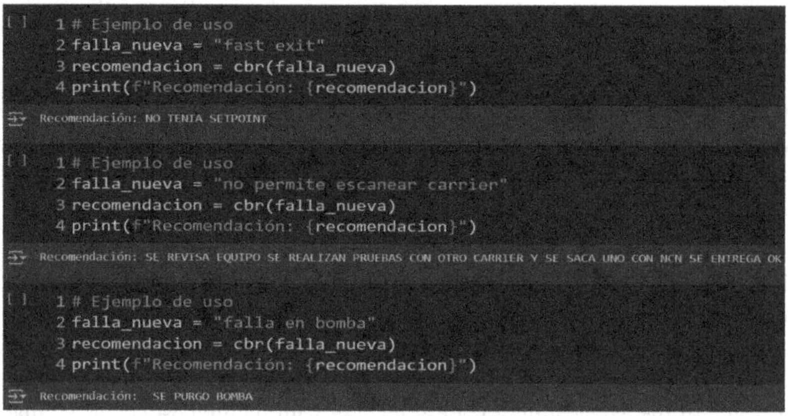

Fig. 12. Examples of CBR System Recommendations.

Implement Continuous Improvement Strategies: The data in the file can serve as the basis for implementing preventive measures, such as operator training, improving maintenance processes, and updating equipment.

5 Discussion of Results

The impact of the system on efficiency and accuracy in detecting and resolving MLCC electroplating failures is analyzed. The limitations of the system and potential future improvements are discussed (Figs. 13, 14, 15 and 16).

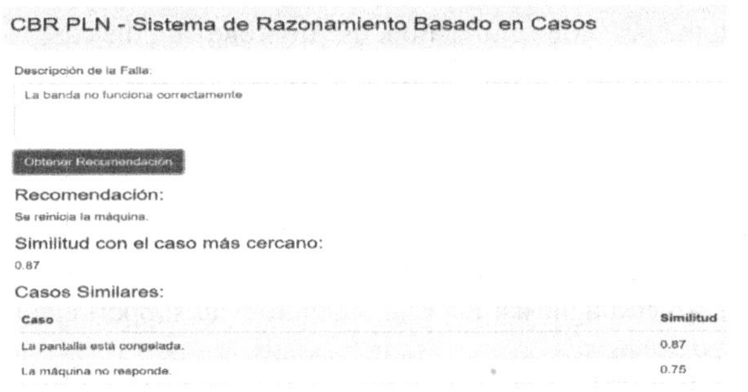

Fig. 13. CBR Failure Recommendation System Interface.

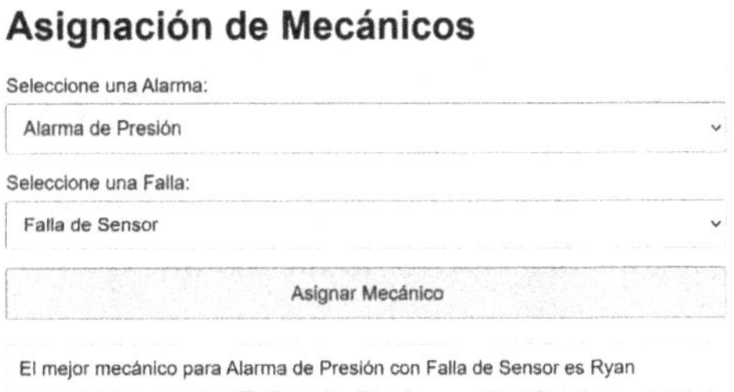

Fig. 14. Mechanic Assignment Interface Based on Failure Type.

Comparison of Intelligent System vs. Technical Maintenance with Tacit Knowledge: Before and After the AI Solution.

Let's analyze and compare both workflows to highlight the benefits of incorporating these technologies in the context of predictive maintenance.

1. **Process Without CBR (Before)**

    In the scenario prior to the implementation of the CBR system, the process begins when a failure is identified on the production line. The production operator, upon detecting the failure, reports it and provides a basic description. Based on this description, an initial decision is made on whether the operator can repair the failure themselves or if technical intervention is required. If it is determined that the

Fig. 15. Time and cost results of repairs performed by our CBR-based Intelligent System.

operator cannot handle the repair, the maintenance supervisor assesses the situation and decides on the best course of action, usually referring the case to a maintenance technician. The technician then diagnoses, repairs the failure, and documents the solution.

Disadvantages:

Dependence on tacit knowledge: The effectiveness of the process relies heavily on the knowledge and experience of the technician and supervisor. Longer response time: Each step requires human intervention, which can lead to delays.

Risk of incorrect diagnosis: Lack of technological support can increase the risk of an incorrect diagnosis.

2. **Process with CBR (After)**

With the integration of the CBR system, the process becomes more refined and automated. The operator continues to report the failure, but now the CBR system plays a crucial role. Upon receiving the failure description, the CBR system searches its database of previous cases and suggests a potential solution to the operator. If the system determines that the repair is beyond the operator's capabilities, the case is automatically escalated to the supervisor and maintenance technician. The technician, armed with recommendations from the CBR system, can address the failure more effectively. Additionally, the technician has the option of closing the failure in the CBR system, enriching the database for future inquiries.

Advantages:

Use of artificial intelligence for diagnosis: CBR technology allows the use of past ex-perience stored in the form of cases to make faster and more accurate diagnoses.

Reduced response time: By automating part of the diagnostic and decision-making pro-cess, failure resolution is accelerated.

Continuous learning: Each new case resolved is added to the database, continuously improving the system's diagnostic capabilities.

Consistency in problem solving: The use of a standardized system reduces variability in responses to similar problems.

The implementation of intelligent systems such as CBR and NLP in predictive mainte-nance not only improves the efficiency and effectiveness of problem solving but also contributes to the development of a more autonomous maintenance system that is less dependent on direct human intervention. This not only reduces production downtime but also improves knowledge management within the organization (Fig. 17).

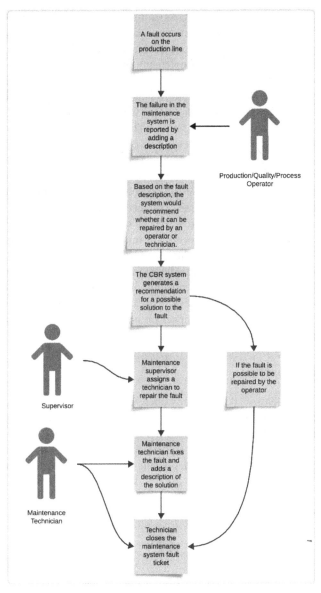

Fig. 16. Production Line Failure Flow Diagram with CBR and NLP Recommendation System.

6 Conclusions

This study has demonstrated the value of an intelligent recommendation system based on case-based reasoning (CBR) and natural language processing (NLP) for alarm management in a multi-layer ceramic capacitor (MLCC) production line. The results highlight the system's ability to improve both efficiency and accuracy in detecting and resolving failures, specifically in the MLCC electroplating process, by providing recommendations that are not only contextualized but also adapt to the specific characteristics of the operators involved. Among the main contributions of this system are:

Intelligent Alarm Prioritization: Implementing automatic alarm prioritization has optimized response times for the maintenance team, minimizing fatigue caused by false alarms or alerts of lesser importance. This facilitates a more efficient approach to addressing critical failures that require immediate attention.

Highly Contextualized Recommendations: The system does not solely rely on identifying alarms but also analyzes the machine's operating parameters to offer accurate suggestions that consider the specific circumstances of each incident. This results in greater effectiveness in operational decisions and reduces downtime.

Personalized Recommendations: A key differentiating factor is the system's ability to adapt recommendations according to the experience level and preferences of each operator, allowing for a personalized approach to failure management. This customization fosters a more dynamic and flexible work environment where maintenance interventions are more precise and tailored to individual needs.

Overall, the system's implementation has significantly optimized maintenance planning, reduced unplanned downtime, and thus increased the overall efficiency of the production line. Additionally, it has been evident that the use of technologies such as CBR and NLP can have a positive impact on an organization's ability to respond to complex failures and manage maintenance more strategically.

7 Future Research

Despite the achievements made, this study opens up various lines of research and development that can further enhance the system's performance in an industrial context. Among the main areas proposed for future work are:

Integration of Real-Time Data: Incorporating real-time data from sensors can lead to greater accuracy in identifying and diagnosing failures. This would allow not only for

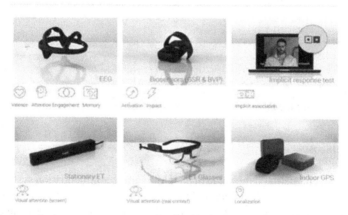

Fig. 17. Neuromarketing components associated with the improvement of a Haptic Interface in Industry 5.0.

anticipation of potential problems but also for more robust predictive monitoring, which could further reduce downtime and improve preventive maintenance planning.

Exploration of Advanced Architectures: The use of deep neural networks and other machine learning techniques could offer significant improvements in the system's ability to analyze complex datasets. This would not only enhance the accuracy of recommendations but also allow for a better interpretation of failure patterns that are more difficult to identify with traditional techniques.

Data Preprocessing: Research is proposed into the use of more sophisticated techniques for noise removal and feature extraction from input data. These techniques can optimize the system's workflow, increasing both the speed and accuracy of recommendations.

Expansion of the Case Database: Including a greater number of historical cases, especially those related to more complex failures, would increase the robustness of the system by providing a greater diversity of reference situations. This would allow the system to improve the quality of recommendations and better adapt to a wider range of failures and operating contexts.

Safety Evaluation: Future research should include a comprehensive evaluation of the system's impact on the operational safety of the production line. This analysis should consider both the identification of risks and the implementation of new safety measures to ensure a safe working environment while optimizing maintenance operations.

Integration with Other Management Systems: It is essential to study the possibility of integrating this recommendation system with other maintenance management and industrial operations platforms, such as CMMS (Computerized Maintenance Management Systems). This integration would facilitate more unified management of resources and processes, which would be especially useful in complex and automated production environments.

Development of an Adaptive Machine Learning System: Research is suggested into the creation of a system capable of learning and continually improving based on user interactions. This adaptive approach could allow the system not only to generate more accurate recommendations over time but also to dynamically adjust to changing production line conditions and user needs.

Incorporation of Haptic Interfaces: To enhance interaction and understanding of recommendations by maintenance personnel, the implementation of haptic interfaces that provide tangible feedback, such as vibrations or physical signals simulating failure symptoms, could be explored. This would not only improve the accuracy of interventions but also enrich the user's experience in interacting with the system.

The opportunities for future research therefore range from improvements in data processing and analysis to the integration of emerging technologies that can further enhance the efficiency and effectiveness of industrial maintenance operations.

References

1. Keleko, A.T., Kamsu-Foguem, B., Ngouna, R.H., et al.: Artificial intelligence and real-time predictive maintenance in industry 4.0: a bibliometric analysis. AI Ethics **2**, 553–577 (2022). https://doi.org/10.1007/s43681-021-00132-6

2. Sezer, E., Romero, D., Guedea, F., Macchi, M., Emmanouilidis, C.: An Industry 4.0-enabled low-cost predictive maintenance approach for SMEs. In: 2018 IEEE International Conference on Engineering, Technology and Innovation (ICE/ITMC), pp. 1–8. Stuttgart (2018). https://doi.org/10.1109/ICE.2018.8436307
3. Koenig, F., Found, P.A., Kumar, M.: Innovative airport 4.0 condition-based maintenance system for baggage handling DCV systems. Int. J. Product. Perform. Manag. **68**(3), 561–577 (2019). https://doi.org/10.1108/IJPPM-04-2018-0136
4. Paprocka, I., Kempa, W.M., Ćwikła, G.: Predictive maintenance scheduling with failure rate described by truncated normal distribution. Sensors (Basel). **20**(23), 6787 (2020). https://doi.org/10.3390/s20236787. PMID:33261083, PMCID:PMC7730090
5. Fink, O., Wang, Q., Svensén, M., Dersin, P., Lee, W., Ducoffe, M.: Potential, challenges and future directions for deep learning in prognostics and health management applications. Eng. Appl. Artif. Intell. **92**, 103678 (2020), https://doi.org/10.1016/j.engappai.2020.103678. ISSN 0952-1976
6. Zonta, T., da Costa, C., da Rosa, R., de Lima, M., da Trindade, E., Pyng Li, G.: Predictive maintenance in the Industry 4.0: a systematic literature review. Comput. Industr. Eng. **150**, 106889 (2020). https://doi.org/10.1016/j.cie.2020.106889. ISSN 0360-8352
7. Guo, C., et al.: Transformer failure diagnosis using fuzzy association rule mining combined with case-based reasoning. IET Gener. Transm. Distrib. **14**, 2202–2208 (2020). https://doi.org/10.1049/iet-gtd.2019.1423
8. Gao Y., et al.: Mechanical equipment health management method based on improved intuitionistic fuzzy entropy and case reasoning technology. Eng. Appl. Artif. Intell. **116**, 105372 (2022). https://doi.org/10.1016/j.engappai.2022.105372. ISSN 0952-1976
9. Dalzochio, J., et al.: Machine learning and reasoning for predictive maintenance in Industry 4.0: current status and challenges. Comput. Ind. **123**, 103298, (2020). https://doi.org/10.1016/j.compind.2020.103298. ISSN 0166-3615
10. Lei, Y., Yang, B., Jiang, X., Jia, F., Li, N., Nandi, A.: Applications of machine learning to machine fault diagnosis: a review and roadmap. Mech. Syst. Sig. Process. **138**, 106587 (2020). https://doi.org/10.1016/j.ymssp.2019.106587. ISSN 0888-3270
11. Montero, J., Schwartz, S., Vingerhoeds, R., Grabot, B., Salaün, M.: Towards multi-model approaches to predictive maintenance: a systematic literature survey on diagnostics and prognostics. J. Manuf. Syst. 56, 539–557 (2020). https://doi.org/10.1016/j.jmsy.2020.07.008. ISSN 0278-6125
12. Psarommatis, F., May, G., Azamfirei, V.: Envisioning maintenance 5.0: insights from a systematic literature review of Industry 4.0 and a proposed framework. J. Manuf. Syst. **68**, 376–399 (2023). https://doi.org/10.1016/j.jmsy.2023.04.009. ISSN 0278–6125
13. Liu, B.: A Case-based reasoning and explaining model for temporal point process. In: Recio-Garcia, J.A., Orozco-del-Castillo, M.G., Bridge, D. (eds.). ICCBR 2024. LNCS, vol. 14775. Springer, Cham (2024). https://doi.org/10.1007/978-3-031-63646-2_9
14. Grumbach, L., Winzig, A., Bergmann, R.: Towards a case-based support for responding emergency calls. In: Recio-Garcia, J.A., Orozco-del-Castillo, M.G., Bridge, D. (eds.) ICCBR 2024. LNCS, vol. 14775, pp. 273–288. Springer, Cham (2024). https://doi.org/10.1007/978-3-031-63646-2_18
15. Wilkerson, K., Leake, D.: On implementing case-based reasoning with large language models. In: Recio-Garcia, J.A., Orozco-del-Castillo, M.G., Bridge, D. (eds.) ICCBR 2024. LNCS, vol. 14775, pp. 414–417. Springer, Cham (2024). https://doi.org/10.1007/978-3-031-63646-2_26
16. Malburg, L., Hotz, M., Bergmann, R.: Improving complex adaptations in process-oriented case-based reasoning by applying rule-based adaptation. In: Recio-Garcia, J.A., Orozco-del-Castillo, M.G., Bridge, D. (eds.) ICCBR 2024. LNCS, vol. 14775, pp. 50–66. Springer, Cham (2024). https://doi.org/10.1007/978-3-031-63646-2_4

17. Jiménez-Diaz, G., Díaz-Agudo, B.: Visualization of similarity models for CBR comprehension and maintenance. In: Recio-Garcia, J.A., Orozco-del-Castillo, M.G., Bridge, D. (eds.). ICCBR 2024. LNCS, vol. 14775, pp. 67–80. Springer, Cham (2024). https://doi.org/10.1007/978-3-031-63646-2_5
18. Mathew, A., Ma, L., Narasimhan, L.: Case-based reasoning for data warehouse schema design. In: 36th International Conference on Computers and Industrial Engineering, ICC and IE 2006 (2006)
19. Al-Najim, A., Al-Amoudi, A., Ooishi, K., Al-Naser, M.: Intelligent Maintenance Recommender System. In: 2022 7th International Conference on Data Science and Machine Learning Applications (CDMA), Riyadh, Saudi Arabia, pp. 212–218 (2022). https://doi.org/10.1109/CDMA54072.2022.00040
20. Darias, J.M., Caro-Martínez, M., Díaz-Agudo, B., Recio-Garcia, J.A.: Using case-based reasoning for capturing expert knowledge on explanation methods. In: Keane, M.T., Wiratunga, N. (eds.) ICCBR 2022. LNCS, vol. 13405, pp. 3–17. Springer, Cham (2022). https://doi.org/10.1007/978-3-031-14923-8_1

Implementation of a Hybrid Tabu Search Algorithm for Solving the Capacitated Vehicle Routing Problem

Carlos Condado-Huerta[1], José Alberto Hernández-Aguilar[1](✉), Martín H. Cruz-Rosales[1], Víctor Pacheco-Valencia[2], and Julio César Ponce-Gallegos[3]

[1] FCAeI, Autonomous University of the State of Morelos, Av. Universidad 1001, 62209 Cuernavaca, Morelos, Mexico
{jose_hernandez,mcr}@uaem.mx

[2] CInC, Autonomous University of the State of Morelos, Av. Universidad 1001, 62209 Cuernavaca, Morelos, Mexico
vhpacval@gmail.com

[3] Autonomous University of Aguascalientes, Av. Universidad 940, Ciudad Universitaria, 20100 Aguascalientes, Mexico
julio.ponce@edu.uaa.mx
http://www.uaem.mx/fcaei, http://cinc.uaem.mx/,
https://www.uaa.mx/portal/

Abstract. The Capacitated Vehicle Routing Problem (CVRP) consists of generating a customer route for each vehicle in which the sum of the customer demands does not exceed the vehicle's capacity. Each vehicle must start from the depot, and when it finishes visiting the last customer, it must return to the depot, thus managing to visit all customers only once by any of the vehicles and have a route solution where the sum of the distances of all the routes is the minimum. This research addressed different algorithms to give a feasible solution with an approximation concerning the best known. Fifty benchmark instances were used, which allowed us to define that the best algorithm to use is Tabu Search, achieving in these fifty instances an approximation close to the optimal, which leads us to conclude that the Tabu Search metaheuristic is easy to adapt for routing problems, allowing to obtain satisfactory results. This research proposed using the randomness and exchange method for an initial solution with a good neighborhood. Work is done with the Tabu Search algorithm to explore the solution space and obtain a global solution, and finally, an intensification process is carried out that improves the solution. This research shows that the proposed algorithms give feasible solutions with an approximation of no more than 10% difference, and in some cases, the best-known solution is obtained.

Keywords: CVRP · Tabu Search · Intensification

1 Introduction

Capacitated Vehicle Routing Problem (CVRP) consists of constructing tours for vehicles with limited capacity, which start and end at a location known as the depot, to transfer goods from the depot to various customers, each with a demand. One salesman must fully satisfy the demand in a single visit.

Vehicle Routing Problem (VRP) presents many variations due to environmental restrictions. CVRP was selected from all the variations of the VRP. CVRP is classified as a combinatorial optimization problem belonging to the class of NP-hard problems; that is, the optimal solution is obtained using algorithms that use non-polynomial time, and it cannot be obtained in a reasonable time. The solutions have different methods to solve the CVRP, divided into exact, heuristic, and metaheuristic methods. In the exact methods, dynamic programming and backtracking search, among others, are more noticeable. In the heuristic methods, constructive algorithms and insertion algorithms, among others, stand out, and in the metaheuristic methods, Simulated Annealing (SA), Tabu Search (TS), and Swarms stand out. We can also find hybridizations of metaheuristics. The problem can be approached with different objectives. Initially, it was solved by assigning clients to vehicles based solely on demand and vehicle capacity without considering the route. Once the subsets were determined, the route for each vehicle was calculated, and the total cost of the routes was summed. A new objective was then considered, which focused on grouping the closest nodes together while ensuring that demand did not exceed vehicle capacity. This research proposes solving a set of capacitated vehicle routing problems with different numbers of nodes and vehicle capacities, using three steps: 1) obtain an initial solution with the method of randomness and nearest neighbor, 2) the TS algorithm according to [1], and 3) An intensification stage.

Many optimization problems need to be solved. The vast majority of these problems have practical implications in science, engineering, economics, and business, which are very complex and challenging to solve [2].

Problems with similar restrictions are analyzed to understand the CVRP problem. They can be used as a starting point, for example, the Traveling Salesman Problem (TSP) [3], which attacks an integral part of the VRP; in this case, the TSP only focuses on making a route in which all the clients are visited only once, and that with the client they start they must end. The shortest path must be considered to make this journey, providing that the route obtained at the end finds an optimal solution (reported in the literature). Over the years, different studies have applied various methodologies to solve the problem, from exact, heuristic, and metaheuristic methods. Among these, the following stand out: implicit enumeration methods, Branch & Bound, Cutting Plane or Dynamic Programming techniques, constructive algorithms, simulated stitching, TS, and Ant Colony (AC), among others. After this problem, the Multiple Traveling Salesman Problem (MTSP) variant was introduced [4]. This problem incorporates a warehouse, unlike the TSP, which does not have one, and some salesmen. The objective is to build exactly one route for each salesperson so that each customer is visited once by one of the salespersons. The route will start and end at the warehouse and can contain, at most, a limited number of customers.

Thus, our research problem is to find a solution to the CVRP [5], which consists of creating routes for vehicles that will store or distribute the demand of a set of customers.

Unlike the TSP and the MTSP, a new restriction or variant has been added: they are no longer sellers and have become vehicles with a limited capacity, and the customers acquire a demand that must be collected or delivered by a single vehicle. The other variants remain the same as in the MTSP.

The CVRP is essential from a theoretical and practical point of view because every company needs to acquire or deliver merchandise to many clients, n, with limited units. In addition, it must consider the time it will take to deliver to each client, the maintenance of each vehicle, and the route it must take, considering that the vehicles must return to the point of origin.

The justification for this project lies in the challenges faced by companies, large or small, in managing the delivery or collection of goods. Planning efficient vehicle routes that ensure each client is visited once within a specific time frame is crucial for reducing operational costs, such as fuel consumption and service time. The objective is to design an algorithm based on the TS metaheuristic to explore and evaluate feasible solutions for the CVRP, aiming to achieve a solution close to the optimal.

Hypothesis (H1) states that if a metaheuristic algorithm efficiently generates and evaluates feasible solutions for the CVRP, then it delivers a solution with a cost close to the optimal one. On the contrary, the null hypothesis (H0) states that if the algorithm is not an effective alternative, then it does not deliver a solution with a cost close to the optimal one.

The algorithm will be evaluated with 40 instances with the following characteristics: 10 units, 200 clients as maximum considering small and medium instances, and 350 clients for large instances of the CVRPLIB [6] library.

The document is organized in the following way. The related work is presented in the second section, and the most relevant works and a timeline are provided. Section three presents the proposed methodology. Section four shows the results obtained, compared with those reported in the literature. Finally, conclusions and future work are presented.

2 Related Work

CVRP has been studied extensively throughout history, and various methods have been implemented to obtain a solution.

Danzig and Ramser (1959) [7], after analyzing a fuel distribution problem, proposed the CVRP and explored how fuel could be distributed to different gas stations, considering the capacity of the vehicles and the demand at each station. Their solution ensures that the company delivers on time and improves the filling times.

Laporte et al. 2000 [8] mention several heuristics that were used to achieve acceptable results for the VRP, commenting that there are heuristics that analyze problems where the number of trucks may or may not be a stopping parameter or where a maximum cost per trip can be defined and that it can also be divided into two phases to reach the solution (first part, obtain k subsets of the original set and then carry out the construction of the route by applying an algorithm for TSP).

Naddef et al. 2002 [9] proposed the Branch and Cut algorithm, mentioning that tests were performed on six instances, of which 2 of them with 135 nodes managed to obtain favorable results for the best-known solution (BKS).

Lori et al. 2006 [10] proposed the TS metaheuristic for solving VRP. Due to computational time issues, it was noted that the algorithm takes longer with more than 50 nodes. It was tested with 27 instances, of which 8 obtained the same results as the BKS, in 11 instances, they were close to the BKS, and in 8 instances, they obtained better results than BKS.

Lori et al. 2007 [10] proposed an exact algorithm (branch and cut). Twelve instances were analyzed, and a maximum solution time of 1 h was considered, obtaining a result that instances of up to 35 nodes can be tested, obtaining excellent results.

Gendreau et al. 2008 [11] proposed the use of TS metaheuristic. Fifty-eight instances were considered with a solution time of 24 h per instance, demonstrating that optimal results were obtained in 33 instances.

Fuellerer et al. 2009 [12] proposed the AC metaheuristic; it was tested with 36 instances executing ten times for each one, and it was compared with the results obtained with TS, demonstrating that the AC is better since it only managed to win TS in 2 instances.

Strodl et al. 2010 [13] proposed variable neighborhood search and exact algorithm (branch and link); it was tested with 36 instances, demonstrating that a time limit must be placed to obtain favorable results and that the algorithm does not enter a cycle.

Leung et al. 2010 [14] proposed SA and packing, tested with 36 instances with an average of 110 min per solution, showing an improvement over the AC algorithm.

Ping Chen, Hou-Kuan Huang, and Xing-Ye Dong (2010) [15] proposed the variable neighborhood descent algorithm, which uses hybrid metaheuristics since accuracy, speed, simplicity, and flexibility are the four criteria to be evaluated of a metaheuristic, combining the strengths of iterated local search and variable neighborhood descent.

Gómez-Atuesta and Rangel-Carvajal (2011) [16] proposed the TS algorithm and SA; using ten instances, they were able to be within 5% of the difference against the best, of which two of them were equal.

Mazin et al. 2012 [17] proposed the Genetic Algorithm (GA) and Cellular-GA; they applied the GA to give a solution with a lower traversal cost; this algorithm is implemented in very large instances, helping to reduce processing time.

Farhanna Mar'i, Wayan Firdaus Mahmudy, and Purnomo Budi Santoso (2018) [18] proposed basic SA and enhanced-SA; they mention that more iterations will increase the computational time; after five iterations, a better result is given for a CVRP, and enhanced-SA gives a better solution to the CVRP.

Wulan Herdianti, Alexander A S Gunawan, Siti Komsiyah (2020) [19]. It is mentioned that the problem expands exponentially but that it is possible to manually make a tour with five nodes. Additionally, the particle swarm optimization method (PSO) algorithm cannot solve the VRP with many locations; the pigeon-inspired optimization (PIO) algorithm requires the position of the pigeon and the speed to be defined, and each interaction is updated when comparing between algorithms the PIO algorithm gave a solution with a lower cost with an approximate difference of 2.5% for the solution of the PSO algorithm.

Hassan Moussa (2021) [20]. Mentions that the K-means algorithm was used to find subsets of the set of nodes using a centroid to associate nodes and then modify this centroid until all the nodes are associated and then generate k centroids once the k subsets are obtained; the Dijkstra algorithm is used to perform the tours of the k subsets.

Pacheco-Valencia et al. (2022, 2023) [21, 22] proposes a 3-phase heuristic algorithm. In phase 1, cities are partitioned into k disjoint subsets. In phase 2, feasible tours are constructed for each subset, and phase 3 improves these tours using a Hill Climbing method.

3 Problem Description

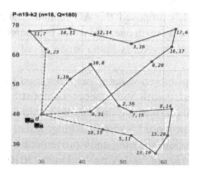

Fig. 1. CVRP instance P-n19-k2

Figure 1 shows a CVRP instance, P-n19-k2, which includes 19 vertices: the depot, labeled as d, and 18 customers with their demands, each identified by a pair of numbers separated by a comma. The instance involves two vehicles, each with a capacity of $Q = 160$.

Let $G = (V, E)$ be a complete, undirected, weighted graph that defines a CVRP, see Fig. 1. Let $i \in V$ be a set of vertices that includes a depot, d, and n customers, each with a demand d_i. Let $(i, j) \in E$ have an associated Euclidean distance in two-dimensional space, $w(i, j) \in E$, and let k be vehicles with a homogeneous capacity Q.

A feasible solution S for a CVRP consists of k tours S^j, for $j = 1, \ldots, k$, which are sequences of customers that a vehicle will visit, starting and ending at the depot. The cost of that solution, $C(S)$ (Eq. 1), is the sum of the distances of the k tours, i.e. $C(S^1) + C(S^2) + \cdots + C(S^k)$, where $C(S^j)$ represents the sum of the distances of the edges included in tour S^j.

$$C(S) = \sum_{j=1}^{k} C(S^j) \qquad (1)$$

The objective is to find a feasible solution S^* whose cost $C(S^*)$ is the minimum, i.e., $C(S^*) = \min_S C(S)$, subject to the following constraints.

- All vehicles must start and end their tours at the depot.
- Each customer must be visited exactly once by one of the vehicles.

- The total demand of the customers included in a tour must not exceed the vehicle's capacity, Q.
- The vehicle must have at least one customer to be considered for its route.

4 Methodology

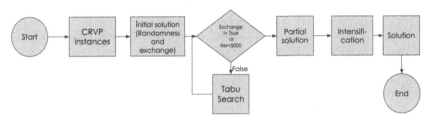

Fig. 2. Flowchart of the proposed methodology

Figure 2 shows the methodology to be used, in which CVRP instances are used, an instance is taken, it goes to the randomness method which gives a different order to the nodes, then to the exchange method a feasible initial solution is created (it considers the accumulated demand which should not exceed the capacity of each vehicle, an exchange of nodes is carried out consecutively until a solution is found that is better than the initial solution, once the solution is obtained, it goes to the simple-TS method which will carry out exchanges between routes regardless of whether it is improved or not considering the capacity of the vehicle once it carries out all the possible exchanges it shows the solution obtained, this solution goes to an intensification process, modifications are made to the routes to find a better one which will allow us to get closer to the BKS or to be very close to the best known one not exceeding 5% difference.

The CVRP data instances were obtained from CVRP Library [6]. Each instance has a file name; its last digits refer to the number of nodes considered. The internal structure of this file has the file name, name of the author, type (CVRP), dimension (number of nodes), capacity, coordinates of the nodes section, demand section, and depot section. The nodes section has the ID number of the node, X, and Y coordinates. The demand section has the number of the node and its demand. The depot section has the depot identifiers. This information was read and stored in a data node structure formed with ID, X, Y, and demand.

The Initial_Solution algorithm (see Algorithm 1 in Fig. 3) generates an initial solution by constructing k tours, where each tour starts and ends at the depot d, and vertices are randomly selected from the instance to be added to the tours. The algorithm begins by initializing an empty solution S and setting $j = 1$, representing the current tour index. For each tour S^j, the depot d is added as the starting point. Then, the algorithm enters a loop where it randomly selects, without repetition, a vertex i from the set $V \setminus \{d\}$, i.e., all vertices except the depot, and append it to the current tour S^j. This process continues until the total demand of the vertices in the tour $\sum_{i \in S^j} d_i$ exceeds the vehicle's capacity Q. Once this condition is met, the depot d is added as the endpoint of the tour, and the tour S^j is included in the solution S. The procedure then moves to the next tour by incrementing j and repeats the process until all k tours are constructed. Finally, the solution S, containing all the tours, is returned.

Algorithm 1: Initial_Solution()
1. $S \leftarrow \emptyset$
2. $j \leftarrow 1$
3. **while** $j \leq k$ **do**
4. $S^j \leftarrow (d)$
5.   **while** $\sum_{i \in S^j} d_i \leq Q$ **do**
6.   Select a vertex i randomly, with no-repetition, from $V \setminus \{d\}$
7. $S^j \leftarrow S^j \cup \{i\}$
8. $S^j \leftarrow S^j \cup \{d\}$
9. $S \leftarrow S \cup \{S^j\}$
10. $j \leftarrow j + 1$
11. **return** S

Fig. 3. Pseudocode to construct an Initial Solution for a CVRP instance

The Neighborhood_By_Swapping algorithm (see Algorithm 2 in Fig. 4) generates a neighborhood of solutions by systematically swapping elements within and between tours. It begins by initializing two indices, j and l, to iterate through the tours S^j, for $j = 1, \ldots, k$ and their vertices S_l^j, for $l = 1, \ldots, |S^j|$. The algorithm checks if the current vertex S_l^j is the last one in the tour S^j; if not, it prepares the next vertex for a swap within the same tour, i.e. $S_{l'}^{j'}$ with $j' = j + 1$ and $l' = l + 1$, otherwise, it moves to the next tour, i.e. with $j' = j$ and $l' = 1$. A nested loop is then used to perform the swaps between the selected vertices, using an auxiliary variable aux to temporarily store values during the exchange. This process continues, swapping elements between all tours until every possible neighboring solution has been generated. The algorithm systematically explores all possible swaps to identify neighboring solutions.

After having both lists, the nodes are grouped for each of the vehicles, considering that in each vehicle, only the number of nodes is added considering the demand, which, when added, does not exceed the capacity of the vehicle. When all the nodes are finished being added and all the vehicles have been occupied, they move on to the following exchange method: exchange one node against all the nodes except the depot to find the best route. This process is carried out for each of the nodes. Suppose a better solution is found than the one being evaluated. In that case, this solution replaces the previous solution, and the process is repeated until no better solution is found than the one being assessed.

The Simple_Tabu_Search algorithm (See Algorithm 3 in Fig. 5) begins with an initial solution S_0, which is assigned to both the current solution S and the best solution S_{best}. An empty tabu list L is initialized to store recently visited attributes. The algorithm then iterates for a fixed number of steps, specifically $100 \times n$. In each iteration, it searches for the best neighboring solution S' of the current solution S. If the attribute $a(S')$ is in the tabu list, that neighbor is skipped; otherwise, if its cost $C(S')$ is lower than that of the current best neighbor $S_{bestNeighbor}$, the best neighbor is updated. After evaluating all neighbors, the current solution S is updated to $S_{bestNeighbor}$, and the tabu list is updated by

Algorithm 2: Neighborhood_By_Swapping(S)

1   $j \leftarrow 1$
2   $l \leftarrow 1$
3   **while** $j \leq k$ **do**
4   **while** $l \leq |S^j|$ **do**
5   **if** $l \neq |S^j|$ **then**
6            $j' \leftarrow j$
7            $l' \leftarrow l + 1$
8   **else**
9            $j' \leftarrow j + 1$
10           $l' \leftarrow 1$
11   **while** $j' \leq k$ **do**
12   **while** $l' \leq k$ **do**
13               $aux \leftarrow S_l^j$
14               $S_l^j \leftarrow S_l^j$
15               $S_{l'}^{j'} \leftarrow aux$
16               $l' \leftarrow l' + 1$
17            $j' \leftarrow j' + 1$
18      $l \leftarrow l + 1$
19   $j \leftarrow j + 1$

Fig. 4. Pseudocode to generate the solutions from the current solution S with neighborhood swapping

Algorithm 3: Simple_Tabu_Search(S_0)

1   $S \leftarrow S_0$
2   $S_{best} \leftarrow S_0$
3   $L \leftarrow \emptyset$
4   **for** $l \leftarrow 1$ **to** $100 \times n$ **do**
5      $S_{bestNeighbor} := \emptyset$
6   **for** each neighbor S' of S **do**
7   **if** $a(S') \in L$ **then**
8   skip
9   **if** $C(S') < C(S_{bestNeighbor})$ **then**
10           $S_{bestNeighbor} \leftarrow v$
11      $S := S_{bestNeighbor}$
12      $L \leftarrow L \cup \{a(S_{bestNeighbor})\}$
13   **if** If the size of L is greater than tabu tenure **then**
14   Delete the oldest item in L
15   **if** $C(S_{bestNeighbor}) < C(S_{best})$ **then**
16         $S_{best} \leftarrow S_{bestNeighbor}$
17 **return** S_{best}

Fig. 5. Pseudocode for the simple tabu search metaheuristic

adding $a(S_{bestNeighbor})$. The oldest attribute is removed if the tabu list exceeds a specified tenure. Additionally, if the cost of the new best neighbor is lower than the current best solution, S_{best} is updated. After completing the iterations, the algorithm returns the best solution S_{best} found during the search.

The best initial solution is obtained at the end of the process, which will be processed with the Simple_Tabu_Search algorithm, see Algorithm 3. The TS algorithm has different processes that must be adjusted, such as short-term memory, short-term memory, and intensification. Small, medium, and large instances were used to adjust the parameters. The adjustments allowed the best solutions to be found.

4.1 Dataset

- Set A (Augerat, 1995) 27 instances.
- Set B (Augerat, 1995) 9 instances.
- Set E (Christofides and Eilon, 1969) 11 instances.
- Set F (Fisher, 1994) 3 instances.

4.2 Hyper-parameters Tunning

- Tabu List: In order not to generalize and after performing the tests, it was found that dividing the number of nodes by 4 (25%) yielded better results.
- Iterations: Tests were carried out from 100 iterations to 8000, in which it was found that in the first 1000 iterations, it was possible to interact with 400, but when continuing with the tests, it was considered that at 5000, better results were obtained.
- Penalty: Tests were carried out from 4 to 12, and 5 was considered.

5 Results and Discussion

The hybrid algorithm developed in this project was coded in Python 3.9 and executed on a DELL Inspiron 15 3000 laptop with the following specifications: Intel(R) Core(TM) i3-1005G1 CPU @ 1.20 GHz, 8 GB RAM, running Windows 11 Home Single Language.

Table 1 shows the results obtained using seven Augerat instances taken from the CVRP library and solved by our proposal. The first column lists the instance name, the second column shows the BKS for the instance, the third column provides the cost obtained using the Random-Insertion (RI) plus Tabu Search (TS), and the fourth column displays the percentage difference between the RI + TS algorithm and the BKS. The fifth column presents the cost obtained with RI + TS + Intensification Stage, and the sixth column shows the percentage difference between the results of this previous approach and the BKS. The average % difference is shown at the end of Table 1, 14.52 for RI + TS, and 5.45 for the proposed methodology.

Table 2 shows the results obtained using the instances taken from CVRP library and solved by [16]. The first column lists the instance name, the second column shows the BKS for the instance, the third column provides the cost obtained using the proposed hybrid algorithm, the fourth column displays the percentage difference between the hybrid algorithm proposed and the BKS. The fifth column presents the cost reported by [16], and finally, the sixth column shows the percentage difference between the results of these authors and the BKS.

Table 1. Comparison BKS versus Random-Insertion(RI) + TS, versus RI + TS + Intensification

Instance	BKS	RI + TS	% difference	RI + TS + Intensification	% difference
A-n32-k5	784.000	839.691	7.103	792.612	1.098
A-n33-k5	661.000	721.092	9.091	685.956	3.775
A-n33-k6	742.000	889.327	19.855	746.026	0.543
A-n34-k5	778.000	868.493	11.631	834.986	7.325
A-n36-k5	799.000	859.872	7.619	854.577	6.956
A-n37-k5	669.000	852.354	27.407	725.681	8.472
A-n37-k6	949.000	1128.775	18.944	1043.943	10.005
		Average %	14.52	Average %	**5.45**

Table 2. Comparison versus Gomez-Atuesta and Rangel-Carvajal [16]

Instance	BKS	Hybrid Algorithm	% difference	Gomez et al. 2011	% difference
E-n16-k3	262	278.726	6.384	284.23	8.485
E-n23-k3	569	568.562	−0.077	568.56	−0.077
E-n33-k4	845	868.644	2.798	844.29	−0.084
P-n50-k10	696	776.565	11.573	732.50	5.244
P-n22-k2	216	221.427	2.513	217.85	0.856
P-n55-k10	694	730.93	5.321	720.76	3.856
M-n200-k17	1373	1559.536	13.586	1579.09	15.009

As Table 2 shows, our hybrid approach obtained two better solutions than the one provided by [16], and obtained the BKS for the instance E-n23-k3. This solution is shown in Fig. 6a. On the other hand, the proposed solution provided solutions very close to BKS, for instance, E-n33-k4 and P-n22-k3.

As is shown in above Fig. 6, the proposed solution (a), for instance E-n23-k3 is equal to the BKS (b).

6 Conclusions and Future Work

The proposed hybrid Tabu Search Algorithm addresses the problem in stages, allowing feasible results to be obtained. In some cases, the difference from the best known is below 5% and even 0%, but in others, it exceeds 10%. Derived from the results obtained, we conclude that H1 is True, and the objective was met.

In future work, combining the exchange and insertion algorithm within the Tabu Search is proposed. We intend to test the algorithm in real-world problems; particularly,

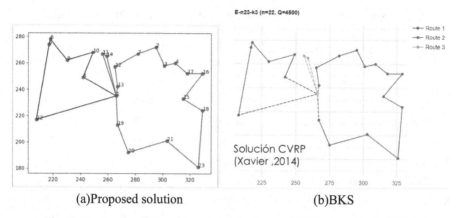

(a)Proposed solution   (b)BKS

Fig. 6. Comparison of the proposed solution versus BKS for instance E-n23-k3

we are interested in testing the methodology in applications of smart cities, like smart waste collection, product delivery by fleet of electric vehicles, and multi-layer transportation systems. The idea is to provide real-time mobile apps to track vehicles and their routes to estimate arrival times for these problems.

Acknowledgments. This study was supported by CONAHCYT-Mexico (scholarship granted to CAC).

Disclosure of Interests. The authors have no competing interests to declare.

References

1. Gómez Atuesta, D., Rangel Carvajal, C.: Formular las metaheurísticas búsqueda tabú y recocido simulado para la solución del cvrp (2011)
2. González-Hernández, I.J., Granillo-Macías, R., Martínez-Flores, J.L., Sánchez-Partida, D., Gibaja-Romero, D.E.: Hybrid model to design an agro-food distribution network considering food quality. Int. J. Ind. Eng. **26**(4) (2019)
3. Flood, M.M.: The traveling-salesman problem. Oper. Res. **4**(1), 61–75 (1956)
4. Bektas, T.: The multiple traveling salesman problem: an overview of formulations and solution procedures. Omega **34**(3), 209–219 (2006)
5. Lei, H., Laporte, G., Guo, B.: The capacitated vehicle routing problem with stochastic demands and time windows. Comput. Oper. Res. **38**(12), 1775–1783 (2011)
6. Reinhelt, G.: TSPLIB: a library of sample instances for the tsp (and related problems) from various sources and of various types (2014). https://comopt.ifi.uniheidelberg.de/software/TSPLIB95
7. Dantzig, G.B., Ramser, J.H.: The truck dispatching problem. Manage. Sci. **6**(1), 80–91 (1959)
8. Laporte, G., Gendreau, M., Potvin, J.Y., Semet, F.: Classical and modern heuristics for the vehicle routing problem. Int. Trans. Oper. Res. **7**(4–5), 285–300 (2000)
9. Naddef, D., Rinaldi, G.: Branch-and-cut algorithms for the capacitated VRP. In: The Vehicle Routing Problem, pp. 53–84. SIAM (2002)

10. Di Lorenzo, D., Bianconcini, T., Taccari, L., Gualtieri, M., Raiconi, P., Lori, A.: Ten years of routist: Vehicle routing lessons learned from practice. In: International Conference on Optimization and Decision Science, pp. 265–276. Springer, Cham (2022)
11. Gendreau, M., Iori, M., Laporte, G., Martello, S.: A tabu search heuristic for the vehicle routing problem with two-dimensional loading constraints. Networks: an Int. J. **51**(1), 4–18 (2008)
12. Fuellerer, G., Doerner, K.F., Hartl, R.F., Iori, M.: Ant colony optimization for the two-dimensional loading vehicle routing problem. Comput. Oper. Res. **36**(3), 655–673 (2009)
13. Strodl, J., Doerner, K.F., Tricoire, F., Hartl, R.F.: On index structures in hybrid metaheuristics for routing problems with hard feasibility checks: an application to the 2-dimensional loading vehicle routing problem. In: Proceedings of the Hybrid Metaheuristics: 7th International Workshop, HM 2010, Vienna, Austria, 1–2 October 2010, pp. 160–173. Springer, Cham (2010)
14. Leung, S.C., Zheng, J., Zhang, D., Zhou, X.: Simulated annealing for the vehicle routing problem with two-dimensional loading constraints. Flex. Serv. Manuf. J. **22**, 61–82 (2010)
15. Chen, P., Huang, H., Dong, X.Y.: Iterated variable neighborhood descent algorithm for the capacitated vehicle routing problem. Expert Syst. Appl. **37**(2), 1620–1627 (2010)
16. Atuesta, D.F.G., Carvajal, C.E.R.: Formular las metaheurísticas búsqueda tabú y recocido simulado para la solución del cvrp (capacitated vehicle routing problem) (2011)
17. Mohammed, M.A., Ahmad, M.S., Mostafa, S.A.: Using genetic algorithm in implementing capacitated vehicle routing problem. In: 2012 International Conference on Computer & Information Science (ICCIS), vol. 1, pp. 257–262. IEEE (2012)
18. Mari, F., Mahmudy, W.F., Santoso, P.B.: An improved simulated annealing for the capacitated vehicle routing problem (CVRP). Jurnal Ilmiah Kursor **9**(3) (2018)
19. Herdianti, W., Gunawan, A.A., Komsiyah, S.: Distribution cost optimization using pigeon inspired optimization method with reverse learning mechanism. Procedia Comput. Sci. **179**, 920–929 (2021)
20. Moussa, H.: Using recursive kmeans and Dijkstra algorithm to solve CVRP. arXiv preprint arXiv:2102.00567 (2021)
21. Pacheco-Valencia, V.H., Vakhania, N., Hernández-Mira, F.Á., Hernández-Aguilar, J.A.: A multi-phase method for euclidean traveling salesman problems. Axioms **11**(9), 439 (2022). https://doi.org/10.3390/axioms11090439
22. Pacheco-Valencia, V.H., Vakhania, N., Hernández-Aguilar, J.A.: An algorithm to solve the euclidean single-depot bounded multiple traveling salesman problem. In: 2023 IEEE World Conference on Applied Intelligence and Computing (AIC), pp. 7–12 (2023). https://doi.org/10.1109/AIC57670.2023.10263920

Intelligent System Associated with a Stochastic RoRo Shipping Problem for a Fleet of Differentiated Vehicles and Collection of Specific Problems

Alberto Ochoa-Zezzatti[1], Irma Hernández-Báez[2(✉)], Axel Bernal[3], and Humberto García-Castellanos[1]

[1] Doctorado en Tecnología, Universidad Autómoma de Ciudad Juárez, Av. Plutarco Elías Calles #1210 Fovisste Chamizal, 32310 Ciudad Juárez, Chihuahua, Mexico
alberto.ochoa@uacj.mx
[2] Universidad Politécnica del Estado de Morelos, Boulevard Cuauhnáhuac 566, Col Lomas del Texcal, 62550 Jiutepec, Morelos, Mexico
ihernandez@upemor.edu.mx
[3] Facultad de Ingeniería, Universidad Anáhuac, Av. Universidad Anáhuac 46, Col. Lomas Anáhuac, 52786 Huixquilucan, State of Mexico, Mexico

Abstract. This research presents the optimization of the vehicle loading process on Roll-on/Roll-off (Ro-Ro) ships through the implementation of a heuristic method. Efficiency in the distribution of vehicles within a Ro-Ro vessel is crucial to maximize cargo capacity and reduce operation times, factors that directly impact the costs and competitiveness of maritime transport companies.

A mathematical model and its corresponding implementation in Python were developed, with the aim of improving the allocation of spaces for different types of vehicles, taking into account the physical restrictions of the ship and the characteristics of each vehicle. The proposed model was validated using real data, demonstrating a significant improvement in charging efficiency compared to traditional methods. The results obtained indicate that the application of heuristic techniques in the loading planning of Ro-Ro vessels can offer optimal solutions in reduced processing times, which translates into a more efficient and profitable operation. This work contributes to the field of logistics and maritime transport, providing a practical tool for decision-making in cargo management on Ro-Ro ships.

Keywords: Stochastic Ro-Ro shipping problem · heuristic method · vehicle loading process optimization

1 Introduction

International maritime transport, a crucial pillar of global trade, has undergone significant transformations in recent decades. In this context, the Roll-on/Roll-off (RoRo) method has emerged as a strategic and innovative solution for the transport of vehicles. This modality, which facilitates the loading and unloading of vehicles by their own means, has revolutionized the sector, offering a superior alternative in terms of efficiency and effectiveness to traditional loading and unloading methods. With its adaptability to a wide range of vehicles, from light cars to heavy industrial equipment, RoRo has gained a prominent position in global maritime logistics [1].

Despite its advantages, RoRo shipping operations face significant challenges. The inherently variable nature of operating conditions, ranging from fluctuating market demand to diverse vehicle dimensions and specifications, constitutes a stochastic challenge of considerable complexity. This stochastic aspect of RoRo transport requires advanced mathematical models and innovative solutions to optimally plan cargo stowage, maximize cargo revenues, and minimize voyage costs [2]. In addition, the pressure for greater efficiency is driving the search for advanced logistics solutions [3].

In the face of these challenges, the implementation of intelligent systems is emerging as a promising solution. The use of advanced technologies, such as artificial intelligence (AI) and machine learning, offers the possibility of significantly improving operational efficiency in the RoRo transport sector. These technologies have the potential to transform decision-making processes, optimizing everything from space allocation on ships to comprehensive route planning. Integrating speed optimization into shipping route planning, for example, can lead to significantly improved solutions, highlighting the importance of considering these variables in an integrated manner [4].

Managing a heterogeneous vehicle fleet in the context of RoRo shipping involves specific challenges [5]. The diversity of vehicles requires detailed management in terms of spatial allocation, security and operational procedures. This management directly impacts the profitability and sustainability of the service, with vehicle stowage efficiency being a critical factor affecting both the operability and profitability of the RoRo shipping service [6]. The increasing complexity of global supply chains and the need for fast and efficient responses to market changes underline the urgency of developing smarter and more adaptive systems. Innovations in vessel design and loading methods, such as those proposed in recent patents, may offer novel solutions to further improve the efficiency of RoRo transport [7].

This study focuses on exploring and developing an intelligent system in the context of RoRo shipping, specifically for the efficient management of differentiated vehicle fleets under stochastic conditions. The aim of this work is to contribute significantly to the existing body of knowledge by providing practical and applicable solutions to the real challenges faced by the RoRo transport industry globally.

Figure 1 illustrates a flow chart of logistics operations in a Ro-Ro port, showing the integration of storage areas, access ramps, and reception and delivery areas, for the efficient handling of semi-trailers and rolling vehicles [8].

2 State of the Art

Stowage planning on RoRo vessels presents major challenges. One of them is the complexity inherent in optimizing cargo space. RoRo vessels carry a wide variety of roll-on/roll-off cargo, including cars, buses, vans, semi-trailers, and project cargo. This heterogeneity in size and type makes it difficult to create optimal stowage plans that maximize revenue while meeting stability and safety constraints [9]. Another challenge is minimizing turnaround time in port. Shifting cargo in port to accommodate loading and unloading of other units consumes valuable time. Stowage planning should seek to minimize or eliminate these shifts to optimize turnaround time in port [9].

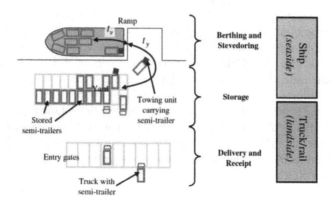

Fig. 1. Flowchart of logistics operations in a Ro-Ro port.

Some of these challenges have been addressed, for example, to facilitate stowage planning, approaches have been developed to discretize the deck area. One method is to divide the deck into rectangular lanes or slots that represent possible cargo locations [13]. Another approach uses a regular grid with cells that merge to create cargo slots [10]. Other authors propose a more refined approach that uses individual grids for each type of cargo on the decks where it can be stowed, ensuring an accurate representation of the available space and optimizing space utilization [9].

Various mathematical models and resolution methods have been proposed. Stowage planning on RoRo vessels has been addressed using operational research techniques, including mixed integer linear programming (MILP), using commercial solvers, recognizing the computational complexity of large-scale problems [11, 13]. Other solutions, such as tabu search and genetic algorithms, have been proposed to find optimal or near-optimal solutions [9, 12].

Another line of research is the evaluation of the cost of shifting. To discourage shifting of cargo in port, methods have been developed to evaluate the cost associated with this operation. One approach is to penalize blocked cargo by assigning a displacement cost proportional to the cargo size, its relative location and the displacement distance [9, 13].

It is recognized that stowage planning must be considered in the broader context of tactical and strategic decisions in maritime transport. This includes cargo selection, route scheduling and vessel allocation to maximize the overall profitability of operations [9].

According to the review, significant progress in stowage planning on RoRo vessels is highlighted, driven by the need to improve efficiency and profitability. Advances in discretization techniques and the development of sophisticated mathematical models and solving methods have contributed to the optimization of stowage plans. However, the literature does not delve into the integration of smart technologies, such as machine learning or artificial intelligence, into the stowage planning process, which presents an opportunity.

3 Methodology

This study employs a heuristic approach to optimize the accommodation of five types of vehicles on different vessels, taking into account the dimensions, weight and characteristics of each vehicle, as well as load balancing to ensure the stability of the vessel during transport. The methodology is structured in four main components, as shown in Fig. 2.

Fig. 2. Methodology used for optimizing the vehicle loading process on Roll-on/Roll-off (Ro-Ro) vessels.

Data Collection and Preparation. At this stage, data collection is carried out, gathering information corresponding to the vessels and vehicles and their corresponding dimensions. In addition, data cleaning and standardization is carried out, including the elimination of missing and duplicate values and the standardization of formats. Finally, an exploratory analysis of the data is carried out.

Table 1 presents the vehicles selected after data collection and preparation, including dimensions and quantities.

Table 2 shows the dimensions and load capacities of the available vessels.

Development of the Optimization Model. Within this stage of the methodology, the optimization technique is selected, defining the heuristic method and establishing the mathematical model. Subsequently, the model is implemented in Python. Finally, an initial validation of the model is carried out through the comparison of results with historical data, making the necessary adjustments.

To maximize the use of space in loading and unloading vehicles on RoRo vessels, a methodology was used based on assigning vehicles to vessels in such a way as to optimize the use of available capacity. First, the volume of each type of vehicle was calculated using its dimensions (length, width, and height). Each vessel has a defined loading capacity in linear meters. This capacity is used as a constraint for load planning. A function was developed to plan the loading of vehicles on a vessel. The function considers the available capacity and assigns vehicles iteratively until there is no more available capacity. The load planning function was applied to each vessel to determine the number and type of vehicles that can be loaded on each one. Three methods were applied to maximize the use of vessel space: the greedy method, the MILP method, and a heuristic method, all with random loading.

Greedy Method: This method selects vehicles from shortest to longest length until the maximum capacity of the vessel is reached.

Table 1. Vehicles and Dimensions

Vehicle Type	Length (m)	Width (m)	Heigth (m)	Units
Compact Vehicle	4.2	1.8	1.6	150
Sedan	4.8	1.9	1.4	200
SUV	5	2	1.8	120
Pick-up Truck	5.5	2.2	1.8	100
Minivan	5.1	1.9	1.8	80
Sports Vehicle	4.5	1.9	1.2	60
Small Truck	6	2.3	2.5	50
Medium Truck	7.5	2.5	3	40
Large Truck	9	2.5	3.5	30
Small Bus	7	2.5	3	25
Large Bus	12	2.5	3.5	15
Motorcycle	2	0.8	1.1	200
Tractor	4	2.5	2.5	10
Utility Vehicle	4	2	1.8	70
Van	5	2	2.2	50
Classic Car	4.5	1.8	1.4	30
Tank Truck	10	2.5	3.5	20
Garbage Truck	7.5	2.5	3	15
Military Vehicle	6	2.5	2.5	20
Ambulance	5.5	2	2.5	10
Police Car	4.8	1.9	1.5	25
Electric Vehicle	4.5	1.8	1.6	50
Autonomous Vehicle	4.6	1.8	1.6	30
Luxury Vehicle	5	1.9	1.4	20
Agricultural Vehicle	4	2.5	2.5	10
Construction Vehicle	6	2.5	3	15

MILP Method: Uses Mixed Integer Linear Programming to find the optimal solution that maximizes the use of available space.

Simple Heuristic Method: This approach allocates vehicles iteratively based on their length and available capacity.

The MILP and Simple Heuristic methods provide more complete and optimized solutions compared to the Greedy method. Vehicles are placed horizontally along the length of the ship to maximize the use of available linear space. The choice of method depends on the availability of computational resources and the execution time required.

Table 2. Vessels and features.

Name	Length (m)	Beam (m)	Tonnage (Tons)	Carrying Cap. (Linear meters)
Ship San Jorge	142.5	23.2	4000	1680
Ship San Guillermo	116.3	17.4	2000	800
Ship Santa Marcela	134.8	17.5	2500	950

Simulation and Evaluation of Scenarios. At this stage, Unity is used to carry out a simulation through the modeling of loading and unloading operations, then an evaluation of loading configurations under different conditions is carried out.

To validate the optimization model and the proposed heuristics, different scenarios were simulated. The main objective is to observe how the model behaves under different load configurations, types of vehicles and ship capacities, thus ensuring the robustness and efficiency of the algorithm in real situations. The simulations allow the validation of the model and the adjustment of parameters to maximize the use of space and load balancing on Ro-Ro ships.

The scenarios are designed to test the model's capacity in situations ranging from a balanced and uniformly distributed load, to an unbalanced load with a high number of heavy vehicles.

Figure 3 shows the algorithm of the heuristic used to solve the problem.

```
1. Initialize Cargo Matrix:
   - Create a 2D matrix of size (Ship Length/50cm) x (Ship Width/50cm).
2. Define Decision Variables:
   - For each vehicle define: length, width, height, and quantity.
3. Sort Vehicles:
   - Sort vehicles by volume, from largest to smallest.
4. For each vehicle in the sorted list:
   - For each cell in the matrix, starting from the top left corner:
   4.1 Check if the vehicle fits in the available cells.
      - Check dimensional and space restrictions between vehicles.
      - If it fits:
        - Place vehicle in the matrix.
        - Record coordinates of its position.
      - If it does not fit, continue with the next cell.
5. Display Results:
   - Display the matrix with the positions of the vehicles.
   - Print list of coordinates for each placed vehicle.
```

Fig. 3. Heuristic Algorithm for Cargo Space Optimization on a Ro-Ro Ship

Analysis of Results and Validation. Finally, in the last stage of the methodology, the results are validated and a sensitivity analysis is performed to evaluate the robustness of the model.

4 Mathematical Model

The development of the mathematical model for this project is based on the combination of two classical optimization problems: The Knapsack Problem and the Bin Packing Problem. Both problems are fundamental in optimization theory and have direct applications in logistics and resource planning.

Bin Packing Problem
This problem deals with how to pack objects of different sizes into fixed-capacity containers in a way that minimizes the number of containers used. In this project, the "containers" are the available spaces on the ship and the "objects" are the vehicles to be loaded. The objective is to maximize the use of the available space on the ship, ensuring that the vehicles are placed so that they do not overlap and that a proper balance of loads is maintained [14, 15].

Knapsack Problem
This problem focuses on maximizing the value of a set of objects that must be placed in a backpack with limited capacity. In our case, "value" can be associated with the ship's payload capacity, while "limited capacity" translates into the dimensions and maximum weight that the ship can support. In addition, the weight of each vehicle is a critical variable that directly affects the balance of loads along and across the ship, which is crucial for stability during transport [16].

By integrating these two approaches, the developed model not only focuses on maximizing the number of loaded vehicles (as in Bin Packing), but also on ensuring that the weight is evenly distributed to avoid stability problems (as in the Knapsack Problem).

Load balancing, which is a critical feature in our model, is not directly addressed in traditional models of the Knapsack Problem or Bin Packing Problem, but is essential in practical applications such as the one we are considering. Therefore, we have extended the mathematical model to include additional constraints to ensure that loads are balanced across the length and width of the ship, using techniques inspired by these classical problems.

Consider the decision variable X_{ijk} as follows:

- X_{ijk}: Binary variable indicating whether vehicle i is placed at position (j, k) on the vessel.

  - $i \in \{1, 2, \ldots, N\}$: Set of vehicle types.
  - $j \in \{1, 2, \ldots, L\}$: Set of possible positions along the length of the ship.
  - $k \in \{1, 2, \ldots, W\}$: Set of possible positions within the width of the ship.

  Consider the following parameters:

- L: Vessel length in terms of 0.1 m (10 cm) cells.
- W: Width of the vessel in terms of 0.1 m (10 cm) cells.
- L_i: Vehicle length i in terms of cells.
- W_i: Vehicle width i in terms of cells.
- P_i: Vehicle weight i.
- n_i: Number of vehicles of the type i.
- s: Additional space (in cells) between vehicles to avoid collisions (in this case, 2 cells representing 20 cm).

The proposed mathematical model is presented below.

$$\text{Max } Z = \sum_{i=1}^{N} \sum_{j=1}^{L} \sum_{k=1}^{W} x_{ijk} \tag{1}$$

s.t.

$$\sum_{j'=j}^{j+l_i-1} \sum_{k'=k}^{k+w_i-1} x_{ijk} \leq 1 \tag{2}$$

$$x_{ijk} + x_{i'j'k'} \leq 1 \tag{3}$$

$$\sum_{j=1}^{L} \sum_{k=1}^{W} x_{ijk} = n_i \tag{4}$$

$$\sum_{j=1}^{L} \sum_{k=1}^{W} x_{ijk} = n_i \tag{5}$$

$$x_{ijk} = 0 \text{ si } \exists x_{i'j'k'} \text{ with distance less than } s \tag{6}$$

$$\sum_{i=1}^{N} \sum_{j=1}^{L/2} \sum_{k=1}^{W} p_i x_{ijk} = \sum_{i=1}^{N} \sum_{j=\frac{L}{2}+1}^{L} \sum_{k=1}^{W} p_i x_{ijk} \tag{7}$$

The objective (1) is to maximize the number of vehicles placed on the vessel, while ensuring balanced weight distribution. Constraint (1) ensures that each vehicle occupies its own space. Constraint (2) prevents vehicles from overlapping. Constraint (3) ensures that each vehicle is placed exactly once. Constraint (4) ensures that the additional space between vehicles is respected. Constraint (5) corresponds to load balance, this restriction ensures that the weight between the left and right side of the ship is balanced.

In (8) the calculation of charging time is presented, where t_{ijk} is the charging time for the vehicle at position (j, k) and v is the charging speed in meters per minute.

$$t_{ijk} = \frac{\sqrt{J^2 + K^2}}{v} \tag{8}$$

5 Results

Below are the results obtained after evaluating three loading scenarios on the Ro-Ro vessel, using different vehicle configurations to analyze the impact on stability, loading time and weight distribution. Each scenario explores a different accommodation strategy—uniform load, heavy load and unbalanced load—allowing to observe how the heuristic algorithm optimizes the use of available space and ensures safety during the voyage.

Scenario 1. Uniform Load

In this scenario, vehicles of all types are uniformly distributed on the selected ship. The objective is to observe how the model optimizes space and balances loads when the distribution is relatively homogeneous.

The main objective was to evenly distribute the different types of vehicles on the ship, prioritizing adequate balancing both along and across the ship. A detailed evaluation of the results obtained is presented below.

The evaluated scenario sought to maximize load balance on the Ro-Ro vessel by distributing different types of vehicles. Large trucks (L) were placed at the lateral ends, pick-up trucks (P) in the center, and SUVs (U) and compact vehicles (C) were distributed laterally, seeking to balance weight. Sedans (S) occupied positions in the central rows. The total load balance showed a slight bias towards the left side of the vessel, with a weight of 72,200 kg, compared to 59,300 kg on the right side. Despite this imbalance, the distribution allowed for efficient occupation of space. The total loading time was 65.58 min.

Figure 4 shows the arrangement of vehicles within the vessel.

Scenario 2. Heavy Load

Mainly large trucks are loaded, with a small number of light vehicles. This scenario simulates situations where heavy loads predominate and it is crucial to assess how the loads are balanced.

In the heavy-lift scenario, vehicle distribution focused on optimizing vessel stability with a majority load of large trucks (L) and pick-up trucks (P). The large trucks were evenly distributed along the vessel, occupying mainly the outer and middle rows, while the pick-ups were placed in the central areas on both sides. The lighter vehicles (SUVs, sedans and compacts) were positioned in the upper rows, ensuring a perfect weight balance between the left and right side of the vessel. This arrangement resulted in an exceptional balance, with 105,000 kg on both sides, and an optimized total loading time of 31.49 min. Table 3 summarizes the distribution and balance of vehicles on the vessel during the heavy load scenario.

Scenario 3. Unbalanced Load

Vehicles are loaded in unequal proportions to simulate an unbalanced situation. This scenario tests the model's ability to adjust the loads and prevent the ship from becoming unbalanced.

The unbalanced loading scenario evaluated how the system handles an uneven distribution of vehicles, highlighting the impact on the stability of the vessel. A single large

```
/_____\
|LLLLLLLLLLLLLL.PPPPPPPPPPPPPPPPPPPLLLLLLLLLL|
|LLLLLLLLLLLLLL.PPPPPPPPPPPPPPPPPPPLLLLLLLLLL|
|LLLLLLLLLLLLLL.PPPPPPPPPPPPPPPPPPPLLLLLLLLLL|
|LLLLLLLLLLLLLL.PPPPPPPPPPPPPPPPPPPLLLLLLLLLL|
|LLLLLLLLLLLLLL.PPPPPPPPPPPPPPPPPPPLLLLLLLLLL|
|LLLLLLLLLLLLLL.PPPPPPPPPPPPPPPPPPPLLLLLLLLLL|
|LLLLLLLLLLLLLL.PPPPPPPPPPPPPPPPPPPLLLLLLLLLL|
|LLLLLLLLLLLLLL.PPPPPPPPPPPPPPPPPPPLLLLLLLLLL|
|LLLLLLLLLLLLLL.PPPPPPPPPPPPPPPPPPPLLLLLLLLLL|
|LLLLLLLLLLLLLL.PPPPPPPPPPPPPPPPPPPLLLLLLLLLL|
|LLLLLLLLLLLLLL....SSSSSSSSSSSSSSSS..LLLLLLLLLL|
|LLLLLLLLLLLLLL....SSSSSSSSSSSSSSSS..LLLLLLLLLL|
|LLLLLLLLLLLLLL....SSSSSSSSSSSSSSSS..LLLLLLLLLL|
|LLLLLLLLLLLLLL....SSSSSSSSSSSSSSSS..LLLLLLLLLL|
|LLLLLLLLLLLLLL....SSSSSSSSSSSSSSSS..LLLLLLLLLL|
|LLLLLLLLLLLLLL....SSSSSSSSSSSSSSSS..LLLLLLLLLL|
|LLLLLLLLLLLLLL....SSSSSSSSSSSSSSSS..LLLLLLLLLL|
|UUUUUUUUUUUUUU...SSSSSSSSSSSSSSSCCCCCCCCCCCC|
|UUUUUUUUUUUUUU...SSSSSSSSSSSSSSSCCCCCCCCCCCC|
|UUUUUUUUUUUUUU..................CCCCCCCCCCCC|
|UUUUUUUUUUUUUU..................CCCCCCCCCCCC|
|UUUUUUUUUUUUUU..................CCCCCCCCCCCC|
|UUUUUUUUUUUUUU..................CCCCCCCCCCCC|
|UUUUUUUUUUUUUU..................CCCCCCCCCCCC|
|UUUUUUUUUUUUUU..................CCCCCCCCCCCC|
|UUUUUUUUUUUUUU..............................|
|UUUUUUUUUUUUUU..............................|
|................UUUU........................|
|................UUUU........................|
|................UUUU........................|
|................UUUU........................|
|................UUUU........................|
|................UUUU........................|
|................UUUU........................|
|................UUUU........................|
|................UUUU........................|
|............................................|
```

Fig. 4. Vehicle distribution for a uniform loading scenario.

Table 3. Balance of vehicles on the vessel during the heavy load scenario

Vehicle Type	Layout	Total Weight Left	Total Weight Right	Charging Time (minutes)
Large Trucks (L)	Outer rows and center	70,000 kg	70,000 kg	31.49
Pick-ups (P)	Center of the ship	25,000 kg	25,000 kg	
SUVs (U), Sedans (S), Compacts (C)	Top rows	10,000 kg	10,000 kg	
Total		105,000 kg	105,000 kg	31.49

truck (L) was placed on the left side of the vessel to compensate for the light load of compact vehicles and SUVs (U) distributed mostly on that same side. Despite the placement of pick-ups (P) towards the right rear, the left side resulted significantly heavier, at 31,900 kg compared to 17,000 kg on the right side. This imbalance could lead to lateral tilting, affecting stability during navigation. The total loading time was 35.11 min, showing moderate efficiency despite the asymmetric distribution.

Table 4 summarizes the uneven distribution of the load on the ship, showing the imbalance that can affect the stability of the ship in this scenario.

Table 4. Balance of vehicles on the vessel during the unbalanced load scenario

Vehicle Type	Layout	Total Weight Left	Total Weight Right	Charging Time (minutes)
Large Trucks (L)	Left side	15,000 kg	0 kg	35.11
SUVs (U)	Central part, with a tendency to the left	12,500 kg	5,000 kg	
Pick-ups (P)	Rear, right side	2,400 kg	10,000 kg	
Compacts (C)	Bottom, right side	2,000 kg	2,000 kg	
Total		31,900 kg	17,000 kg	35.11

6 Conclusions

This research presents the development of a mathematical model for efficient stowage on RoRo ships, as well as its validation through real data and simulations.

The results obtained indicate that the heavy load scenario turned out to be the most stable, with an almost perfect balance of loads on both sides of the vessel. The unbalanced load scenario, on the other hand, presented a significant risk to stability, while the uniform load scenario showed a moderate imbalance.

The heavy load scenario also turned out to be the most efficient in terms of time, followed by the unbalanced load scenario, and finally the uniform load scenario. This suggests that simplicity in the variety of vehicles (as in the heavy load scenario) facilitates faster and more effective loading.

To optimize both the stability and efficiency of the vessel, it is recommended to adopt a strategy similar to that of the heavy load scenario, where the weight is evenly distributed on both sides of the vessel. Although unbalanced loading can be faster in some cases, the risks associated with the tilting of the vessel outweigh the potential benefits of faster loading.

In conclusion, the balance between stability and efficiency must guide loading decisions on Ro-Ro vessels. The heavy lift scenario offers a model to follow to keep a vessel balanced and operational in a reasonable time.

7 Future Research

Building upon the mathematical model for RoRo ship stowage developed in this research, there are several avenues for future work that could enhance both the accuracy and applicability of the findings: Using machine learning algorithms, such as reinforcement learning, could further optimize stowage plans by predicting the most efficient and stable configurations based on historical data and real-time input. By training models on various loading scenarios, the system could autonomously propose stowage plans that balance speed and stability based on specific vehicle types and weights.

Future research could focus on integrating Internet of Things (IoT) devices to monitor real-time data on load shifts, vehicle conditions, and environmental factors during the stowage process. This would allow the system to dynamically adjust the loading configuration to maintain optimal stability throughout the voyage, especially in cases of unforeseen events such as shifting cargo or mechanical failures.

While this study focused on a specific set of vehicle types, future research could expand the scope to include a broader variety of cargo, including oversized or irregularly shaped vehicles, and even non-vehicle cargo. This would test the robustness of the model and potentially highlight additional variables that need to be considered for effective stowage in more diverse scenarios.

References

1. Parvasi, S.P., Main, A.R., Pacino, D.: RoRo ships stowage planning: using a novel MIP model and stability constraints. In: 2022 8th International Conference on Control, Decision and Information Technologies (CoDIT), pp. 1309–1311 (2022)
2. Zhang, Y., Tian, H., He, L., Ma, S., Yang, L.: A two-phase stowage approach for passenger-cargo RoRo ship based on 2D-KP: coping with complex rotation and safe navigation constraints. IEEE Access (2020)
3. Wijnolst, N., Wergeland, T.: Shipping Innovation. IOS Press (2009)
4. Andersson, H., Fagerholt, K., Hobbesland, K.: Integrated maritime fleet deployment and speed optimization: case study from RoRo shipping. Comput. Oper. Res. **55**, 233–240 (2015)
5. Ro-Ro (roll-on roll-off) ship. Oxford Reference. https://www.oxfordreference.com/view/10.1093/oi/authority.20110803100428566. Accessed 19 Mar 2024
6. Øvstebø, B.O., Hvattum, L.M., Fagerholt, K.: Optimization of stowage plans for RoRo ships. Comput. Oper. Res. **38**(10), 1425–1434 (2011)
7. Strang, R.W.A.: Method of loading wheeled vehicles in a container. PCT/AU2009/001340 (2010)
8. Henesey, L., Lizneva, Y., Philipp, R., Meyer, C., Gerlitz, L.: Improved load planning of RoRo Vessels by adopting Blockchain and Internet-of-Things (2020). https://doi.org/10.46354/i3m.2020.hms.009
9. Puisa, R.: Optimal stowage on Ro-Ro decks for efficiency and safety. J. Mar. Eng. Technol. **20**(1), 17–33 (2021). https://doi.org/10.1080/20464177.2018.1516942
10. Fancello, G., Serra, P., Mancini, S.: A network design optimization problem for Ro-Ro freight transport in the tyrrhenian area. Transp. Probl. **14**, 4 (2019). https://doi.org/10.20858/tp.2019.14.4.6
11. Schrotenboer, A.: Exact and heuristic methods for optimization in distributed logistics. [Thesis fully internal (DIV), University of Groningen]. University of Groningen, SOM research school (2020). https://doi.org/10.33612/diss.112911958
12. Dragan, D., Kramberger, T., Popovic, V.: Optimization Methods and heuristics and their role in supply chains and logistics. Quant. Methods Logist. (2020). https://doi.org/10.37528/FTTE/9786673954196.008
13. Hansen, J., Fagerholt, K., Stålhane, M., Rakke, J.: An adaptive large neighborhood search heuristic for the planar storage location assignment problem: application to stowage planning for Roll-on Roll-off ships. J. Heuristics **26**, 885–912 (2020). https://doi.org/10.1007/s10732-020-09451-z

14. Coffman, E.G., Garey, M.R., Johnson, D.S.: Approximation Algorithms for Bin Packing: A Survey. Springer, Cham (1996)
15. Garey, M.R., Johnson, D.S.: Computers and Intractability: A Guide to the Theory of NP-Completeness. W.H. Freeman (1979)
16. Martello, S., Toth, P.: Knapsack Problems: Algorithms and Computer Implementations. Wiley (1990)

Author Index

A

Acosta-Mesa, Héctor-Gabriel II-99, II-112, II-125, II-135, II-147, II-155, II-167
Acosta-Roman, Jose Luis I-84
Alberto, Ochoa-Zezzatti I-199
Alonso, Guerrero-Osuna Héctor I-199
Angeles-Hernadez, J. C. II-190
Aquino-Bolaños, Elia-Nora II-112, II-167
Aramburo-Aguilar, Jorge II-147
Arellano-Vazquez, Magali II-73
Avendaño-Garrido, Martha-Lorena II-112

B

Barradas-Palmeros, Jesús-Arnulfo II-125, II-135
Barrientos, Antonio Muñoz I-55
Barrón-Estrada, María Lucía II-37
Bautista, Humberto Muñoz I-55
Bernal, Axel I-284

C

Campaña, José Ismael Ojeda I-216
Canizales-Martinez, Juan Pablo I-248
Canul-Reich, Juana II-179
Castaneda-Diaz, Rafael I-121, I-133
Castillo, Oscar I-3
Castillon, Manuel Omar Meranza II-15
Castillo-Viveros, Nemesio I-180
Castro-Espinoza, F. A. II-190
Ceballos-Cancino, Héctor Gibrán II-179
Condado-Huerta, Carlos I-272
Cortes-Antonio, Prometeo I-3
Cruz-Chávez, Marco Antonio II-179
Cruz-Rosales, Martín H. I-272

D

de Dios González Torres, Juan II-53
De la Cruz Hernández, Erick II-179
De los Santos, José I-28
de Santiago, José Roberto Escamilla I-233
Díaz, Alejandro Padilla I-55

Diego, Carlos Iván Ramón I-216
Domínguez-Gómez, Daniel II-53

E

Escalante-Ramírez, Boris II-73
Escobar, David Alonso Carranza I-163
Escobar, Rafael Torres I-16
Espadas-Baños, Gilberto I-67

F

Fernandes, A. M. II-190
Ferreira-Escutia, Rogelio II-87
Flota-Bañuelos, Manuel I-67
Fuentes-Tomás, José-Antonio II-155

G

Garcia, Mario I-97
García-Castellanos, Humberto I-284
García-Gorrostieta, Jesús Miguel II-61
Gil-Núñez, José Miguel II-37
Gómez-Jiménez, Salvador I-133
Gonzalez, Claudia I. I-97, I-109
González-Franco, Nimrod II-45
González-López, Samuel II-15, II-61
González-Serna, Gabriel II-45
Guajardo, Hector M. I-3
Guerrero-Mendez, Carlos I-121, I-133
Guevara, L. II-190
Gutiérrez-Gnecchi, José A. II-87

H

Hernández, E. G. Salgado II-190
Hernández, Maylin I-28
Hernández, Yasmín II-53
Hernández-Aguilar, José Alberto I-272
Hernández-Báez, Irma I-284
Hernandez-Mendez, Sergio II-147
Hernández-Sánchez, Atalia-Yael II-147
Herrera-Sánchez, David II-155

I

Ibañez-Espiga, María Blanca II-37
Lizeth, Castañeda-Miranda Celina I-199
Lopez-Betancur, Daniela I-121, I-133
López-Herrera, Carlos-Alberto II-125, II-135
López-Lobato, Adriana-Laura II-99, II-112
López-López, Aurelio II-61
López-Sánchez, Máximo II-45

M

Madrid-Monteverde, José David II-61
Marin-Hernandez, Antonio II-147
Marquez, Maximiliano Ponce II-15
Martinez-Ytuza, Luis I-121
Mata-Rivera, Miguel F. II-73
Melin, Patricia I-3
Méndez-Patiño, Arturo II-87
Mezura-Montes, Efrén II-99, II-125, II-135, II-155, II-179
Minutti-Martinez, Carlos II-73
Montes-Rivera, Martin I-84
Morales, Rafael II-25
Morales-Reyes, José-Luis II-112, II-167

N

Nacarati-da-Silva, I. II-190
Núñez-Vieyra, Adrián II-87

O

Ochoa-Ortiz, Alberto I-180
Ochoa-Zezzatti, Alberto I-16, I-28, I-67, I-84, I-216, I-233, I-248, I-284
Octavio, Solís-Sánchez Luis I-199
Olivares-Rojas, Juan C. II-87
Oliveira, T. II-190
Olveres, Jimena II-73
Ortíz, Ángel I-28
Ortiz-Hernandez, Javier II-53

P

Pacheco-Valencia, Víctor I-272
Padilla, Cynthia Cristina Martinez I-143

Ponce-Gallegos, Julio César I-272
Ponce-Mendoza, Ulises II-61
Puma-Ttito, Flossi I-121, I-133

Q

Quej-Solís, César I-67

R

Ramírez, Efrén González I-133
Ramiro, Esquivel-Félix I-199
Ramos Herrera, Iván Alejandro I-43
Rentería, Jesús Raúl Cruz II-15
Reyes, Joshuar I-28
Reyes-Archundia, Enrique II-87
Reyes-Escalante, Aida-Yarira I-233
Rivera-López, Rafael II-179
Rodríguez, Francisco Javier Álvarez I-55
Rodríguez-Hernández, Rogelio I-180
Romero-Ramos, Luis II-45

S

Sánchez-Jiménez, Eduardo II-53
Sánchez-Sánchez, Christian II-3
Santos, Ayrton I-97
Saucedo-Anaya, Tonatiuh I-121
Soto, Gilberto Borrego II-15
Sucar, L. Enrique II-25

T

Torres, Cesar I-109
Trejo, Francisco I-16

V

Valdez, Fevrier I-3
Valenzuela-Robles, Blanca II-45
Villar-Patiño, Carmen I-248

Z

Zamora, Guillermina Muñoz II-15
Zatarain, Virginia Berenice Niebla I-216
Zataraín-Cabada, Ramón II-37

Printed in the United States
by Baker & Taylor Publisher Services